1945 Wolfgang Pauli (*Austria*): The exclusion principle.

1946 Percy Williams Bridgman (*U.S.A.*): High-pressure physics.

1947 Sir Edward Appleton (*England*): Discovery of ionospheric reflection of radio waves.

1948 Patrick Maynard Stuart Blackett (*England*): Discoveries in cosmic radiation.

1949 Hideki Yukawa (*Japan*): Theoretical prediction of the meson.

1950 Cecil Frank Powell (*England*): Photographic method of studying atomic nuclei; discovery of the charged pion.

1951 Sir John Douglas Cockcroft (*England*); Ernest Thomas Sinton Walton (*Ireland*): Transmutation of atomic nuclei by accelerated atomic particles.

1952 Felix Bloch (*U.S.A.*); Edward Mills Purcell (*U.S.A.*): Measurement of magnetic fields in atomic nuclei.

1953 Fritz Zernike (*Netherlands*): Development of phase contrast microscopy.

1954 Max Born (*Germany*): Contributions to quantum mechanics. Walther Bothe (*Germany*): Studies of cosmic radiation by use of the coincidence method.

1955 Willis E. Lamb, Jr. (*U.S.A.*); Polykarp Kusch (*U.S.A.*): Atomic measurements.

1956 John Bardeen (*U.S.A.*); Walter H. Brattain (*U.S.A.*); William B. Schockley (*U.S.A.*): Invention and development of the transistor.

1957 Chen Ning Yang (*China, U.S.A.*); Tsung Dau Lee (*China, U.S.A.*): Overthrow of principle of conservation of parity.

1958 Pavel A. Cerenkov (*U.S.S.R.*); Ila M. Frank (*U.S.S.R.*); Igor Y. Tamm (*U.S.S.R.*): Interpretation of radiation effects and development of a cosmic-ray counter.

1959 Owen Chamberlain (*U.S.A.*); Emilio Gino Segrè (*U.S.A.*): Discovery of the antiproton.

1960 Donald A. Glaser (*U.S.A.*): Invention of bubble chamber.

1961 Robert L. Hofstadter (*U.S.A.*): Electromagnetic structure of nucleons from high-energy electron scattering.

Rudolph L. Mössbauer (*Germany*): Discovery of recoilless resonance absorption of gamma rays in nuclei.

1962 Lev D. Landau (*U.S.S.R.*): Theories of condensed matter.

1963 Eugene P. Wigner (*U.S.A.*); Marie Goeppert-Mayer (*U.S.A.*); J. Hans D. Jensen (*Germany*): Nuclear shell structure.

1964 Charles H. Townes (*U.S.A.*); Nikolai G. Basov (*U.S.S.R.*); Alexander M. Prokhorov (*U.S.S.R.*): Amplification by laser — maser devices.

1965 Richard P. Feynman (*U.S.A.*); Julian S. Schwinger (*U.S.A.*); Sin-Itiro Tomonaga (*Japan*): Quantum electrodynamics.

1966 Alfred Kastler (*France*): Atomic energy levels.

1967 Hans A. Bethe (*Germany*): Nuclear theory and stellar energy production.

1968 Luis W. Alvarez (*U.S.A.*): Elementary particles.

1969 Murray Gell-Mann (*U.S.A.*): Theory of elementary particles.

1970 Hannes Alfvén (*Sweden*): Astrophysics.

1971 Dennis Gabor (*U.S.A.*): Holography.

1972 John Bardeen (*U.S.A.*); Leon Cooper (*U.S.A.*); John Schrieffer (*U.S.A.*): Theory of superconductivity.

1973 Leo Isaki (*Japan*); Ivar Giaever (*U.S.A.*): Discoveries in semiconductor physics. Brian Josephson (*England*): Prediction of "Josephson effects" in superconductivity.

1974 Antony Hewish (*England*): Discovery of pulsars. Martin Ryle (*England*): Radioastronomy.

1975 Aage Bohr (*Denmark*); Ben Mottelson (*Denmark*): The collective model of the nucleus. James Rainwater (*U.S.A.*): Measurements of nuclear electric quadrupole moments.

1976 Burton Richter (*U.S.A.*); Samuel C. C. Ting (*China, U.S.A.*): Discovery of the psi/J particle.

1977 Phillip W. Anderson (*U.S.A.*); Sir Neville Mott (*England*); John H. Van Vleck (*U.S.A.*): Theoretical investigations of the electronic structure of magnetic and disorders systems.

MODERN PHYSICS

for scientists and engineers

modern physics

for scientists and engineers

S. K. Kim and E. N. Strait

Macalester College

Macmillan Publishing Co., Inc.
New York

Collier Macmillan Publishers
London

Macmillan Publishing Co., Inc.
866 Third Avenue, New York, New York 10022

Collier Macmillan Canada, Ltd.

Library of Congress Cataloging in Publication Data

Kim, Sung Kyu, (date)
  Modern physics for scientists and engineers.

  Includes bibliographies and index.
  1. Physics.   I. Strait, E. N., joint author.
II. Title.
QC21.2.K49   530   77–5365
ISBN 0–02–363780–3

Printing: 1 2 3 4 5 6 7 8   Year: 8 9 0 1 2 3 4

Preface

This book is intended as an introduction to modern physics for students aspiring to careers in science and engineering. It is presumed that students will have had a calculus-based course in introductory classical physics, and thus be familiar with Newtonian mechanics, electricity and magnetism using Maxwell's equations in integral form, interference effects with light, and the geometrical optics of image formation. Facility with elementary calculus and exposure to partial differentiation are also assumed.

The present text begins with the theory of special relativity. The discussion of relativistic momentum and energy leads naturally to the photon model of light, and this in turn leads to the treatment of the wave nature of matter. Wave equations are then constructed and their solutions are given a probability interpretation. The fundamental ideas of the Schrödinger theory are applied to the hydrogen atom with emphasis on physical insights into the meanings and uses of quantum mechanics. The purely quantum mechanical ideas of indistinguishable particles and the exchange effect are also discussed. A description of x-ray diffraction then follows, leading into studies of crystal structure and crystal imperfections. In our treatment of the electronic properties of solids, emphasis is given to the concept of energy bands and an understanding of semiconductor devices.

Our discussion of the nucleus is wider in coverage than most modern physics textbooks at the introductory level. It is our firm conviction that an understanding of basic nuclear phenomena will be expected of every scientist and engineer as the question of energy production becomes more urgent. Our longer than average chapter on the technology of detectors and accelerators gives recognition to the profound and essential role this technology has played in the development of modern physics. Our work concludes with a discussion of elementary particles. By stressing the fundamental ideas of quantum field theory in the description of the strong, electromagnetic, and weak interactions, we have attempted to remind the student that physics remains an open intellectual discipline with its need for vigorous and imaginative thought undiminished.

The sequence of topics in this book is similar to that found in many established course syllabi. This particular order of presentation seems to preserve clarity of thought and logical unity that, admittedly, are characteristic of physics. Historical references, interesting and important as they are, have been confined to those cases in which they would not detract from the main flow of ideas.

The SI (MKS) system of units is familiar to most students. Thus no definitions are given when the volt is used in connection with such measurements as pulse heights and acceleration of charged particles and the ampere in those examples that call for numerical values of current. For numerical values of energy, we use the electron volt, which is a derivative unit employed in both SI and other systems of units. Despite the favored use of SI units in dealing with macroscopic phenomena, the cgs system of units is more "natural" and is thus adopted here for the description of atomic and subatomic phenomena. The mixture of units, however, is not likely to cause confusion. The student, for instance, will almost never need to know the charge of the electron in the cgs unit because of our nearly exclusive use of the dimensionless fine structure constant. In any case the student will encounter different systems of units in the literature, and early exposure to them must be regarded as a valuable learning experience.

Our reviewers have been very helpful, and the incorporation of their suggestions has improved as well as enriched our presentation here. They are Professor Malcolm Goldberg of Westchester Community College, Professor David Jenkins of Virginia Polytechnic Institute and State University, Professor Thomas T. Thwaites of The Pennsylvania State University, and Professor Walter D. Wilson of California Polytechnic State University at San Luis Obispo. To them and to our wise and vigorous editor, Mr. James L. Walsh, we owe a great debt of gratitude.

Finally, we wish to express our deepest appreciation to our families for their encouragement and understanding, without which this project would not have reached completion.

S. K. K.
E. N. S.

Contents

3

4

Electrons and the Exclusion Principle 157

X Rays and Crystallography 179

Electrical Properties of Solids 213

Basic Properties of the Nucleus   243

Nuclear Reactions   271

Nuclear Models   299

Radioactivity and Alpha Decay   325

Beta Decay and Gamma Decay   353

15   Detectors and Accelerators   379

16   Elementary Particles   425

Galilean relativity and the
Michelson-Morley experiment

1.1   The Law of Inertia

Man has always been curious about the nature of motion. If the heavens and the motions of heavenly bodies held a fascination for ancient man, the arrow and the cart and their motions were of more immediate concern to him. He had probably long wondered about the causes of motion, and found it to be quite reasonable when it was suggested that an object in motion required the continuous application of an external force. Such a notion of motion was formalized by Aristotle (384–322 B.C.), and, amazingly, went largely unchallenged for some 2000 years.

The first successful challenge to the Aristotelian view of motion was proferred by Galileo Galilei (1564–1642). Concerning uniform motion, for example, Aristotle asserted that an object in uniform motion on a flat surface required the continuous application of an external force. Galileo, on the other hand, argued that an object on a flat surface would remain in uniform motion only in the absence or cancellation of all external forces.

Observe, Galileo argued, the motion of a block on an essentially frictionless plane. On an inclined plane a block, left alone following an initial upward push, will eventually come to rest. It must therefore be pushed upward continuously if it is to be in uniform motion. On a downward slope a block must be retarded continuously by an external agent to be in uniform motion. On a horizontal plane, then, which is neither upwardly nor downwardly inclined, an object will require neither a push nor a retarding force to remain in uniform motion. Similarly, an object at rest will remain at rest only in the absence of all external forces.

Galileo believed these assertions about the nature of uniform motion and rest to be a natural abstraction from observation, and hence to constitute a law of motion. Referred to as the law of inertia, it was later generalized by Isaac Newton (1642–1721) to apply to all bodies, terrestrial as well as celestial. Newton incorporated it as his first law of motion:

> Every body continues in its state of rest, or of uniform motion in a straight line, unless it is compelled to change that state by forces impressed on it.

1.2   Uniform Motion and the Frame of Reference

Aristotle asserted that uniform motion was inherently different from the state of rest. As a way of establishing the difference, he suggested a simple procedure: Throw an object vertically upward and determine where it returns. Return of such an object to the thrower implies a state of rest for the thrower. According to Aristotle, then, a vertically thrown object would not return to a moving thrower. This ancient notion of absolute motion was occasionally questioned but not overthrown until the middle of the seventeenth century.

In *Two New Sciences,* published in 1638, Galileo exposed the falsity of Aristotle's claims about the nature of motion. He argued: Imagine a boat in uniform motion on a calm sea. Let a sailor throw an object vertically upward. The object will certainly return to the sailor. According to Aristotle, the sailor would then have to ascribe a state of rest to himself and the boat. A shorebound observer, however, will ascribe a state of uniform motion to the boat. The distinction between rest and uniform motion, which is absolute in Aristotelian physics, thus recedes in Galilean physics. Galileo asserted that the sailor will, in fact, experience no effects of the motion that the shorebound observer ascribes to him. Whether the shorebound observer ascribes a state of rest or a state of uniform motion to the boat, the sailor will in no way experience any difference. Galileo consequently proposed that there is no inherent difference between rest and uniform motion; the difference is only of a relative nature. The boat is, at once, in a state of rest relative to the sailor and in a state of motion relative to the shore. Only the points of reference are different. Thus the frame of reference emerges as a fundamental notion in the Galilean description of motion. It is, in fact, meaningless to speak of motion and rest apart from a frame of reference.

1.3   Galilean Transformation Equations

The description of motion entails the measurements of position and time, and these must be referred to a frame of reference. Observations of motion can be made in any frame of reference; however, let us consider only those frames of reference in which the law of inertia holds. These are called **inertial frames of reference.**

In order to examine how the measurements of position and time in different frames are related, let us introduce two inertial observers, one of whom employs an unprimed Cartesian coordinate system, and the other a primed system. The axes of the coordinate system are oriented so that the corresponding axes will be parallel.

Suppose now that an object is located at the position (x, y, z) in the unprimed system, and at the position (x', y', z') in the primed system. Then the relationship between (x, y, z) and $(x', y'\ z')$, according to Figure 1.1, is

$$
\begin{aligned}
x &= x' + x_{O'} \\
y &= y' \\
z &= z'
\end{aligned}
\tag{1.1}
$$

where the position of the primed origin O' in the unprimed system is $(x_{O'}, 0, 0)$. On the other hand, the position of the unprimed origin O in the primed system is $(x'_O, 0, 0)$, so that we have

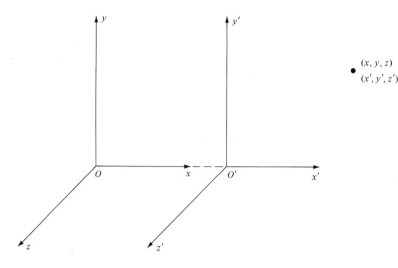

Figure 1.1 Cartesian coordinate systems.

$$x' = x + x'_0$$
$$y' = y \qquad\qquad (1.2)$$
$$z' = z$$

Equations (1.1) and (1.2) are consistent with ordinary experience and follow from the tacit assumption that space is absolute, homogeneous, and isotropic.

The measurement of time requires a clock. Since each observer must use his own clock, let us place a clock at each of the coordinate origins, O and O', and synchronize the clocks to read zero when the origins O and O' coincide. To be physically meaningful, the clocks must advance at the same rate when they are at rest relative to each other; however, we shall make no assumptions about how clock rates change, if at all, with motion.

Now let the origin O' of the primed system move uniformly at velocity v along the x axis in the unprimed system, as shown in Figure 1.2. This velocity must clearly be measured by an unprimed clock so that, from the definition of velocity, we have

$$x_{0'} = vt$$

Equation (1.1) can then be cast in the form

$$x = x' + vt$$
$$y = y'$$
$$z = z'$$

This can be rewritten as

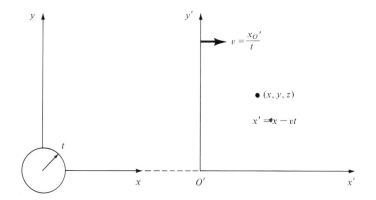

Figure 1.2 The unprimed clock reads zero when the origins of the coordinate systems coincide. The primed system moves along the x axis at velocity **v**.

$$x' = x - vt$$
$$y' = y \qquad\qquad (1.3)$$
$$z' = z$$

Similarly, the origin O of the unprimed system moves at velocity v' along the negative x' axis in the primed system. This is shown in Figure 1.3. If v' denotes the magnitude of the velocity of O, then the position of O is given by

$$x'_0 = -v't'$$

Substitution in Eq. (1.2) yields

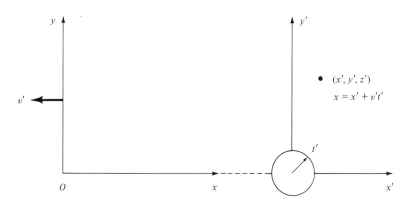

Figure 1.3 The primed clock reads zero when the origins of the coordinate systems coincide. The unprimed system moves in the negative x' direction at velocity **v'**.

$$x' = x - v't'$$
$$y' = y$$
$$z' = z$$

which leads to

$$x = x' + v't'$$
$$y = y' \tag{1.4}$$
$$z = z'$$

Note that there exists a symmetry between the unprimed and the primed coordinate systems. More specifically, if the unprimed observer ascribes a motion to the primed system, then the primed observer must also ascribe a similar motion to the unprimed system. The velocities, thus ascribed to each other, are opposite in direction but identical in magnitude, that is, $v = v'$. As a result of this important observation, Eq. (1.4) may be cast in the form

$$x = x' + vt'$$
$$y = y' \tag{1.5}$$
$$z = z'$$

Equations (1.3) and (1.5) are the space part of the Galilean coordinate transformation equations: They constitute the transformation equations for the coordinates of an absolute, homogeneous, and isotropic space.

When the Galilean transformation equations for x and x' are added, a relationship between the unprimed and the primed time coordinates emerges, namely,

$$t = t' \tag{1.6}$$

This result implies that two clocks, which keep time at the same rate when at rest relative to each other, must keep time at the same rate when in relative uniform motion. There is no reason to expect that the ticking rate of the clock in the unprimed system would have any relation to the motion of the primed system. Therefore Eq. (1.6) predicts that the time-keeping behavior of the clock in the primed system is independent of its motion relative to the unprimed system. This equation further implies that clock readings in an absolute space do not depend on the positions of the clocks. Equation (1.6) is thus equivalent to the statement that time is absolute and homogeneous.

Equations (1.3), (1.5), and (1.6) constitute the so-called **Galilean coordinate transformation equations:**

$$
\begin{array}{ll}
x = x' + vt' & x' = x - vt \\
y = y' & y' = y \\
z = z' & z' = z \\
t = t' & t' = t
\end{array} \tag{1.7}
$$

An immediate consequence of the Galilean coordinate transformation is the invariance of time intervals: Time intervals are the same in all inertial frames of reference.

$$\Delta t = t_2 - t_1 = t'_2 - t'_1 = \Delta t' \tag{1.8}$$

Equation (1.8) describes Newton's often quoted statement: "Absolute, true, mathematical time, of itself, and from its own nature, flows equably without relation to anything external."

Similarly, space intervals also constitute an invariant in classical mechanics. Suppose we transform a space interval $\Delta x'$ in the primed system to a corresponding space interval Δx in the unprimed system. The Galilean transformation equations yield

$$\begin{aligned}
\Delta x' &= x'_2 - x'_1 \\
&= (x_2 - vt_2) - (x_1 - vt_1) \\
&= \Delta x - v(t_2 - t_1) \\
&= \Delta x
\end{aligned} \tag{1.9}$$

provided

$$t_2 = t_1$$

Thus measurements of positions x_2 and x_1 must be made simultaneously, if the notion of a space interval is to be meaningful.

1.4   The Invariance of Classical Mechanics

Velocity is not an invariant quantity. From the definition of velocity.

$$\dot{\mathbf{r}} = \frac{d\mathbf{r}}{dt}$$

$$\dot{\mathbf{r}}' = \frac{d\mathbf{r}'}{dt'} \tag{1.10}$$

It is clear that an object having a velocity $\dot{x}'$ in the primed system would have a velocity $\dot{x}$ in the unprimed system, which is related to $\dot{x}'$ as follows:

$$\dot{x} = \dot{x}' + v \tag{1.11}$$

In general, the components of the velocity vector $\dot{\mathbf{r}}$ transform as

$$\dot{x} = \dot{x}' + v$$
$$\dot{y} = \dot{y}'$$
$$\dot{z} = \dot{z}' \tag{1.12}$$
$$\dot{x}' = \dot{x} - v$$
$$\dot{y}' = \dot{y}$$
$$\dot{z}' = \dot{z} \tag{1.13}$$

Equations (1.12) and (1.13) are called the **Galilean velocity transformation equations.**

Let us now apply the Galilean velocity transformation equations to the law of momentum conservation in a simple collision problem as shown in Figure 1.4. In classical mechanics, the definition of momentum, which is the product of mass and velocity, assumes the invariance of mass. The expression for the conservation of momentum in the unprimed system is

$$m\dot{x}_1 + M\dot{X}_1 = m\dot{x}_2 + M\dot{X}_2 \tag{1.14}$$

where the subscripts 1 and 2 identify the velocities before and after the collision. Equation (1.14), under the Galilean velocity transformation, takes the form

$$m(\dot{x}_1' + v) + M(\dot{X}_1' + v) = m(\dot{x}_2' + v) + M(\dot{X}_2' + v)$$

or

$$m\dot{x}_1' + M\dot{X}_1' = m\dot{x}_2' + M\dot{X}_2' \tag{1.15}$$

Equation (1.15) expresses the conservation of momentum in the primed system. Note that Eqs. (1.14) and (1.15) have the same mathe-

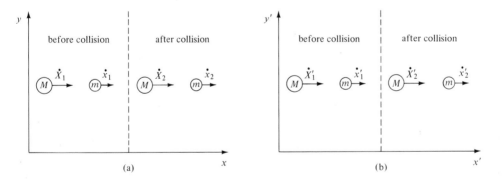

before collision   after collision   before collision   after collision

(a) x (b) x'

Figure 1.4 The principle of momentum conservation is invariant under the Galilean velocity transformation. Momentum conservation implies (a) $M\dot{X}_1 + m\dot{x}_1 = M\dot{X}_2 + m\dot{x}_2$ in the unprimed system and (b) $M\dot{X}_1' + m\dot{x}_1' = M\dot{X}_2' + m\dot{x}_2'$ in the primed system.

matical form. In general, the law of momentum conservation has the same mathematical structure in all frames of reference. The law of momentum conservation is thus said to be invariant under the Galilean velocity transformation.

Acceleration constitutes an invariant in classical mechanics:

$$\ddot{x} = \frac{d\dot{x}}{dt} = \frac{d}{dt}(\dot{x}' + v) = \ddot{x}'$$

$$\ddot{y} = \ddot{y}' \tag{1.16}$$

$$\ddot{z} = \ddot{z}'$$

Newton's equation of motion, $\mathbf{F} = m\ddot{\mathbf{r}}$, is then invariant under the Galilean transformation. In fact, Newton's laws of motion, and thus the laws of classical mechanics, are invariant under the Galilean transformation. The invariance of mechanical laws is referred to as the **Galilean principle of relativity.**

Galilean relativity is a profound principle — simple in content yet far-reaching in scope. It has its origin in the observation that the effects of uniform motion are indistinguishable from the effects of rest. It recognizes the fact that there exists no physical distinction between the states of rest and of uniform motion; there only exists a relative distinction. Thus no mechanical phenomenon would distinguish one inertial observer from another; all inertial frames of reference are equivalent. Galilean relativity serves as the guiding principle for the formulation of mechanical laws; all laws of mechanics must be so formulated as to be invariant under the Galilean transformation.

Classical mechanics is consistent with the principle of Galilean relativity. The notions of absolute rest and absolute uniform motion are then untenable in classical mechanics. It may be said, however, that Newton did not completely part with the notions of absolute rest and absolute motion, but conceived of space as being something absolute. In fact, the Newtonian conception of an absolute space, if not Newtonian mechanics, admitted the notions of absolute rest and absolute motion, and the hope that an absolute frame of reference might someday be identified was kept alive by subsequent generations of physicists.

1.5   Electromagnetic Ether

Electromagnetic phenomena are not described by Newtonian mechanics. The science of electricity and magnetism developed mostly during the nineteenth century through the experimental discoveries of physicists such as Coulomb, Oersted, Ampère, and Faraday. In the 1860s, James Clerk Maxwell (1831–1879) discovered that the laws of electromagnetism, as they then existed, were not internally consistent. The removal of the inconsistency, he found, required the introduction of a hypothetical term, which in turn led to the possible existence of a new phenomenon.

The new phenomenon predicted by Maxwell's equations of electromagnetism was the electromagnetic wave. The predicted velocity of propagation of the electromagnetic wave in empty space turned out to be identical to the observed velocity of light. This identity led Maxwell to the revolutionary conclusion that light must be an electromagnetic wave.

Since the early 1800s, the wave nature of light had been well established by the works of Young and Fresnel. However, light was viewed as a mechanical wave, like a wave on a string, and its propagation, therefore, required a physical medium, which was called ether. The polarization of light, for example, was pictured in terms of the transverse vibrations of the ether, and the amplitudes of reflected and transmitted light were successfully described from a mechanistic model of ether vibrations. The concept of ether was thus an integral part of the wave theory of light of Young and Fresnel.

In such a historical context it is not surprising that Maxwell incorporated the ether idea in his electromagnetic theory of light. He envisioned the "empty space" of his theory to be filled with the so-called electromagnetic ether. He wrote:

> The vast interplanetary and interstellar regions will no longer be regarded as waste places in the universe, which the Creator has not seen fit to fill with the symbols of the manifold order of His kingdom. We shall find them to be already full of this wonderful medium. . . . It extends unbroken from star to star; and when a molecule of hydrogen vibrates in the Dog-Star, the medium receives the impulse of these vibrations; and after carrying them in its immense bosom for three years, delivers them in due course, regular order, and full tale into the spectroscope of Mr. Huggins at Tulse Hill.

Maxwell's equations are not invariant under the Galilean transformation. For electromagnetic phenomena, therefore, the inertial frames of reference are not equivalent, implying that there must exist a unique frame of reference. It is only in this frame that an electromagnetic wave would propagate at the predicted speed, $c = 3 \times 10^{10}$ cm/sec. In any other frame of reference in relative motion the speed of light would differ from the predicted value c. Observation of such a difference would then establish the existence of an absolute frame of reference, which would in turn be identified with the ether frame of reference. Maxwell's electromagnetic theory thus provided a way of detecting the electromagnetic ether.

1.6 The Michelson-Morley Experiment

Imagine a sea of electromagnetic ether pervading space. Since it is unlikely that the earth would remain stationary in the ether, an experiment could conceivably be devised to detect the earth's absolute motion, that is, its motion in the ether. Various assumptions might be made about this motion. If the sun is at rest in the ether, then the earth's velocity through the ether would be that of its orbital motion, approximately 10^{-4} times the speed of light. If the sun is in motion through the ether, then, at some point in its orbit, the velocity of the earth in the ether would exceed the orbital velocity. The factor of 10^{-4}, however, led Maxwell to doubt that any terrestrial experiment would reveal the earth's absolute motion.

In 1881, despite the apparent difficulty of the measurement, Albert A. Michelson (1852–1931) embarked on an experiment to detect the earth's motion in the ether. In 1887, he and E. W. Morley obtained results that bore decisively on the question of the reality of the ether.

Their experiment was conceptually simple: It consisted of measuring the speed of light relative to the earth. Their apparatus was made of a light source, a beam splitter, two mirrors, and a viewing telescope. The mirrors were placed at the ends of two mutually perpendicular arms of equal lengths. The distance between the beam splitter and the mirrors was about 120 cm. Light from the source was then split by the beam splitter, and the beams, upon reflection from the mirrors, were rejoined by the beam splitter to produce an interference pattern of parallel bright and dark fringes. A schematic of the apparatus is given in Figure 1.5.

The experiment can most simply be analyzed in the earth frame of reference. Suppose that the ether current is flowing in the negative

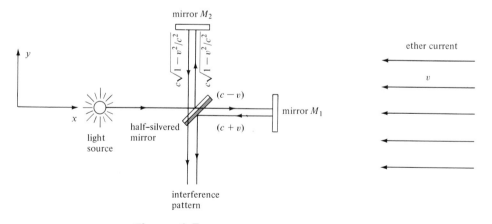

Figure 1.5 A schematic of the Michelson-Morley apparatus.

x direction at a velocity v. The beam, which travels in the x direction (i.e., parallel to the direction of the ether current), would then have a velocity $(c - v)$ on the way toward the mirror M_1 and a velocity $(c + v)$ on its return trip. The total transit time is

$$t_x = \frac{2L_x}{c(1 - v^2/c^2)}$$ (1.17)

where L_x is the distance between M_1 and the beam splitter.

The beam, which travels in the y-direction (i.e., perpendicular to the direction of the ether current), would appear to have a speed u, given by

$$u = c \sqrt{1 - v^2/c^2}$$ (1.18)

If the distance between the mirror M_2 and the beam splitter is L_y, then the total transit time for the beam is

$$t_y = \frac{2L_y}{c \sqrt{1 - v^2/c^2}}$$ (1.19)

The transit times given by Eqs. (1.17) and (1.19) are different for $v \neq 0$. Their difference is

$$\Delta t = t_x - t_y = \frac{2}{c}\left[\frac{L_x}{1 - v^2/c^2} - \frac{L_y}{\sqrt{1 - v^2/c^2}}\right]$$ (1.20)

which, in the approximation $v \ll c$, reduces to*

$$\Delta t \approx \frac{2}{c}\left[(L_x - L_y) + \frac{v^2}{c^2}\left(L_x - \frac{L_y}{2}\right)\right]$$ (1.21)

If the apparatus is rotated through 90°, then the difference in the transit times is

$$\Delta t' = t_x' - t_y' = \frac{2}{c}\left(\frac{L_x}{\sqrt{1 - v^2/c^2}} - \frac{L_y}{1 - v^2/c^2}\right)$$ (1.22)

reducing, in a similar approximation, to

$$\Delta t' \approx \frac{2}{c}\left[(L_x - L_y) + \frac{v^2}{c^2}\left(\frac{L_x}{2} - L_y\right)\right]$$ (1.23)

---

* $\dfrac{1}{\sqrt{1 - x^2}} \approx 1 + \tfrac{1}{2}x^2 \qquad x \ll 1$

Notice that Δt and $\Delta t'$ are not equal, which implies a path difference for the two beams of light, yielding a shift in the interference pattern. The path difference is given by

$$c(\Delta t - \Delta t') = (L_x + L_y)\frac{v^2}{c^2} \tag{1.24}$$

In the Michelson-Morley experiment of 1887, the effective distance between the beam splitter and the mirrors was increased to about 1100 cm by means of multiple reflections. If the orbital velocity of the earth is assumed to be comparable to the earth's velocity in the ether (i.e., $v = 3 \times 10^6$ cm/sec), the expected path difference in the two beams would be

$$c(\Delta t - \Delta t') = 2200 \times 10^{-8} \text{ cm}$$

A shift in the interference pattern is given by

$$\delta = \frac{c(\Delta t - \Delta t')}{\lambda} \tag{1.25}$$

where λ is the wavelength of the light. Michelson and Morley used a sodium light beam of $\lambda = 5900$ Å and thus could expect a shift of about 0.4 fringe. Since the detection limit of the interferometer was a shift of 0.005 fringe, such a shift would have been readily observed if it existed. However, no fringe shift in the interference pattern was detected.

1.7 Summary

By demolishing the Aristotelian notion of absolute motion and absolute rest, Galileo provided a foundation for Newtonian mechanics. Classical mechanics is consistent with the Galilean principle of relativity which has its origin in the relative character of rest and uniform motion. The mathematical content of Galilean relativity is the Galilean transformation equations, and the laws of classical mechanics are invariant under the Galilean transformation.

The laws of electricity and magnetism are not invariant under the Galilean transformation, and this fact implies that there may exist an absolute frame of reference. The known speed of light in this absolute frame gives rise to the possible measurement of absolute motion by means of light, and the Michelson-Morley experiment is an attempt to measure the absolute motion of the earth. The null result of the Michelson-Morley experiment deals yet another blow to the notion of absolute motion, and casts a serious doubt on the universal validity of the Galilean principle of relativity.

Problems

1.1 (a) If the Galilean sailor of Section 1.2 throws a ball vertically upward, give a mathematical description of the path of the ball in the sailor's frame of reference.

(b) If the boat is in uniform motion relative to the shore, obtain, by application of the Galilean transformation, the path of the ball in the shore-based frame of reference.

1.2 Show that a reference frame in uniform motion to an inertial frame is also an inertial frame.

1.3 If the principle of conservation of kinetic energy is invariant under the Galilean transformation, show that such invariance leads necessarily to the conservation of linear momentum.

1.4 Show that the equation of motion for a simple harmonic oscillator is invariant under a Galilean transformation.

1.5 Derive Eq. (1.18).

1.6 Verify Eq. (1.21).

1.7 As an attempt to explain the null result of the Michelson-Morley experiment, it was suggested that the earth drags the ether with it. How does this "ether-drag" hypothesis help explain Michelson's observation, and how reasonable is this hypothesis?

1.8 Lorentz and FitzGerald proposed, independently, that the null result of the Michelson-Morley experiment could be explained by a hypothesis of length contraction. Formulate a length-contraction hypothesis that would be consistent with the observation of Michelson and Morley, that is, how the dimensions of an object would have to change by virtue of its motion through the ether.

1.9 (a) Given $L_x = L_y = 120$ cm, $v = 3 \times 10^6$ cm/sec, and $\lambda = 5900$ Å, calculate the amount of the shift in the interference pattern.

(b) If the detection limit is a shift of 0.005 fringe, calculate the upper limit of the earth's velocity relative to the ether from the null result of the experiment using $L_x = L_y = 1100$ cm.

Suggestions for Further Reading

Ford, K. W. *Classical and Modern Physics*, vol. 3. Lexington, Mass.: Xerox College Publishing, 1974. Chapter 19.

French, A. P. *Special Relativity*. New York: W. W. Norton & Company, Inc., 1968. Chapter 2.

Smith, J. H. *Introduction to Special Relativity*. New York: W. A. Benjamin, Inc., 1965. Chapters 1 and 2.

Taylor, E. F., and J. A. Wheeler. *Spacetime Physics*. San Francisco: W. H. Freeman and Company, 1966. Sections 1.1–1.3.

Special relativity: space and time

2.1   Einstein's Principle of Relativity

The notion of absolute rest is without meaning in Newtonian mechanics. The laws of mechanics are invariant under the Galilean transformation, and the notion of invariance underlies the entire structure of classical mechanics.

The theory of electricity and magnetism developed by James Clerk Maxwell in the 1860s is not invariant under the Galilean transformation. In 1904, H. A. Lorentz discovered a set of transformation equations that rendered Maxwell's equations invariant. H. Poincaré recognized the importance of the invariance idea in electrodynamics and showed that all the equations of electrodynamics can be made invariant under the Lorentz transformation, provided the charge and current densities are suitably transformed.

In 1905, Albert Einstein (1879–1955) formulated a new version of the relativity principle, which contained the suggestion of Poincaré and the null result of the Michelson-Morley experiment. In Einstein's own words:[1]

> the unsuccessful attempts to discover any motion of the earth relative to the "light medium" suggest that the phenomena of electrodynamics as well as of mechanics possess no properties corresponding to the idea of absolute rest. They suggest rather that . . . the same laws of electrodynamics and optics will be valid for all frames of reference for which the equations of motion hold good.

Einstein thus recognized the possibility that the principle of relativity may be applicable to all physical phenomena. He continues:

> We will raise this conjecture (the purport of which will hereafter be called the "Principle of Relativity") to the status of a postulate, which is only apparently irreconcilable with the former, namely, that light is always propagated in empty space with a definite velocity c which is independent of the state of motion of the emitting body.

The principle of relativity is associated with Einstein, because it was Einstein who, more than any of his contemporaries, appreciated its profound and pervasive nature. His relativity may be stated: Physical laws must be the same in all inertial frames of reference. This implies that not only the mathematical forms of the physical laws but also the physical constants must be the same in all reference frames. Einstein's relativity principle thus introduces the notion of invariance to all of physics.

The principle of relativity is equivalent to the existence of a universal set of transformation equations and the invariance of all physical laws under the transformation. Such a set of transformation equations was first discovered by Lorentz and are referred to as the **Lorentz transformation equations.** Here we shall construct the Lorentz transformation equations from the **principle of relativity** and the **constancy of the speed of light.**

Consider, in Figure 2.1(a), a transformation equation of the form

$$x = ax' + bt' \tag{2.1}$$

[1] A. Einstein, *Annalen der Physik,* **17**:891 (1905); trans. W. Perrett and G. B. Jeffery in *The Principle of Relativity* (New York: Dover, 1923).

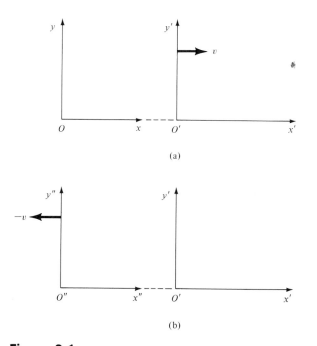

Figure 2.1 (a) The primed system is in motion at velocity *v* in the *x* direction of the unprimed system. (b) The double primed system is in motion at velocity −*v* in the *x'* direction of the primed system.

where a and b are arbitrary coefficients. If the primed coordinate system moves at velocity v relative to the unprimed system, then the unprimed origin, $x_0 = 0$, has a velocity $-v$ in the primed system, that is, $\dot{x}'_0 = -v = -b/a$. Hence the coefficients a and b are related according to

$$b = va \qquad (2.2)$$

Equation (2.1) can then be rewritten

$$x = a(x' + vt') \qquad (2.3)$$

Consider, now, a double primed system, which is moving at velocity $-v$ relative to the primed system, as in Figure 2.1(b). The transformation from the double primed system to the primed system, according to Eq. (2.3), is

$$x' = a[x'' + (-v)t''] \qquad (2.4)$$

The double primed system, however, is identical to the unprimed system, that is, $x'' = x$ and $t'' = t$. Thus Eq. (2.4) takes the form

$$x' = a (x - vt) \qquad (2.5)$$

which is the inverse transformation of Eq. (2.3).

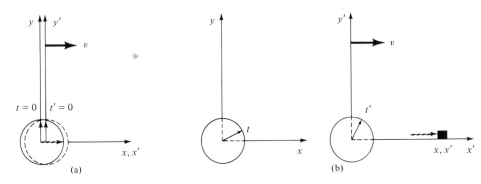

Figure 2.2 (a) When a flash is emitted at x = 0, the origins of both coordinate systems coincide and both clocks read zero. (b) When the flash is absorbed by a detector whose position coordinates are (x, 0, 0) and (x′, 0, 0), the clocks read t and t′, respectively.

The coefficient a in the transformation equations can be determined by invoking the constancy of the speed of light. Suppose a light source at $x = 0$ emits a flash when the coordinate origins O and O' coincide. Suppose further that clocks, located at the origins, are started by the light emission. Then the propagation of the flash in the horizontal direction is, in each system, described by

$$x = ct$$

and

$$x' = ct' \tag{2.6}$$

respectively. Now the substitution of Eq. (2.6) for t' and t in Eqs. (2.3) and (2.5) yields

$$a = \left(1 - \frac{v^2}{c^2}\right)^{-1/2} \tag{2.7}$$

whence the transformation equations for the x and x' coordinates take the form

$$x = \frac{x' + vt'}{\sqrt{1 - v^2/c^2}} \tag{2.8}$$

and

$$x' = \frac{x - vt}{\sqrt{1 - v^2/c^2}} \tag{2.9}$$

The y and z coordinates transform according to

$$y' = y$$
$$z' = z$$

(2.10)

which follow directly from the principle of relativity (see Section 2.4).

The transformation equation for the time coordinate is now easy to obtain. Solve for x in Eq. (2.9), and substitute it for x in Eq. (2.8). This leads to

$$t = \frac{t' + vx'/c^2}{\sqrt{1 - v^2/c^2}}$$

(2.11)

Similarly, solving for x' in Eq. (2.8) and substituting it for x' in Eq. (2.9) yields

$$t' = \frac{t - vx/c^2}{\sqrt{1 - v^2/c^2}}$$

(2.12)

The coordinate transformation equations thus constructed are identical to the transformation equations that Lorentz discovered in connection

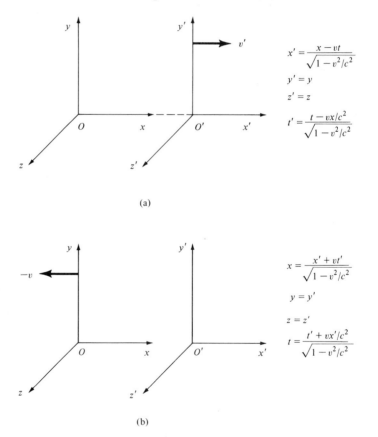

(a)

(b)

Figure 2.3 The Lorentz equations for the transformation (a) from the unprimed to the primed system and (b) from the primed to the unprimed system.

with Maxwell's equations, and are here summarized (see also Figure 2.3):

$$x' = \frac{x - vt}{\sqrt{1 - (v^2/c^2)}} \qquad\qquad x = \frac{x' + vt'}{\sqrt{1 - v^2/c^2}}$$

$$y' = y \qquad\qquad y = y'$$
$$z' = z \qquad\qquad z = z'$$

$$t'. = \frac{t - vx/c^2}{\sqrt{1 - v^2/c^2}} \quad (2.13) \quad t = \frac{t' + vx'/c^2}{\sqrt{1 - v^2/c^2}} \qquad (2.14)$$

The principle of relativity requires all the laws of physics to be invariant under the Lorentz transformation.

2.2 Lorentz Velocity Transformations

The frame of reference is an integral part of the concept of velocity. Velocity is the time rate of change of the position vector, where time and position are measured in the same frame. That is,

$$\mathbf{V} = \dot{\mathbf{r}} = \frac{d\mathbf{r}}{dt} \qquad \text{(in unprimed system)}$$

$$\mathbf{V}' = \dot{\mathbf{r}}' = \frac{d\mathbf{r}'}{dt'} \qquad \text{(in primed system)} \qquad (2.15)$$

The Lorentz transformation equations for velocity follow directly from the definition of velocity. The differential form of the Lorentz coordinate transformation equations is

$$dx' = \frac{dx - v\,dt}{\sqrt{1 - v^2/c^2}}$$

$$dy' = dy$$
$$dz' = dz$$

$$dt' = \frac{dt - v\,dx/c^2}{\sqrt{1 - v^2/c^2}} \qquad (2.16)$$

which leads to a new velocity transformation

$$\dot{x}' = \frac{dx'}{dt'} = \frac{dx - v\,dt}{dt - (v/c^2)dx} = \frac{\dot{x} - v}{1 - v\dot{x}/c^2} \qquad (2.17)$$

The y and z components of velocity transform according to

$$\dot{y}' = \frac{dy'}{dt'} = \frac{dy \sqrt{1 - v^2/c^2}}{dt - (v/c^2)\, dx} = \frac{\dot{y}\sqrt{1 - v^2/c^2}}{1 - v\dot{x}/c^2}$$

$$\dot{z}' = \frac{\dot{z}\sqrt{1 - v^2/c^2}}{1 - v\dot{x}/c^2} \tag{2.18}$$

Equations (2.17) and (2.18) are the **Lorentz velocity transformation equations.**

An immediate consequence of the Lorentz velocity transformation is the constancy of the speed of light c. Consider a light flash traveling in the x direction. If its speed is measured to be $\dot{x} = c$ in the unprimed system, its speed in the primed system would be, according to the relativistic velocity addition law, Eq. (2.17),

$$\dot{x}' = \frac{\dot{x} - v}{1 - v\dot{x}/c^2} = c$$

Thus, if the Lorentz transformation is postulated and the principle of relativity is stated in terms of invariance under the Lorentz transformation, then the constancy of the speed of light need not be given explicitly as a separate postulate for the theory of special relativity.

2.3 Relative Time

The constancy of the speed of light clashes violently with the Newtonian concepts of absolute space and absolute time. The notions of absolute space and absolute time are thus incompatible with Einstein's principle of relativity.

Suppose a light flash, emitted at the origin O', propagates along the y' axis, as in Figure 2.4(a). If it is in transit for a time t', as measured by the primed clock, the position of the light flash is given by

$$y' = \dot{y}'t' \tag{2.19}$$

where $\dot{y}' = c$. Viewed in the unprimed system (see Figure 2.4b), the flash follows a diagonal path, and the y component of the position of the flash at any time t, measured by the unprimed clock, is simply

$$y = \dot{y}t$$

which, by means of the velocity transformation, takes the form

$$y = \left(\frac{\dot{y}'\sqrt{1 - v^2/c^2}}{1 + \dot{x}'v/c^2}\right)t = \dot{y}'(\sqrt{1 - v^2/c^2})t, \quad \dot{x}' = 0 \tag{2.20}$$

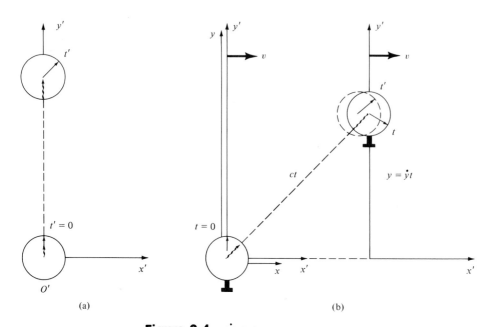

Figure 2.4 (a) A light flash emitted at O' propagates along the y' axis for a time interval t'. (b) The same flash follows a diagonal path according to the unprimed observer and is in transit for a time interval t. The y-component of the velocity of the flash is denoted by $\dot{y}$.

This can be rewritten as

$$y = \left(\frac{y'}{t'}\right)(\sqrt{1 - v^2/c^2})t$$

which, because $y = y'$, leads to a relationship between t and t':

$$t = \frac{t'}{\sqrt{1 - v^2/c^2}} \tag{2.21}$$

This is a remarkable result. It says that time in the primed system is slower than time in the unprimed system: Time is relative.

The relative nature of time can be deduced in a straightforward way by applying the Lorentz transformation to time-measuring devices, namely, clocks. For example, place a clock at the origin of the primed system, as in Figure 2.5, and two other clocks in the unprimed system, one at the origin and the other at an arbitrary position x. Let the time lapse on the primed clock as it moves from the position of one unprimed clock to the other be $(t' - t_0')$. According to the Lorentz transformation

$$t = \frac{t' + x'/c^2}{\sqrt{1 - v^2/c^2}}$$

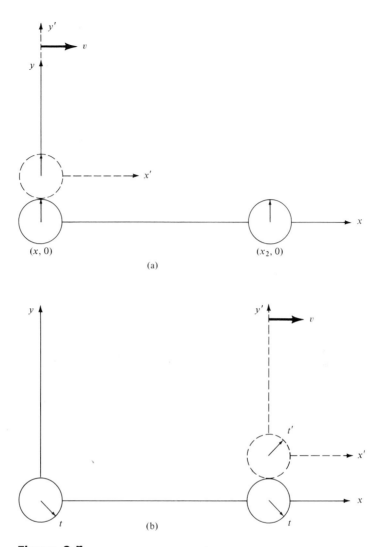

Figure 2.5 The x and x' axes are intended to coincide. (a) All three clocks read zero. (b) The moving clock keeps time at a slower rate than the stationary clocks, that is, $\Delta t' = \Delta t \sqrt{1 - v^2/c^2}$.

the time lapse on the unprimed clocks must be

$$t - t_o = \frac{t' - t_0'}{\sqrt{1 - v^2/c^2}} \tag{2.22}$$

If we set $t_0 = t_0' = 0$ (the clocks positioned at the coordinate origins read zero when they coincide), the subsequent clock readings compare according to

$$t = \frac{t'}{\sqrt{1 - v^2/c^2}}$$

implying that the moving clock keeps time at a slower rate than the stationary clock.

To illustrate the relative nature of time, it is interesting to compare the pulse rates of an astronaut and a ground-based observer. Suppose their pulse rates are identically 70 beats per minute when they are both at rest. Then if the astronaut is placed in a spaceship moving at constant velocity v, say $v = 0.99c$, his heart would pump only 70 times during the ground-based observer's some 500 beats. This is so, because 1 min on the spaceship clock corresponds to 7.1 min on the ground-based clock:

$$t = \frac{t'}{\sqrt{1 - v^2/c^2}} = 7.1 \text{ min}$$

Thus the stationary observer would attribute a slower pulse rate to the astronaut. The time interval between the pulses of the astronaut is simply longer than the observer's own, and we say that time in the spaceship is dilated relative to time on the ground. Time is relative, because motion dilates time, and Eq. (2.21) is therefore referred to as Einstein's **time-dilation equation.**

The frame of reference attached to an object under observation is called a rest frame. Measurements in such a frame are referred to as proper measurements. The **proper** time of an object, then, would be the time measured in a frame that is at rest relative to the object. The proper lifetime of a radioactive particle, for example, is its lifetime measured in its rest frame.[2] If we denote the proper time by t_0 and the nonproper time by t, then we may write the time-dilation equation in the form

$$t = \frac{t_0}{\sqrt{1 - v^2/c^2}} \tag{2.23}$$

The principle of relativity thus renders untenable the familiar notion of absolute time. Time does not, as Newton put it, "flow equably without any relation to anything external . . . ." Rather, it flows unequably in different frames of reference, and exists only in intimate relation to a frame of reference.

2.4 Relative Space

Consider two hollow cylinders with mirrors at the ends, as in Figure 2.6. Place one along the x' axis and the other along the y' axis,

[2] See the film *Time Dilation—An Experiment with Mu-Mesons*, by D. H. Frisch and J. H. Smith, Education Development Center, Newton, Mass., 1963; Frisch and Smith. *American Journal of Physics*, **31**:342–355 (1963).

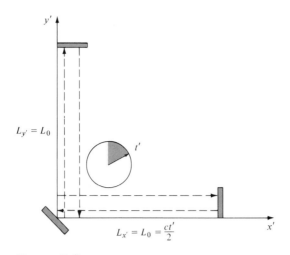

Figure 2.6 Two cylinders of equal lengths in the primed system.

with both cylinders having one mirror at the origin O'. Let the mirrors at the origin emit pulses of light at the same instant of time. Adjust the lengths of the cylinders so that the light pulses, upon reflection by the far-end mirrors, will arrive at the origin at the same moment. Then the lengths of the cylinders are equal, that is, $L_{x'} = L_{y'} = L_0$. If a lapse of time t' is recorded on the clock in the primed system, then the length of the cylinder is

$$L_0 = \frac{ct'}{2} \tag{2.24}$$

Let us now view the cylinders relative to the unprimed system. The Lorentz coordinate transformation immediately yields that the length of the vertical cylinder remains invariant, that is, $L_y = L_{y'} = L_0$. This fact can also be deduced directly from the principle of relativity. Suppose we have two rods A and B, as in Figure 2.7(a), that are of equal lengths when at rest relative to each other. Attach a brush to the end of rod B, and place each rod along the y axis in its own frame. Assume that when the rod moves in the x direction, its y dimension or vertical length undergoes contraction. In order to deduce consequences of such an assumption, let us suppose that rod B moves relative to rod A [see Figure 2.7(b)]. Upon passing by rod A, rod B would then leave a mark on rod A, because rod B is in motion in the rest frame of rod A. In the rest frame of rod B, on the other hand, rod A is in motion and its vertical length would contract [see Figure 2.7(c)]. Upon passing by rod B, rod A would therefore miss the brush of rod B. The result is that rod A ends up with a brush mark in one frame and no brush mark in another frame. Such, however, contradicts the principle of relativity. The initial assumption must, therefore, be false, and

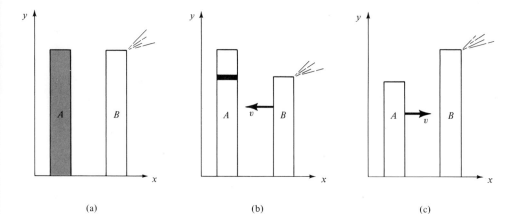

(a)   (b)   (c)

Figure 2.7   Assume that rods moving in the x direction contract in their vertical lengths. (a) Two rods of equal lengths are both stationary. (b) Rod **B** is in motion, and leaves a mark on rod **A**. (c) Rod **A** is in motion, and hence misses the brush.

we are led to the general conclusion that the vertical length of a rod moving horizontally remains the same.

Figure 2.8 pictures the vertical cylinder in motion relative to the unprimed system. Viewed in the unprimed system, the light pulse in the vertical cylinder follows a diagonal path. If its total time of flight is measured to be t_y by the unprimed clock, then the pulse in the vertical cylinder traverses a total distance ct_y. The path of the light forms

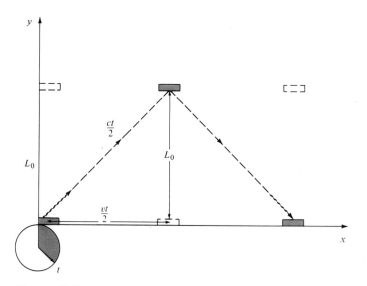

Figure 2.8   The path of a light flash trapped in a vertical cylinder is diagonal when viewed in the unprimed system.

the hypotenuse of a right triangle in Figure 2.8, and the Pythagorean theorem yields

$$\left(\frac{ct_y}{2}\right)^2 = L_0^2 + \left(\frac{vt_y}{2}\right)^2$$

or

$$t_y = \frac{2L_0}{c\sqrt{1 - v^2/c^2}} \qquad\qquad (2.25)$$

The observation of the light pulse in the horizontal cylinder is best analyzed in two parts. When the pulse is in propagation in the $+x$ direction, it travels a distance ct_1, and Figure 2.9(a) gives

$$ct_1 = L_x + vt_1 \qquad\qquad (2.26)$$

Here t_1 is the time elapsed on the unprimed clock during the transit of the light from one mirror to the other. On the return trip, the pulse

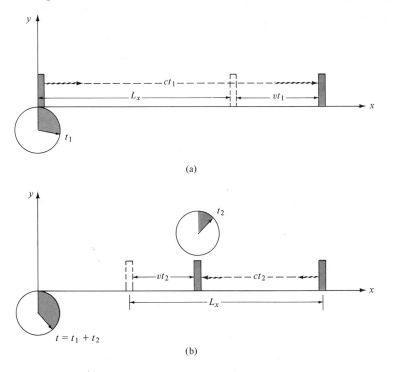

(a)

(b)

Figure 2.9 The moving horizontal cylinder is viewed in the unprimed system. (a) The outbound flash is in transit for a time interval t_1. (b) The inbound flash is in transit for a time interval t_2.

spends a time t_2 in transit, thus traversing a distance ct_2, and from Figure 2.9(b) we have

$$ct_2 = L_x - vt_2 \tag{2.27}$$

The total time of transit of the light pulse in the horizontal cylinder is

$$t_x = t_1 + t_2 = \frac{2L_x}{c(1 - v^2/c^2)} \tag{2.28}$$

Now the principle of relativity dictates that the transit times of the light pulses in both cylinders must be equal ($t_x = t_y$), yielding the result

$$L_x = L_0 \sqrt{1 - v^2/c^2} \tag{2.29}$$

This implies that, viewed in the unprimed system, the horizontal cylinder is shorter in length than the vertical cylinder. Since the length of the vertical cylinder equals the length of the horizontal cylinder at rest, the length of the horizontal cylinder is shorter in motion than at rest. This result is referred to as **length contraction:** a moving rod contracts in the direction of motion.

It is instructive to obtain the length-contraction equation directly from the Lorentz coordinate transformation. Suppose a rod is at rest in the primed system with its ends at x_1' and x_2' so that its length is

$$L' = x_2' - x_1' \tag{2.30}$$

The Lorentz transformation then gives

$$x_2' - x_1' = \frac{x_2 - x_1 - v(t_2 - t_1)}{\sqrt{1 - v^2/c^2}} \tag{2.31}$$

If $(x_2 - x_1)$ is to represent the length of the rod in the unprimed system, the ends of the rod must be observed at the same instants of time, namely, $t_1 = t_2$. Hence

$$L = L' \sqrt{1 - v^2/c^2} \tag{2.32}$$

where

$$L = x_2 - x_1 \tag{2.33}$$

Notice that it is always the nonproper length that contracts. Proper length is the length of an object in its rest frame.

The contraction of a moving rod implies that space cannot be viewed in absolute terms. A space interval is different in different frames, because a meter stick in one frame differs from a meter stick in another frame. Thus the Newtonian statement that space exists "without relation to anything external, remains always similar and immovable," is no longer valid. In the theory of relativity, space exists only in intimate relation to a reference frame: Space is a relative concept.

2.5   Relative Simultaneity

It should come as no surprise to the reader that simultaneity must also be a relative concept. Consider two simultaneous events in the primed system. An event is specified by a set of space and time coordinates, and constitutes a point in the so-called space-time continuum. An example of two events is illustrated in Figure 2.10, where two detectors positioned at the ends, x_1' and x_2', of a hollow cylinder register the arrivals of light signals emitted at the midpoint of the cylinder. Since both signals must arrive at the detectors at the same time, that is, $t_1' = t_2'$, they constitute two simultaneous events in the primed system. Moreover, the space interval between the two events gives the proper length of the cylinder, $L' = x_2' - x_1'$.

The observation of the two events in the unprimed system is

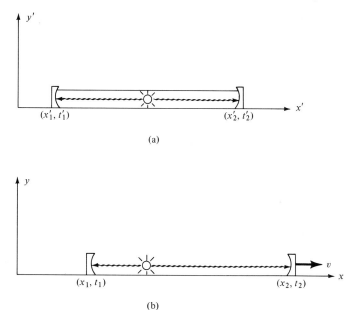

(a)

(b)

Figure 2.10   (a) Two light signals emitted at the midpoint of the cylinder register at the detectors positioned at x_1' and x_2' at the same time, that is, $t_1' = t_2'$. (b) The cylinder is moving at v, and the light signal registers at the detector at x_1 first, that is, $t_1 < t_2$.

described by the Lorentz transformation. According to the unprimed clock, the detectors register the arrivals of the light signals at different times, and the time interval between them is

$$t_2 - t_1 = \frac{v(x_2' - x_1')/c^2}{\sqrt{1 - v^2/c^2}} \tag{2.34}$$

The light signal reaches the detector positioned at x_1 ahead of the detector positioned at x_2, and the two events are not simultaneous. The space interval between the detectors is

$$x_2 - x_1 = \frac{x_2' - x_1'}{\sqrt{1 - v^2/c^2}} \tag{2.35}$$

Here $(x_2 - x_1)$ does not give the length of the cylinder in the unprimed system but represents the total distance traversed by the light signals.

A notion related to simultaneity is clock synchronization. Let the detectors of Figure 2.10 be replaced with clocks as in Figure 2.11, and have them synchronized in the primed system so that their readings will be the same. Their spatial separation $(x_2' - x_1')$, then, gives the proper distance between the clocks. We know, however, that the space interval $(x_2 - x_1)$ between the clocks, as viewed in the unprimed system, does not represent the length of the cylinder. Moreover, the clocks at

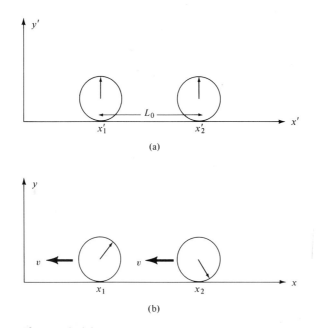

(a)

(b)

Figure 2.11 (a) Two clocks, separated by L_0, are synchronized in the primed system. (b) In the unprimed system, the two clocks are in motion in the $-x$ direction, and the rear clock at x_2 is ahead of the front clock at x_1.

x_1' and x_2', at any time t in the unprimed system, will, according to the Lorentz transformation for the time coordinate, not read the same:

$$t = \frac{t_1' + vx_1'/c^2}{\sqrt{1 - v^2/c^2}} \tag{2.36}$$

$$t = \frac{t_2' + vx_2'/c^2}{\sqrt{1 - v^2/c^2}} \tag{2.37}$$

or, equating the right-hand sides of Eqs. (2.36) and (2.37),

$$t_2' = t_1' - \frac{(x_2' - x_1')v}{c^2}$$

$$= t_1' - \frac{L'v}{c^2} \tag{2.38}$$

Thus if two clocks are synchronized in the primed system, they are out of synchronization by $L'v/c^2$ in the unprimed system, and the clock at x_1' (the rear clock) is ahead of the clock at x_2' (the front clock) by $L_0 v/c^2$ where L_0 denotes the proper distance between the clocks.

Example 2.1   Two clocks, separated by 5 light-hr, are synchronized according to an observer at rest relative to the clocks. How much are they out of synchronization according to a spaceship observer who is in motion at $v = 0.8c$ from clock 1 to clock 2?

Solution.   According to the spaceship observer clocks 1 and 2 are in motion, and clock 1 constitutes the front clock and clock 2 the rear clock. Clock 2 is thus ahead of clock 1 by

$$\frac{L_0 v}{c^2} = \frac{(5c)(0.8c)}{c^2} \text{ hr} = 4 \text{ hr} \quad \bullet$$

If the principle of relativity is accepted, no absolute significance can be attached to the notion of simultaneity. The notion of relative simultaneity is rooted in the fact that, in relativity, space and time are mixed in an inseparable way. A simple example will illustrate this point. Consider two simultaneous events that are spatially separated in the primed system. If the primed system is in motion, as is customary in our discussions, then the events will clearly not be simultaneous in the unprimed system. They will thus be separated both in space and in time. In general, a space interval in one frame will appear in another frame as a mixture of space and time intervals. Space and time, therefore, can no longer be treated independently of each other. Rather, in relativity, space and time must be placed on an equal footing; indeed, they are interrelated. Thus arises the notion of space-time, and physical events must henceforth be described in the so-called space-time continuum.

2.6 Space-Time Intervals

Although neither the space interval nor the time interval constitute invariant quantities, an invariant interval in space-time can be defined:

$$I^2 = c^2(t_2 - t_1)^2 - (x_2 - x_1)^2 - (y_2 - y_1)^2 - (z_2 - z_1)^2 \tag{2.39}$$

As a way of exploring the physical content of the notion of the space-time interval, consider first the case where the value of the space-time interval is zero, that is, $I = 0$. If we take the first event to be $(0, 0, 0, 0)$ and the second event (x, y, z, t), then the expression for the space-time interval becomes in the unprimed and the primed systems, respectively,

$$(ct)^2 - x^2 - y^2 - z^2 = 0 \tag{2.40}$$

and

$$(ct')^2 - x'^2 - y'^2 - z'^2 = 0 \tag{2.41}$$

Equations (2.40) and (2.41) describe the propagation of light from the origins of the unprimed and the primed systems. The invariance of a zero space-time interval is thus equivalent to the invariance of the speed of light.

Consider, secondly, two events in the primed system with a zero space interval, that is, $I^2 > 0$. An example of such events is two separate readings on a clock stationary in the primed system. Such an interval is called **timelike,** and its invariance implies

$$c^2(\Delta t')^2 = c^2(\Delta t)^2 - (\Delta x)^2$$

or

$$(\Delta t') = (\Delta t) \sqrt{1 - v^2/c^2} \tag{2.42}$$

where $v = \Delta x/\Delta t$. Equation (2.42) is the time-dilation equation, and $\Delta t'$ is a proper time interval. Proper time is thus a Lorentz invariant.

Consider, finally, two events on the x' axis in the primed system, separated only by space, that is, $I^2 < 0$. Such an interval is called **spacelike.** Spacelike intervals are also invariant; therefore, if an interval is spacelike in one frame, it must be spacelike in all frames. In the unprimed system, the two events have an interval

$$I^2 = c^2(t_2 - t_1)^2 - (x_2 - x_1)^2 = -(x_2' - x_1')^2 \tag{2.43}$$

implying that

$$c(t_2 - t_1) < x_2 - x_1 \tag{2.44}$$

Since the speed of light is a maximum speed attainable in nature, events at x_1 and x_2 are not causally connected. Causality holds only for those events that have a timelike separation in space-time.

Space-time constitutes a four-dimensional continuum. Space and time must now be regarded as the position and the time coordinates of space-time. Space-time is thus more fundamental than the separate notions of space and time. As pointed out by V. Weisskopf, "relativity theory is really a theory of the absolute, of the quantities that remain unchanged from any standpoint (invariants)."

2.7 Summary

The special theory of relativity is founded upon the postulates of the relativity principle and of the constancy of the speed of light. The principle of relativity states that all the laws of physics must be invariant under the Lorentz transformation. The Lorentz coordinate transformation equations can be constructed from the postulates of special relativity. The Lorentz velocity transformation equations are obtained from the definition of velocity, and the relativistic velocity addition law is

$$v = \frac{v_1 + v_2}{1 + v_1 v_2 / c^2}$$

Time and space are no longer absolute. Moving clocks keep time at a slower rate than stationary clocks, and the time-dilation effect is described by

$$t = \frac{t'}{\sqrt{1 - v^2/c^2}}$$

A moving object contracts in the direction of its motion, and the length-contraction effect is described by

$$L = L_0 \sqrt{1 - v^2/c^2}$$

Simultaneity is also a relative concept: Two events simultaneous to one observer are not simultaneous to another observer, and the time interval between the nonsimultaneous events is given by

$$\Delta t = \frac{v\Delta x'/c^2}{\sqrt{1 - v^2/c^2}}$$

If two clocks, separated by L_0, are synchronized to an observer at rest relative to them, they are out of synchronization to another observer moving relative to the clocks, who will find the rear clock to be ahead of the front clock by $L_0 v/c^2$.

Because space and time intervals are relative, an invariant interval in space-time is defined, and is considered to be more fundamental. To be consistent with causality, two events must have a timelike separation; events having a spacelike separation cannot be causally related.

Problems

2.1 Argue that the unprimed system of Figure 2.1 and the double-primed system of Figure 2.2 are identical.

2.2 Verify Eq. (2.7).

2.3 Verify Eq. (2.11).

2.4 Suppose a primed observer is moving at $v = 0.5c$ in the x direction relative to an unprimed observer. If the primed observer finds the velocity of an object to be $\dot{x}' = 0.75c$, what would the unprimed observer find the object's velocity to be?

2.5 If a double-primed system moves at v_2 relative to a primed system, and the primed system moves at v_1 relative to an unprimed system, show that the double-primed system moves at v relative to the unprimed system, which is given by

$$v = \frac{v_1 + v_2}{1 + v_1 v_2/c^2}$$

2.6 Suppose a light signal is emitted at the origin O' of a primed system and propagates along the y' axis. If the primed system moves at velocity v in the x direction of an unprimed system, calculate the x and y components of the velocity vector of the light signal in the unprimed system.

2.7 Suppose a primed clock that is set to read zero at the origin of an unprimed coordinate system moves at velocity v along the x axis.
  (a) Express the position of the moving clock as a function of unprimed time, that is, as recorded in a clock fixed in the unprimed system.
  (b) If a stationary clock is fixed at the unprimed coordinate origin and is synchronized to the moving clock when both are at the origin of the unprimed system, show that the moving clock keeps time at a slower rate than the stationary clock by a factor of $\sqrt{1 - v^2/c^2}$.

2.8 Suppose the mean life of a muon at rest is observed to be 2.2×10^{-6} sec in a spaceship moving at $v = 0.99c$ relative to the ground. What is its mean life as measured on the ground-based clock?

2.9 Verify Eq. (2.28).

2.10 A light signal propagates a distance of 300 cm along that x' axis, as measured in the primed system. If the primed system is in motion at $v = 0.99c$ in the $+x$ direction of the unprimed system, calculate the distance of transit of the light signal as measured in the unprimed system.

2.11 Verify Eq. (2.34).

2.12 Suppose the earth is in contact with a stellar object 25 light-years away. If the earth- and the star-based clocks are synchronized according to the earth-based observer, how much will they be out of synchronization according to the observer in a spaceship in transit from the earth to the star at $v = 0.8c$?

2.13 An astronaut journeys to a distant object 25 light-years away.
   (a) If his spaceship maintains a constant velocity of $v = 0.9998c$, show that he would age only 6 months during the outbound trip. (Neglect any effects of acceleration.)
   (b) How much will the earth-based people age according to the earth-based clock during the astronaut's outbound trip?
   (c) How much time will the astronaut observe to elapse on the earth- and the star-based clocks during his outbound trip?
   (d) Assuming that the earth- and the star-based clocks are synchronized to the earth-based observer, how much will the earth-based people have aged according to the astronaut at the moment of his landing on the star?

2.14 A 200-m train passes through a 100-m tunnel.
   (a) How fast must the train move in order for it to be just contained within the tunnel according to a ground-based observer?
   (b) What is the length of the tunnel according to a train-based observer?
   (c) The ground-based observer observes that lightning strikes the ends of the tunnel simultaneously precisely at the moment the train is just contained within the tunnel. Will the train-based observer observe that the train got hit by the lightning bolts? Explain.

2.15 Show that the space-time interval defined by

$$I^2 = c^2t^2 - x^2 - y^2 - z^2$$

is invariant under the Lorentz transformation.

2.16 Consider two clocks in the primed system, which are separated by a space interval s'. If t_1' and t_2' denote readings on clock 1 and clock 2, respectively, show that the quantity $[s'^2 + c^2(t_2' - t_1')^2]$ is a relativistic invariant.

Suggestions for Further Reading

Feynman, R. P., R. B. Leighton, and L. M. Sands. *The Feynman Lectures on Physics,* vol. 1. Reading, Mass.: Addison-Wesley Publishing Company, 1963. Chapters 15, 18, and 20.

Ford, K. W. *Classical and Modern Physics,* vol. 3. Lexington, Mass.: Xerox College Publishing, 1974. Chapter 20.

French, A. P. *Special Relativity*. New York: W. W. Norton & Company, Inc., 1968. Chapters 3 and 4.

Mermin, N. D. *Space and Time in Special Relativity*. New York: McGraw-Hill Book Company, 1968.

Smith, J. H. *Introduction to Special Relativity*. New York: W. A. Benjamin, Inc., 1965.

Weisskopf, V. F. "The Visual Appearance of Rapidly Moving Objects," *Physics in the Twentieth Century: Selected Essays*. Cambridge, Mass.: The MIT Press, 1972. Pp. 238–247.

Special relativity: energy and momentum

3.1 Relativistic Momentum

Momentum, defined as the product of mass and velocity, constitutes one of the fundamental concepts of classical mechanics. Its conservation is well established and underlies its great utility. However, momentum as defined in classical mechanics is not conserved in relativistic mechanics and therefore must be redefined if it is to be useful in relativity. One way to extend momentum conservation to

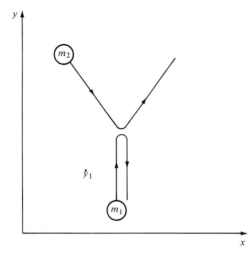

Figure 3.1 An elastic collision between two particles is viewed in the unprimed system. Their masses are equal when they are at rest.

relativity is to retain the mathematical form of its definition as the product of mass and velocity. But such a procedure immediately implies that inasmuch as velocity must now transform according to the Lorentz transformation, the classical notion of mass as an invariant quantity can no longer remain valid. This is precisely what happens when the principle of momentum conservation is generalized to relativistic mechanics: Mass emerges as a relative concept.

The relative nature of mass can be explored by means of a collision problem, such as one depicted in Figure 3.1. Assume that in this unprimed system two colliding particles have identical masses when they are at rest relative to each other, that is, $m_1(0) = m_2(0)$. Before the collision, particle 1 moves in the positive y direction and particle 2 approaches particle 1 at an angle with the x axis. After the collision, which is perfectly elastic, particle 1 reverses its direction and moves in the negative y direction with no change in the magnitude of its velocity $\dot{y}_1$. Particle 2, likewise, reverses its direction such that the y component of its velocity vector remains unchanged in magnitude. Then, in terms of their momenta, whose conservation we wish to retain in relativity, such a collision is described by

$$m_2 \dot{x}_2 = \text{constant}$$
$$m_1 \dot{y}_1 + m_2 \dot{y}_2 = 0 \tag{3.1}$$

where m_1 and m_2 denote the masses of the particles in motion, and are yet to be determined. To this end, Eq. (3.1) can be put in a more useful form

$$\frac{m_2}{m_1} = -\frac{\dot{y}_1}{\dot{y}_2} \tag{3.2}$$

The relationship between $\dot{y}_1$ and $\dot{y}_2$ in Eq. (3.2) is most simply established if the collision is viewed in the primed system moving at

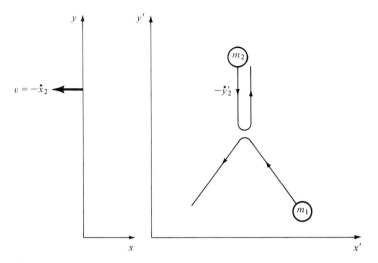

Figure 3.2   The collision of Figure 3.1 is viewed in the primed system moving at $v = \dot{x}_2$ relative to the unprimed system.

$v = \dot{x}_2$ along the x axis (Figure 3.2). In the primed system, particle 2 approaches particle 1 along the negative y' axis and particle 1 approaches particle 2 at an angle with the x' axis. The collision viewed in the primed system is identical in character to that viewed in the unprimed system except for the fact that the role of each particle is interchanged. Thus, for example, the velocity of particle 1 in the unprimed system is equal to the negative of the velocity of particle 2 in the primed system, that is,

$$\dot{y}_1 = -\dot{y}_2' \tag{3.3}$$

The speed $\dot{y}_2'$ of particle 2 in the primed system, on the other hand, is related to the y component $\dot{y}_2$ of its velocity in the unprimed system according to the Lorentz velocity transformation

$$\dot{y}_2' = \frac{\dot{y}_2\sqrt{1 - v^2/c^2}}{1 - v\dot{x}_2/c^2}$$

which, upon substitution of $v = \dot{x}_2$, reduces to

$$\dot{y}_2' = \frac{\dot{y}_2}{\sqrt{1 - \dot{x}_2^2/c^2}} \tag{3.4}$$

Then, the combination of Eqs. (3.3) and (3.4) leads to a relationship between $\dot{y}_1$ and $\dot{y}_2$, which is

$$-\frac{\dot{y}_1}{\dot{y}_2} = \frac{1}{\sqrt{1 - \dot{x}_2^2/c^2}} \tag{3.5}$$

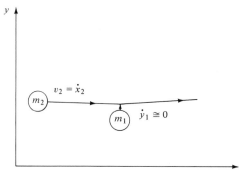

Figure 3.3   The limiting case of the collision of Figure 3.1.

This result, in turn, yields, via Eq. (3.2), a relationship between the masses of the particles in motion:

$$\frac{m_2}{m_1} = \frac{1}{\sqrt{1 - \dot{x}_2^2/c^2}} \qquad (3.6)$$

Note at once that the masses of the particles in motion are not identical, although they were identical when they were at rest relative to each other. This is an important result, and is an inescapable consequence of the principle of momentum conservation.

Now, to explore this strange result about the nature of mass, consider a limiting case of the collision: Let particle 1 be at rest, that is, $\dot{y}_1 = 0$. The motion of particle 2 must then be revised such that it makes a grazing collision with particle 1. In such a limit (see Figure 3.3), the components of the velocity $\mathbf{v}_2$ of particle 2 are $\dot{y}_2 = 0$ and $\dot{x}_2 = v_2 = v$. Furthermore, the mass m_1 of particle 1 in motion approaches in magnitude its mass at rest, which is, in turn, equal to the rest mass of particle 2:

$$m_1 \to m_1(0) = m_2(0) \qquad \dot{y}_1 \to 0 \qquad (3.7)$$

Hence Eq. (3.6) takes the form

$$m_2(v) = \frac{m_2(0)}{\sqrt{1 - v^2/c^2}} \qquad (3.8)$$

Equation (3.8) states that the mass of particle 2 in motion at velocity v is greater than its mass at rest by a factor $(1 - v^2/c^2)^{-1/2}$.

We have thus shown that if momentum is defined as the product of mass and velocity, that is, $\mathbf{p} = m\mathbf{v}$, and the principle of momentum conservation is retained in relativity, the notion of invariant mass must be discarded. Rather, the mass of an object must increase with velocity according to

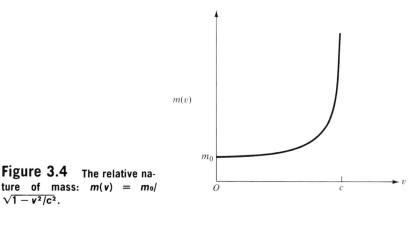

Figure 3.4 The relative nature of mass: $m(v) = m_0/\sqrt{1-v^2/c^2}$.

$$m = \frac{m_0}{\sqrt{1-v^2/c^2}} \tag{3.9}$$

where m and m_0 denote the mass at velocity v and at rest, respectively (Figure 3.4). The quantity m_0 is also referred to as the rest, or proper, mass of a particle. Finally, the explicit form of relativistic momentum is

$$\mathbf{p} = \frac{m_0\mathbf{v}}{\sqrt{1-v^2/c^2}} \tag{3.10}$$

3.2 Relativistic Energy

Suppose we square Eq. (3.9) and invoke the definition of relativistic momentum, $\mathbf{p} = m\mathbf{v}$. Then we are led, after minor rearrangement, to an interesting result, namely,

$$(mc^2)^2 - (pc)^2 = (m_0c^2)^2 \tag{3.11}$$

If we introduce the definitions $E = mc^2$ and $E_0 = m_0c^2$, Eq. (3.11) takes the form

$$E^2 - (pc)^2 = E_0^2 \tag{3.12}$$

Let us now examine Eq. (3.12) in the nonrelativistic approximation, that is, $v \ll c$. Putting Eq. (3.12) in the form

$$(E - E_0)(E + E_0) = p^2c^2$$

and noting that, for a free particle moving at a nonrelativistic speed, the moving mass deviates little from the rest mass, that is,

$$E + E_0 \approx 2\,E_0$$

Equation (3.12) reduces to

$$E - E_0 \approx \frac{p^2}{2m_0} \tag{3.13}$$

Notice that the right-hand side of Eq. (3.13) is precisely the Newtonian expression for the kinetic energy of a particle of mass m_0. Hence the quantity E is the sum of kinetic energy and the quantity $E_0 = m_0 c^2$, whereupon we are tempted to interpret the latter as the rest, or intrinsic, energy of a particle. Then it is natural to interpret the quantity $E = mc^2$ as the total energy of a particle. Such an interpretation, indeed, turns out to be correct, and gives rise to the important result that mass represents the energy content of a physical entity.

3.3 Lorentz Transformations for Energy and Momentum

Let us now examine how the relativistic energy, $E = m_0 c^2 / \sqrt{1 - (v^2/c^2)}$, and the relativistic momentum, $\mathbf{p}\, m_0 \mathbf{v} / \sqrt{1 - (v^2/c^2)}$, transform under the Lorentz velocity transformation. A particle of rest mass m_0 and velocity $\dot{\mathbf{r}}$ in the unprimed system has the following values for energy and momentum:

$$p_x = \frac{m_0 \dot{x}}{\sqrt{1 - \dot{r}^2/c^2}}$$

$$p_y = \frac{m_0 \dot{y}}{\sqrt{1 - \dot{r}^2/c^2}}$$

$$E = \frac{m_0 c^2}{\sqrt{1 - \dot{r}^2/c^2}} \tag{3.14}$$

Viewed in a primed system moving at velocity v along the x axis, the particle has primed values $\mathbf{p}'$ and E' for momentum and energy (Figure 3.5). A relationship between the primed and unprimed values of energy and momentum may be obtained by noting that the energy of the particle in the primed system is

$$E' = \frac{m_0 c^2}{\sqrt{1 - (\dot{r}'/c)^2}}$$

$$= \frac{m_0 c^2}{\sqrt{1 - (\dot{x}'^2 + \dot{y}'^2)/c^2}} \tag{3.15}$$

The radicant in Eq. (3.15), whose parts, under the Lorentz velocity transformation, become

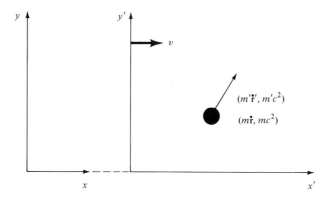

Figure 3.5 Momentum and energy of a particle: (p, E) in the unprimed system and (p', E') in the primed system.

$$1 - \left(\frac{\dot{x}'}{c}\right)^2 = 1 - \frac{(\dot{x} - v)^2}{c^2(1 - \dot{x}v/c^2)^2} \tag{3.16}$$

and

$$\left(\frac{\dot{y}'}{c}\right)^2 = \frac{(\dot{y}\sqrt{1 - v^2/c^2})^2}{c^2(1 - \dot{x}v/c^2)^2} \tag{3.17}$$

takes the form:

$$1 - \left(\frac{\dot{r}'}{c}\right)^2 = \frac{(1 - \dot{r}^2/c^2)(1 - v^2/c^2)}{(1 - \dot{x}v/c^2)^2} \tag{3.18}$$

Substitution of Eq. (3.18) into Eq. (3.15), then, yields the Lorentz transformation for energy:

$$E' = \frac{m_0 c^2/\sqrt{1 - \dot{r}^2/c^2} - m_0\dot{x}v/\sqrt{1 - \dot{r}^2/c^2}}{\sqrt{1 - v^2/c^2}}$$

or

$$E' = \frac{E - vp_x}{\sqrt{1 - v^2/c^2}} \tag{3.19}$$

The Lorentz transformation for the x' component of momentum emerges when Eq. (3.18) is applied to the definition of momentum. That is,

$$p'_x = \frac{m_0\dot{x}'}{\sqrt{1 - (\dot{r}'/c)^2}}$$

$$= \frac{m_0(\dot{x} - v)}{\sqrt{1 - \dot{r}^2/c^2}\,\sqrt{1 - v^2/c^2}}$$

or

$$p'_x = \frac{p_x - vE/c^2}{\sqrt{1 - v^2/c^2}}$$ (3.20)

The transformation of the y' component of momentum is similarly obtained:

$$p'_y = \frac{m_0 \dot{y}'}{\sqrt{1 - (\dot{r}'/c)^2}}$$

$$= \frac{m_0 \dot{y} \sqrt{1 - v^2/c^2}}{\sqrt{1 - (\dot{r}'/c)^2} \, (1 - \dot{x}v/c^2)}$$

or

$$p'_y = p_y = \frac{m_0 \dot{y}}{\sqrt{1 - \dot{r}^2/c^2}}$$ (3.21)

Notice that the transformations of energy and momentum are of exactly the same form as those of the time and space coordinates, which are

$$t' = \frac{t - vx/c^2}{\sqrt{1 - v^2/c^2}}$$

$$x' = \frac{x - vt}{\sqrt{1 - v^2/c^2}}$$ (3.22)

$$y' = y$$

Furthermore, there exists a Lorentz invariant E_0 for energy and momentum, which is defined by

$$E_0^2 = E^2 - (pc)^2 = E'^2 + (p'c)^2$$
$$= (m_0 c^2)^2$$ (3.23)

Thus relativistic energy and momentum are linked intimately so as to constitute the so-called four-vector momentum, and energy and momentum are referred to as the time and space components of the four-vector momentum. Accordingly, if momentum is conserved, energy must also be conserved. Now the four-vector momentum is conserved in all real processes, and this fact gives rise to the principle of conservation of energy-momentum.

3.4   Energy and Momentum of Radiation

The electromagnetic wave transports both energy and momentum. To establish this fact, consider a particle of charge q, and expose it to an electromagnetic wave propagating in the x direction. Such a particle experiences a Lorentz force:

$$\mathbf{F} = q\mathcal{E} + \frac{q\mathbf{v} \times \mathbf{B}}{c} \qquad (3.24)$$

In Eq. (3.24) $\mathcal{E}$ and $\mathbf{B}$ are the electric and magnetic fields of incident radiation, and $\mathbf{v}$ is the velocity of the charged particle. If the particle is initially at rest, and the radiation is polarized such that $\mathcal{E} = \mathcal{E}\mathbf{e}_y$ and $\mathbf{B} = B\mathbf{e}_z$, the Lorentz force gives rise to a velocity $\dot{y}\mathbf{e}_y$, which is parallel to the electric field $\mathcal{E}$. (Here $\mathbf{e}$ is a unit vector.) Once in motion, the particle experiences a magnetic force in the direction of wave propagation, and thus gains a small velocity component $\dot{x}\mathbf{e}_x$. As a result, the velocity of a charged particle in response to radiation has both x and y components:

$$\mathbf{v} = \dot{x}\mathbf{e}_x + \dot{y}\mathbf{e}_y \qquad (3.25)$$

Hence, the Lorentz force can be written in the form

$$\mathbf{F} = q\mathcal{E}\mathbf{e}_y + \frac{q\dot{y}B}{c}\mathbf{e}_x - \frac{q\dot{x}B}{c}\mathbf{e}_y \qquad (3.26)$$

Consider now the time average of the Lorentz force over one cycle. The longitudinal velocity component $\dot{x}$ of the particle is small and increases very slowly compared to its transverse velocity component $\dot{y}$. Hence the third term in Eq. (3.26), which contains $\dot{x}$, may be neglected. The electric field $\mathcal{E}$ is a sinusoidal function, and consequently the time average of $\mathcal{E}$ over one cycle is zero, that is,

$$\frac{\int_0^{2\pi} \mathcal{E}\, dt}{2\pi} = 0 \qquad (3.27)$$

which implies that the first term of Eq. (3.26) also averages to zero. Therefore Eq. (3.26), upon time-averaging over one cycle, reduces to

$$\langle \mathbf{F} \rangle = \frac{q}{c}\langle \dot{y}B \rangle \mathbf{e}_x \qquad (3.28)$$

Equation (3.28) implies that the time-averaged motion of the charged particle is described by

$$\left\langle \frac{d\mathbf{p}}{dt} \right\rangle = \frac{q}{c}\langle \dot{y}B \rangle \mathbf{e}_x \qquad (3.29)$$

where $\mathbf{p}$ is the momentum of the particle.

Now the rate at which electromagnetic energy is delivered to a charged particle is given by

$$\frac{dE}{dt} = \mathbf{v} \cdot \mathbf{F}$$

which, by use of Eqs. (3.24) and (3.25), reduces to

$$\frac{dE}{dt} = q\dot{y}\mathscr{E} \tag{3.30}$$

For a traveling wave, the electric and the magnetic fields are equal in magnitude, that is, $|\mathscr{E}| = |\mathbf{B}|$, so that Eq. (3.30) can be rewritten as

$$\frac{dE}{dt} = q\dot{y}B \tag{3.31}$$

or, upon time-averaging,

$$\left\langle \frac{dE}{dt} \right\rangle = q\langle \dot{y}B \rangle \tag{3.32}$$

Substitution of Eq. (3.32) into Eq. (3.29), then, yields

$$\left\langle \frac{d\mathbf{p}}{dt} \right\rangle = \frac{1}{c} \left\langle \frac{dE}{dt} \right\rangle \mathbf{e}_x \tag{3.33}$$

which says that if the charged particle removes energy E from the incident radiation during some finite interval of time, the particle must necessarily remove a linear momentum $\mathbf{p}$ from the radiation during the same time interval. It is thus natural to ascribe energy and momentum to electromagnetic radiation. Furthermore, Eq. (3.33) gives rise to the important result that the energy and momentum delivered by radiation obey the relation

$$\mathbf{p}c = E\mathbf{e}_x$$

or

$$pc = E \tag{3.34}$$

Notice now that the energy-momentum invariant E_0 for electromagnetic radiation is zero. That is, from Eq. (3.34):

$$E_0^2 = E^2 - p^2c^2 = 0 \tag{3.35}$$

It is thus tempting to view the electromagnetic wave as a physical entity having zero rest mass, that is, $E_0 = 0$. Such a view is, at least, consistent with relativity.

3.5   The Photoelectric Effect

Observed first by H. Hertz in 1887, the photoelectric effect refers to a phenomenon of electron emission from metal surfaces by irradiation with light. The identification of the emitted particles as electrons was made by P. Lenard in 1900.

The interpretation of the phenomenon was initially sought in the electromagnetic theory of light. The electromagnetic description of the photoelectric effect proceeded as follows: Atomic electrons in the metal are set in oscillatory motion by incident radiation, and the amplitude of the electron's oscillation is determined by the amplitude $\mathscr{E}_0$ of the electric field. Now the square of the field amplitude, $\mathscr{E}_0^2$, defines the intensity of radiation. The square of the amplitude of the electron's oscillation, on the other hand, gives a quantity proportional to the kinetic energy of the electron. Hence the kinetic energy of the electrons emitted by irradiation with light is proportional to the intensity of incident radiation.

Another prediction of the electromagnetic theory is that the absorption rate of energy by an atom in the photoelectric effect is also proportional to the intensity of incident radiation. This is easily seen by noting that if an atom exposes an area A to radiation of intensity I, then the product of A and I gives the rate at which energy is absorbed by the atom. Thus the more intense the radiation, the quicker the electron absorbs the energy required for ejection. Put another way, the delay time between initial irradiation and final ejection is inversely proportional to the intensity of radiation.

Neither prediction of the electromagnetic theory agrees with observation. It was established by Lenard in 1902 that the maximum kinetic energy of the electrons emitted upon irradiation is independent of the intensity of incident radiation. It was also noticed early in the study of the photoelectric effect that there was no appreciable time lag between irradiation and electron emission, and that the time delay was roughly independent of light intensity. (In 1928, E. O. Lawrence and J. W. Beams established an upper limit of 3×10^{-9} sec on the delay time in the photoelectric effect.) These were serious discrepancies, and although there were some aspects of the phenomenon that were successfully described by the electromagnetic theory, they pointed to some fundamental flaws in the theory.

In 1905, Einstein proposed a quantum theory of the photoelectric effect. The quantum hypothesis was first enunciated in 1900 by Max

Planck as a way of explaining blackbody radiation. The quantum hypothesis represented a radical departure from classical physics in that it replaced the classical notion of the continuum with the notion of discreteness. Specifically, Planck assumed that the energy of the atomic oscillator does not constitute a continuum of states but takes on a discrete set of values given by

$$E_n = nh\nu, \quad n = 0, 1, 2, 3, \ldots \tag{3.36}$$

where h is a proportionally constant, and ν is the frequency of oscillation. The proportionality constant h is now called **Planck's constant.**

Einstein proposed that not only the oscillator but also the radiation it emits and absorbs be quantized. He envisioned, for example, that a transition from the $nh\nu$ state to the $(n - 1)h\nu$ state in the oscillator would be achieved by the emission of a burst of radiation whose energy is $h\nu$. He thus introduced the notion of "graininess" to electromagnetic radiation. He pictured the electromagnetic field transporting energy not in a continuous manner as in Maxwell's theory but rather in "lumps," that is, in the form of particles. These localized packets of radiant energy are called photons. The energy E of a photon is then determined by its frequency ν:

$$E = h\nu \tag{3.37}$$

where h is Planck's constant with a numerical value

$$h = 6.626 \times 10^{-27} \text{ erg sec}$$
$$= 4.136 \times 10^{-15} \text{ eV sec}$$

In the photon view of radiation, the photoelectric process is pictured as follows. A photon imparts all of its energy to an electron in the material under irradiation. The ejected electron then carries a kinetic energy given by

$$\tfrac{1}{2}mv^2 = h\nu - W \tag{3.38}$$

where W is the binding energy of the electron. Electrons are bound to the surface with different energies, and the lowest value of the binding energies is called the **work function** W_0 of the material. The photoelectric, equation can then be expressed in the form:

$$\tfrac{1}{2}mv_{max}^2 = h\nu - W_0 \tag{3.39}$$

where v_{max} is the maximum velocity that is reached by the photoelectrons. It is thus clear that the kinetic energy of the photoelectron is a function of the frequency of the incident radiation and not of its intensity. Furthermore, the time lag between irradiation and ejection of an electron is also independent of the intensity of radiation, and is negligible because the interaction between the photon and the electron can be likened to a collision. Finally, the field amplitude of radiation is interpreted as a measure of the photon flux; specifically, the square of the field amplitude is proportional to the photon flux, and the intensity of radiation is identical to the product of the photon flux and the energy of the photons.

Einstein's photoelectric equation was experimentally verified by R. A. Millikan in 1916. Millikan measured the maximum kinetic energies of photoelectrons as a function of the frequency of incident radiation, and found the experimental points to fall on a straight line in agreement with Einstein's prediction. Thus the existence of the photon, that is, the quantization of radiant energy, was beyond doubt. Furthermore, the slope of the straight line yielded a numerical value of h, which agreed within 0.5% with the value of h determined by Planck from an analysis of the blackbody radiation spectrum. This was, of course, a profound result, for Planck's constant could now be elevated to the status of a universal constant.

3.6   Energy and Momentum of the Photon

Let us now ask how the energy of a photon transforms from one inertial frame to another. Since Planck's constant remains invariant in all frames, our question bears on the transformation of frequency.

Consider Figure 3.6 which shows a light source at the origin O' of the primed system. The primed system, as usual, moves at velocity v in the x direction of the unprimed system. Let it emit a light pulse in

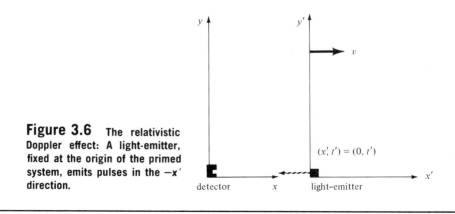

Figure 3.6 The relativistic Doppler effect: A light-emitter, fixed at the origin of the primed system, emits pulses in the $-x'$ direction.

the negative x' direction at $t_1' = 0$. The event of light emission, then, has the following unprimed coordinates:

$$x_1 = \frac{x_1' + vt_1'}{\sqrt{1 - v^2/c^2}} = 0$$

$$t_1 = \frac{t_1' + vx_1'/c^2}{\sqrt{1 - v^2/c^2}} = 0 \qquad \textbf{(3.40)}$$

Let a second pulse be emitted at $t_2' = t_1' + \tau'$ where τ' denotes the period of oscillation of the light pulse in the primed system. The unprimed coordinates of the event are

$$x_2 = \frac{x_2' + vt_2'}{\sqrt{1 - v^2/c^2}} = \frac{v\tau'}{\sqrt{1 - v^2/c^2}}$$

$$t_2 = \frac{t_2' + vx_2'/c^2}{\sqrt{1 - v^2/c^2}} = \frac{\tau'}{\sqrt{1 - v^2/c^2}} \qquad \text{.3.41)}$$

Note that t_2 is the time, in the unprimed system, at which the second pulse is emitted.

Suppose now that a detector of the light pulses is placed at the origin O of the unprimed system. The first pulse arrives at the detector at t_1, because the pulse spends no time in transit, that is, $x_1/c = 0$. The second pulse, which is emitted at t_2, traverses a distance x_2 before reaching the detector, and thus spends a time, $\Delta t = x_2/c$, in transit. Consequently, the time lapse τ between the arrival of the first pulse and that of the second pulse at the detector is

$$\tau = (t_2 - t_1) + \Delta t \qquad \textbf{(3.42)}$$

which, from use of Eqs. (3.40) and (3.41), reduces to

$$\tau = \frac{\tau'(1 + v/c)}{\sqrt{1 - v^2/c^2}} \qquad \textbf{(3.43)}$$

The time interval τ is, of course, the period of oscillation of the pulses as observed in the unprimed system. Now the frequency of oscillation is the reciprocal of the period, that is, $\nu = 1/\tau$, and thus:

$$\nu = \nu' \frac{\sqrt{1 - v^2/c^2}}{1 + v/c}$$

or

$$\nu = \nu' \frac{1 - v/c}{\sqrt{1 - v^2/c^2}} \qquad \textbf{(3.44)}$$

Equation (3.44) describes the logitudinal Doppler shift that arises from the motion of the light source away from the observer. A similar analysis of a light pulse, propagating in the positive x' direction, leads to an equation for a Doppler shift that occurs when the light source approaches the detector:

$$\nu = \nu' \frac{1 + v/c}{\sqrt{1 - v^2/c^2}} \tag{3.45}$$

Now the energy of a photon is proportional to its frequency, that is, $E = h\nu$. Consequently, the Doppler shift equations, namely, Eqs. (3.44) and (3.45), describe, in essence, the transformation of the energy of the photon from one frame to another frame. Specifically,

$$E = \frac{E' + vE'/c}{\sqrt{1 - v^2/c^2}}$$
$$E' = \frac{E - vE/c}{\sqrt{1 - v^2/c^2}} \tag{3.46}$$

Notice that Eq. (3.46) is in the same form as Eq. (3.19), and we are thus led to a relation between the energy and momentum of a photon:

$$p = \frac{E}{c}$$
$$p' = \frac{E'}{c} \tag{3.47}$$

Equation (3.47), then, implies that the energy-momentum invariant E_0 for a photon is zero:

$$E^2 - p^2c^2 = E'^2 - p'^2c^2 = 0 \tag{3.48}$$

Recall now that the energy-momentum invariant for a material particle gives its rest energy. The photon is thus referred to as a particle of zero rest mass, although a photon at rest is, strictly speaking, without meaning.

It is interesting to note that the momentum of a photon, given by Eq. (3.47), can be expressed in terms of its wavelength via the well-known relation for electromagnetic radiation, $\lambda\nu = c$. It is

$$p = \frac{h}{\lambda} \tag{3.49}$$

Equation (3.49) is a natural result of Einstein's photoelectric equation and relativity.

3.7 Compton Scattering

A decisive experimental proof of the photon nature of radiation was provided by A. H. Compton in 1923. Compton bombarded a graphite target with x rays from a molybdenum tube, and measured the angle-dependent energies of the scattered radiation. The scattered radiation was observed to consist of two components for a given angle; the first component had the same wavelength as the incident radiation, and the second component had a longer wavelength. The results were then analyzed by treating radiation as photons.

Suppose a photon of frequency ν collides with an electron at rest. Upon collision, the electron recoils at an angle ϕ, and the scattered photon of frequency ν' emerges at an angle θ, relative to the incident direction (Figure 3.7). The collision conserves energy and momentum and thus

$$E + m_0c^2 = E' + mc^2 \tag{3.50}$$
$$\mathbf{p} - \mathbf{p}' = \mathbf{p}_e \tag{3.51}$$

If Eq. (3.50) is squared in the form

$$(E + m_0c^2 - E')^2 = (mc^2)^2 \tag{3.52}$$

and the square of Eq. (3.51), multiplied by c^2, is subtracted from Eq. (3.52), the result is

$$[(E + m_0c^2) - E']^2 - (\mathbf{p} - \mathbf{p}')^2c^2 = (mc^2)^2 - p_e^2c^2$$
$$= (m_0c^2)^2 \tag{3.53}$$

which can be cast in the form

$$(E^2 - p^2c^2) + (E'^2 - p'^2c^2) - 2E'(E + m_0c^2) + 2\mathbf{p} \cdot \mathbf{p}'c^2$$
$$+ 2Em_0c^2 = 0 \tag{3.54}$$

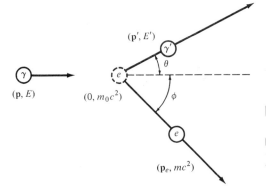

Figure 3.7 A schematic diagram of the Compton effect: A photon undergoes an elastic collision with a stationary electron. The photon scatters and the electron recoils.

Note that the first two terms in Eq. (3.54) are zero. Now, if the dot product is written out explicitly, that is, $\mathbf{p} \cdot \mathbf{p}'c^2 = pp'c^2 \cos \theta$, Eq. (3.54) can be put in the form

$$\frac{E - E'}{pp'c^2} = \frac{1}{m_0 c^2} (1 - \cos \theta)$$

or

$$\frac{1}{p'} - \frac{1}{p} = \frac{1}{m_0 c} (1 - \cos \theta) \tag{3.55}$$

It is more useful to express Eq. (3.55) in terms of the wavelengths of the incident and the scattered photons, and this can be done by use of Eq. (3.49), which gives

$$\lambda' - \lambda = \frac{h}{m_0 c} (1 - \cos \theta) \tag{3.56}$$

The quantity $h/m_0 c$ is called the Compton wavelength of the electron, and has the numerical value of

$$\frac{h}{m_0 c} = 2.43 \times 10^{-10} \text{ cm} = 0.024 \text{ Å}$$

Equation (3.56) described successfully the increase in the wavelength of the radiation scattered by a free electron (Figure 3.8). The radiation scattered by the carbon nucleus, on the other hand, suffered no change in its wavelength, because the Compton wavelength of a nucleus is much smaller than 0.7 Å, which was the wavelength of the photons in Compton's original experiment.

Note that Compton treated the photon as a whole particle. Just as the incident photon is a whole photon, so also is the scattered photon; it is not a fraction of a photon. If a fraction of a photon were scattered, the conservation laws would give a result quite different from the Compton result. In 1950 Cross and Ramsey[1] reported that when the direction of the scattered photons is specified, the Compton electrons recoil in the direction required by the conservation laws within ±1 degree. The view that the Compton effect is a result of a direct interaction between the photon and the electron is further supported by the observation of Hofstadter and McIntyre[2] that the time lag between

[1] W. G. Cross and N. F. Ramsey, "The Conservation of Energy and Momentum in Compton Scattering," *Physical Review,* **80:**929–936 (1950).
[2] R. Hofstadter and J. A. McIntyre, "Simultaneity in the Compton Effect," *Physical Review,* **78:**24–28 (1950).

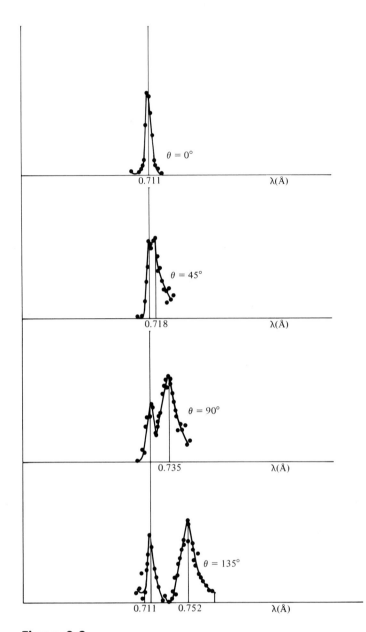

Figure 3.8 Compton's experimental results [*Phy. Rev.*, **22**:409 (1923)]: Intensity of scattered radiation versus wavelengths of scattered radiation at four different scattering angles.

the emission of the electron and that of the scattered photon is less than 1.5×10^{-8} sec. The Compton effect can, therefore, be viewed as a decisive phenomenon in support of the quantization of radiant energy.

3.8 Equivalence of Mass and Energy

Consider a light source in the primed system (Figure 3.9). If it emits two pulses of equal frequencies in opposite directions along the x' axis, the total energy of the pulses is

$$E' = h\nu'_+ + h\nu'_- = 2h\nu' \tag{3.57}$$

In the unprimed system, the same pulses are not of equal frequencies because of the Doppler effect, and the total energy of the pulses is

$$E = h\nu_+ + h\nu_- \tag{3.58}$$

which, upon substitution of Eqs. (3.44) and (3.45), reduces to

$$E = \frac{E'}{\sqrt{1 - v^2/c^2}} \tag{3.59}$$

Equation (3.59) simply says that the light emitter in the primed system emits more energy to the unprimed observer than to the primed observer. It is thus an equation that describes the relative character of the energy of radiation.

In order to explore the implications of the notion of relative energy, let us calculate the quantity $\Delta E = E - E'$ in the nonrelativistic approximation, that is, $v \ll c$. The result is

$$\Delta E = E' \left(\frac{v^2}{2c^2} \right) \tag{3.60}$$

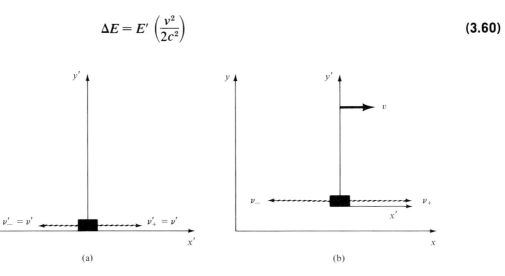

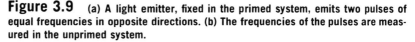

Figure 3.9 (a) A light emitter, fixed in the primed system, emits two pulses of equal frequencies in opposite directions. (b) The frequencies of the pulses are measured in the unprimed system.

The quantity ΔE, which is the extra energy observed in the unprimed system, has its origin in the motion of the light emitter, and must therefore be associated with the kinetic energy of the emitter. But kinetic energy depends only on two quantities, namely, mass and the square of velocity. Thus the origin of energy can be traced either to mass or velocity. However, because an emitter at rest can still radiate, the origin of energy must be traced to mass. That is, energy must ultimately arise from the conversion of mass. Now if E' comes from the conversion of mass m', then the extra energy observed in the unprimed system must come from the kinetic energy of m', that is,

$$\Delta E = \tfrac{1}{2} m' v^2 \tag{3.61}$$

The precise relationship between energy and mass, then, emerges from Eqs. (3.60) and (3.61):

$$E' = m' c^2 \tag{3.62}$$

Equation (3.62) establishes the equivalence of energy and mass.

An example of the conversion of mass into energy is the annihilation of an electron-positron pair. A positron is the antimatter counterpart of an electron, and is equal in mass and opposite in charge to the electron. The conversion of mass into radiant energy occurs when the particles come in close proximity. Annihilation at rest produces two photons moving in opposite directions, a consequence of the conservation of momentum. Also, the sum of the photon energies is, as required by the conservation of mass-energy, precisely equal to the sum of the rest energies of the particles.

The creation of an electron-positron pair constitutes an example of conversion of radiant energy into mass. Pair creation occurs when an energetic photon enters the Coulomb field of a nucleus. Pair creation from a photon does not occur in empty space because such a process cannot simultaneously conserve mass-energy and momentum, and is thus forbidden.

Most elementary particles are created from energy. And all unstable particles decay into stable particles by converting mass into energy. Neither mass nor energy is now separately conserved; it is mass-energy that is conserved in all processes.

Example 3.1 Calculate the energies of the gamma rays produced by the annihilation of an electron-positron pair at rest.

Solution. The principle of momentum conservation dictates that since the total momentum of an electron-positron pair at rest is zero, the total momentum of gamma rays must also be zero. Hence there must be produced at least two gamma rays moving in opposite directions with equal momenta. Since $E = pc$ for a photon, the two gamma rays must carry equal energies, and each gamma ray carries an energy equal to the rest energy of the electron, that is,

$$E_\gamma = m_0 c^2 = 0.51 \text{ MeV} \quad \bullet$$

Example 3.2 Suppose that an electron collides with another electron at rest. Calculate the minumum kinetic energy of the incident electron required for such a reaction to produce a pair of electron and positron.

Solution. The conservation of the so-called electron family number, which we merely state here without further explanation, requires that the products of such a reaction consist of three electrons and one positron

$$e + e \rightarrow e + e + e + \bar{e}$$

where $\bar{e}$ denotes a positron. The total momentum of the electrons before the collision is zero in the center-of-mass system, and so the total momentum of the four particles after the collision must also be zero in the center-of-mass system. Furthermore, the four product particles will be at rest in the center-of-mass system inasmuch as the incident electron provides only the minimum kinetic energy required for such a reaction. In the laboratory system, then, the four product particles will all have the same velocity and may thus be regarded as a "particle" of rest mass $4m_0$. The momentum p of the incident electron will then equal the momentum of this collective "particle." Consequently, the conservation of energy for the reaction, which says that the total energy of the two electrons before the collision must equal the total energy of the collective "particle" after the collision, can now be written as

$$\sqrt{p^2 c^2 + E_0^2} + E_0 = \sqrt{p^2 c^2 + (4E_0)^2}$$

Squaring both sides and rearranging terms, we obtain

$$\sqrt{p^2 c^2 + E_0^2} = 7E_0$$

which implies that the total energy of the incident electron must be $7E_0$. Hence the minimum kinetic energy of the incident electron required for the reaction is

$$K = E - E_0 = 6E_0 \quad \bullet$$

Example 3.3 Although the conversion of a 1.02 MeV gamma ray into a pair of electron and positron is energetically possible, explain why such a process does not take place in empty space.

Solution. The rest energy of an electron is 0.51 MeV, and hence the threshold energy for pair creation is 1.02 MeV. At the threshold energy the electron and the positron are created with zero kinetic energies and consequently with zero momenta. However, the momentum of the incident gamma ray is $p = 1.02$ MeV/c, and thus such a process violates the conservation of momentum. As a result, pair creation at the threshold energy does not occur in empty space. In order for such a process to occur, there must exist a massive charged particle, such as a nucleus, which will carry the momentum of the gamma ray without any appreciable kinetic energy. Thus pair creation processes from gamma rays take place in the neighborhoods of nuclei.

3.9 Summary

The requirement that the principle of momentum conservation be valid in relativity leads to the result that mass must be a relative quantity, namely,

$$m = \frac{m_0}{\sqrt{1 - v^2/c^2}}$$

The definition of momentum in relativity is

$$\mathbf{p} = m\mathbf{v} = \frac{m_0\mathbf{v}}{\sqrt{1 - v^2/c^2}}$$

An important relation among the total energy E, the rest energy E_0, and the momentum $\mathbf{p}$ of a particle is

$$E^2 = p^2c^2 + E_0^2$$

The kinetic energy of a relativistic particle is given by

$$K = E - E_0$$

The total energy E and the momentum $\mathbf{p}$ of a particle transform in a manner similar to the time and the space coordinates, and give rise to a Lorentz invariant E_0 defined by

$$E_0^2 = E^2 - p^2c^2$$

The energy-momentum relation for a photon is

$$E^2 - p^2c^2 = 0 \quad \text{or} \quad E = pc$$

Hence a photon may be regarded as a physical entity of zero rest mass. The energy of a photon is determined by its frequency ν

$$E = h\nu$$

where the proportionality constant h is Planck's constant

$$h = 6.626 \times 10^{-27} \text{ erg-sec}$$
$$= 4.136 \times 10^{-15} \text{ eV-sec}$$

The momentum of a photon is determined by its wavelength

$$p = \frac{h}{\lambda}$$

The photoelectric effect refers to a phenomenon in which a photon imparts its entire energy to an electron bound to a metal surface, thus releasing it. Einstein's photoelectric equation is

$$\tfrac{1}{2}mv^2_{\text{max}} = h\nu - W_0$$

where W_0 is called the work function of a metal.

In the Compton effect a photon scatters off of a free electron, imparting only a part of its energy to the electron. The wavelength of the scattered photon is given by

$$\lambda' - \lambda = \frac{h}{m_0 c}(1 - \cos\theta)$$

and the kinetic energy of the recoiling electron is given by

$$K = h\nu \left(1 - \frac{\lambda}{\lambda'}\right)$$

The numerical value of the so-called Compton wavelength of the electron is

$$\frac{h}{m_0 c} = 0.0243 \text{ Å}$$

The relativistic Doppler equations are

$$\nu = \nu' \frac{1 \pm v/c}{\sqrt{1 - v^2/c^2}}$$

By use of these equations Einstein derived the formula, $E = mc^2$, establishing the equivalence of mass and energy. Examples of the total conversion of mass into energy and energy into mass are the annihilation of an electron-positron pair and the creation of an electron-positron pair by a gamma ray in the neighborhood of a massive charged particle.

Problems

3.1 Calculate the energy in electron volts that would be required to accelerate an electron from $v = 0$ to $v = 0.90c$ and from $v = 0.90c$ to $v = 0.99c$.

3.2 It is often convenient to express momentum in units of MeV/c. What is the kinetic energy of an electron whose momentum is **(a)** 1.0 MeV/c and **(b)** 10.0 MeV/c.

3.3 Calculate the nonrelativistic and the relativistic kinetic energies of an electron at **(a)** $v = 0.1c$, **(b)** $v = 0.5c$, and **(c)** $v = 0.9c$.

3.4 Show that the rest energy of a particle is a relativistic invariant.

3.5 Show that the velocity of a particle having a kinetic energy K is given by

$$v = c\left[1 - \frac{E_0^2}{(E_0 + K)^2}\right]^{1/2}$$

where E_0 is the rest energy of the particle.

3.6 What is the total energy of a photon whose momentum is 5 MeV/c?

3.7 The red light of the helium-neon laser has a wavelength of 6328 Å. What is the energy of the photons making up the laser beam?

3.8 The work function of potassium is 2.20 eV. If the surface of potassium is irradiated with ultraviolet radiation of wavelength 3500 Å, calculate the maximum kinetic energy of the emitted photoelectrons.

3.9 The rest energy of a neutral pion is 134.98 MeV.
  (a) If a neutral pion at rest decays into two gamma rays, what are the energies of the photons?
  (b) If the pion decays in flight at $v = 0.9c$, what are the energies of the two gamma rays?

3.10 Calculate the momentum of a photon whose wavelength is **(a)** 6000 Å, **(b)** 1.0 Å, and **(c)** 1.4×10^{-13} cm.

3.11 **(a)** If a 100 MeV photon is scattered by a proton, calculate the energy of the photon scattered through 90° from the incident direction.
  (b) Calculate the velocity of the recoiling proton. (The rest energy of the proton is 938.26 MeV.)

3.12 An x-ray photon of wavelength 0.711 Å is scattered by a free electron at an angle 135° from the incident direction. What is the direction of the recoiling electron?

3.13 A charged pion decays into a muon and a neutrino. If it decays at rest, calculate the energy of the emitted neutrino. (The rest energies of the pion, the muon, and the neutrino are 140 MeV, 106 MeV, and 0, respectively.)

3.14 Show that a free electron cannot absorb a photon.

3.15 Suppose that a photon, upon striking an electron at rest, converts into a pair of electron and positron. What minimum energy would such a photon require in order to produce an electron-positron pair?

3.16 An antiproton can be created simultaneously with a proton in a collision where a proton at rest is struck by an energetic proton. In such a reaction, one starts with two protons and ends up with three protons and one antiproton, that is,

$$p + p \rightarrow p + p + p + \bar{p}.$$

If the rest mass of the antiproton is equal to the rest mass of the proton, calculate the minimum kinetic energy of the incident proton that would be required to create an antiproton.

Suggestions for Further Reading

Feynman, R. P., R. B. Leighton, and L. M. Sands. *The Feynman Lectures on Physics,* vol. I. Reading, Mass.: Addison-Wesley Publishing Company, 1963. Chapters 15, 16, and 17.

Ford, K. W. *Classical and Modern Physics,* vol. 3. Lexington, Mass.: Xerox College Publishing, 1974. Chapters 20, 21, and 23.

French, A. P. *Special Relativity.* New York: W. W. Norton & Company, Inc., 1968.

Wichmann, E. H. *Quantum Physics: Berkeley Physics Course,* vol. 4. New York: McGraw-Hill Book Company, 1967. Chapters 1, 2, and 4.

The wave nature of radiation and matter

4.1   The Wave Nature of Light

The wave theory of light had its origin in the work of Huygens in the late 1600s, and its confirmation came in the early 1800s through Young's and Fresnel's demonstrations of interference and diffraction phenomena. The experimental works of Coulomb, Ampère, and Faraday on seemingly unrelated electromagnetic phenomena, then, laid the foundation for Maxwell's profound discovery in 1865 that light waves are electromagnetic in nature—that light is a propagating electromagnetic field satisfying a wave equation.

Fundamental to the wave nature of light is its interference property, and this property arises from the linear character of Maxwell's wave equation. The linearity of Maxwell's wave equation implies that if $\mathscr{E}_1$ and $\mathscr{E}_2$ are solutions of the wave equation, then their sum $\mathscr{E}_1 + \mathscr{E}_2$ must also be a solution. In physical terms, this means that if $\mathscr{E}_1$ and $\mathscr{E}_2$ describe two light waves, then $\mathscr{E}_1 + \mathscr{E}_2$ must also describe a physically realizable situation. This fact gives rise to the important conclusion that light waves obey the superposition principle. And the interference pattern of waves is a direct consequence of the superposition principle that underlies all wave phenomena.

An electromagnetic wave of angular frequency $\omega = 2\pi\nu$ and wave number $k = 2\pi/\lambda$, propagating in the x direction, is described by the electric and magnetic fields of the form

$$\mathscr{E} = \mathscr{E}_0 \cos(\omega t - kx + \phi)$$
$$B = B_0 \cos(\omega t - kx + \phi) \tag{4.1}$$

where $\mathscr{E}_0$ and B_0 denote the field amplitudes and ϕ is the phase constant. When all the electromagnetic fields of the same relative phase from a source superpose so as to reach a constructive interference maximum, such gives rise to a beam of monochromatic radiation. The intensity of the beam is determined by the time-averaged square of the resultant electric field

$$I\left(\frac{\text{erg}}{\text{cm}^2 \text{ sec}}\right) \propto \langle \mathscr{E}^2 \rangle \tag{4.2}$$

The superposing character of waves is so important that it is worth reviewing the essence of the familiar two-slit experiment. Let light waves of angular frequency ω and wave number k illuminate an opaque screen with two narrow slits, as in Figure 4.1. The slit separation d is comparable to the wavelength λ of the waves. If the light waves are observed at a point far away from the slits, that is, $r \gg d$, then the light waves at the point of observation can be described by the electric fields of the form

$$\mathscr{E}_1 = \mathscr{E}_0(r_1) \cos(\omega t - kr_1 + \phi_1)$$
$$\mathscr{E}_2 = \mathscr{E}_0(r_2) \cos(\omega t - kr_2 + \phi_2) \tag{4.3}$$

where ϕ_1 and ϕ_2 are the phase constants. Since the relative phase constant does not change with time, the slits may be regarded as coherent sources of light.

In the far-field approximation of the superposition of $\mathscr{E}_1$ and $\mathscr{E}_2$, the distances from the slits to the point of observation r_1 and r_2 are nearly equal, so that the field amplitudes can be evaluated at the aver-

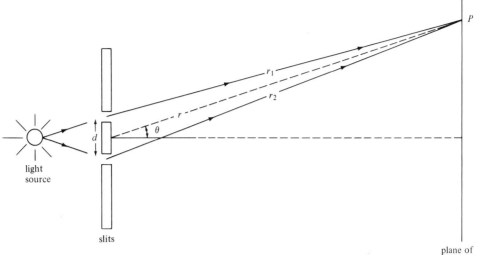

Figure 4.1 A schematic diagram of a two-slit interference experiment.

age value of r_1 and r_2, namely, at r. However, the path difference $(r_2 - r_1)$ is not zero, and is given by

$$r_2 - r_1 = d \sin \theta \qquad (4.4)$$

If the path difference is an integral multiple of the wavelength, that is,

$$d \sin \theta = m\lambda \qquad m = 0, 1, 2, 3, \ldots \qquad (4.5)$$

$\mathscr{E}_1$ and $\mathscr{E}_2$ interfere constructively. If the path difference is an integral multiple of half a wavelength,

$$d \sin \theta = (m + \tfrac{1}{2})\lambda \qquad m = 0, 1, 2, \ldots \qquad (4.6)$$

$\mathscr{E}_1$ and $\mathscr{E}_2$ interfere destructively.

The calculation of the intensities of the resultant fields at the points of observation is facilitated by setting the phase constants to be zero, that is, $\phi_1 = \phi_2 = 0$. The total field, $\mathscr{E}$, at the point of observation is simply the superposition of $\mathscr{E}_1$ and $\mathscr{E}_2$:

$$\begin{aligned} \mathscr{E} &= \mathscr{E}_1 + \mathscr{E}_2 \\ &= \mathscr{E}_0(r) \left[\cos(\omega t - kr_1) + \cos(\omega t - kr_2) \right] \end{aligned} \qquad (4.7)$$

The expression for the total field, by use of the trigonometric identity

$$\cos a + \cos b = 2 \cos \tfrac{1}{2}(a + b) \cos \tfrac{1}{2}(a - b)$$

simplifies to:

$$\mathscr{E}(r,\theta,t) = \mathscr{E}_0(r)\ 2\cos(\omega t - kr)\cos\tfrac{1}{2}k(r_2 - r_1)$$
$$= \mathscr{E}_0(r,\theta)\cos(\omega t - kr) \tag{4.8}$$

where the angle-dependent amplitude, $\mathscr{E}_0(r,\theta)$, is given by

$$\mathscr{E}_0(r,\theta) = 2\mathscr{E}_0(r)\cos\tfrac{1}{2}k(r_2 - r_1)$$
$$= 2\mathscr{E}_0(r)\cos\frac{1}{2}\left(\frac{2\pi\ d\ \sin\theta}{\lambda}\right) \tag{4.9}$$

The observed intensity of light at the point of observation is given by the time-averaged value of the square of the resultant field:

$$I(r,\theta) \propto \langle \mathscr{E}^2(r,\theta,t)\rangle = \frac{\mathscr{E}_0^2(r,\theta)}{2}$$

or

$$I(r,\theta) = 4I_0(r,\theta)\cos^2\left(\frac{\pi d\ \sin\theta}{\lambda}\right) \tag{4.10}$$

where $I_0(r,\theta)$ is the intensity of the light from a single slit. This result is in complete agreement with the observed two-slit interference pattern of light waves.

4.2   The Probability Interpretation of the Wave

In 1887, Heinrich Hertz made two important experimental observations. The first observation, which consisted of the generation and the detection of electromagnetic waves, constituted a decisive confirmation of Maxwell's electromagnetic theory. The second observation was the photoelectric effect, which, as discussed in Section 3.5, is not fully explained by Maxwell's electromagnetic theory. The successful theory of the photoelectric effect, which Einstein proposed in 1905, is based on the corpuscular view of light. Einstein's photon model of light was decisively confirmed by the Compton effect. In both phenomena light is treated as a particle, specified by the dynamical variables of energy and momentum. In the interference phenomena of light, on the other hand, light is treated as a wave and is specified by a wavelength. In view of these facts light is said to exhibit both wave and particle properties. Wave and particle constitute a duality, and such is the origin of the notion of the dual nature of light.

The photon is described by an electromagnetic field, which, in turn, is a solution of Maxwell's wave equation. It thus becomes necessary to give a photon interpretation to Maxwell's electromagnetic wave. Consider, for this purpose, a beam of monochromatic radiation, which, as a constructive interference maximum of all waves from a source with the same relative phase, can be described by an electromagnetic field of the form

$$\mathscr{E} = \mathscr{E}_0 \cos (\omega t - kx) \tag{4.11}$$

The square of the amplitude then gives the intensity of the beam. In the photon view of radiation, the intensity of the beam must be determined by the photon flux N; more specifically,

$$I = N\hbar\omega \tag{4.12}$$

where $\hbar = h/2\pi$. Thus the amplitude of the electromagnetic wave turns out to be a measure of the photon flux, that is,

$$N \propto \mathscr{E}_0^2 \tag{4.13}$$

Suppose now that a photon is emitted by a single atom. Such a photon can be described by an electromagnetic field spreading outwardly like a spherical shell. As well known, the amplitude of such a wave decreases inversely as the distance away from the source. The photon flux, then, decreases inversely as the square of the distance from the source. But the notion of the photon flux in this case is clearly without precise meaning. A more meaningful interpretation of the square of the field amplitude is to discard the notion of the photon flux and replace it with the notion of the probability of detecting the photon. That is, let the square of the field amplitude give a physical quantity that is proportional to the probability of detecting the photon. The decreasing field amplitude in the present example, then, implies that the probability of detecting the photon with a detector would decrease inversely as the square of the distance from the source. This new interpretation is consistent with the observation that no matter where it is detected, the photon delivers a full energy packet of $\hbar\omega$ into the detector and thus registers as a particle.

The probabilistic interpretation of the square of the field amplitude provides a consistent picture of electromagnetic radiation. The classical energy flux, for example, is now interpreted as giving the probability of detecting a photon of energy $\hbar\omega$ per unit area per unit time. For a large number of photons, then, the probability flux can be identified with the photon flux, and the product of the photon flux and the photon energy can, in turn, be identified with the classical energy flux. If the number of photons is large enough, the concept of the photon flux can obviously be replaced with the more familiar concept of the radiation flux. The classical interpretation of the intensity of radiation may thus be regarded as a macroscopic view of radiation. Therefore we can conclude that, in the context of the probabilistic interpretation of the square of the field amplitude, the corpuscular photon and the electromagnetic wave are one and the same thing.

4.3 The de Broglie Hypothesis of Matter Waves

The photon is a particle with wave properties. Its particle aspects are specified by energy E and momentum p, and its wave aspects by frequency ω and wave number k. As the energy-momentum invariant of the photon is zero,

$$E^2 - (pc)^2 = 0 \tag{4.14}$$

so is the frequency-wave number invariant of the photon also zero

$$\omega^2 - (kc)^2 = 0 \tag{4.15}$$

Clearly, the four-vector momentum of Eq. (4.14) and the four-dimensional wave vector of Eq. (4.15) describe the same physical entity, namely, the photon. It is thus not surprising that their corresponding components are related in the same way;

$$E = \hbar\omega$$
$$p = \hbar k \tag{4.16}$$

It is significant that the linkage between energy and frequency and between momentum and wave number is Planck's constant $\hbar$. It suggests the quantum basis of duality.

In 1923, Louis-Victor de Broglie proposed that this dual nature may not be a property unique to the photon but may apply to all physical entities. Now the energy-momentum invariant for a material particle of rest mass m_0 is

$$E^2 - (pc)^2 = (m_0c^2)^2 \tag{4.17}$$

Analogy with the photon, then, suggests that the wave property of a material particle would be described by a frequency ω and a wave number k, which must, according to relativity, satisfy a dispersion relation of the form

$$\omega^2 - (kc)^2 = \omega_0^2 \tag{4.18}$$

where ω_0 is a relativistic invariant and may be called the "rest frequency" of a particle. De Broglie's bold idea of duality was to demand that both Eqs. (4.17) and (4.18) describe one and the same material entity, and that radiation and matter be viewed in a unified way so that the relations

$$E = \hbar\omega \qquad \text{(4.19)}$$
$$p = \hbar k \qquad \text{(4.20)}$$

hold for a material particle as well. Eq. (4.20) may now be rewritten so as to give the wavelength of a material particle explicitly

$$\lambda = \frac{h}{p} \qquad \text{(4.21)}$$

A profound implication of Eq. (4.21) is that the product of the wavelength and the momentum of matter and radiation defines a universal constant, namely, Planck's constant. Thus in the quantum world the de Broglie wavelength and the momentum of a physical entity are absolutely inseparable, suggesting that its wave and particle aspects must likewise be inseparable. We have already seen that such is indeed the case for the photon. An experimental verification of the de Broglie relation for a material particle will then establish a fundamental unity underlying radiation and matter.

4.4 The Observation of Electron Waves

The wave property of the photon is exhibited in the interference and diffraction effects that it produces. Interference and diffraction are not two distinct phenomena. The intensity pattern of light waves is referred to as an interference pattern, when it is produced by the superposition of waves from a discrete set of coherent sources such as two "narrow" slits. An intensity pattern is referred to as a diffraction pattern, when it is produced by the superposition of waves from a large number of coherent sources such as a grating or from a continuous distribution of sources such as a "wide" slit. Both phenomena are readily observable, and a consistent explanation for them exists only within the framework of a wave theory.

Despite the great theoretical appeal of the de Broglie hypothesis, the notion of material particles producing interference effects is totally foreign to ordinary experience. In fact, classical physics does not admit the possibility of material particles producing interference effects, because classical particles do not obey the principle of superposition. However, if the de Broglie hypothesis is valid, all physical entities must display wave properties; therefore it is worth exploring why ordinary particles do not seem to exhibit interference effects. To this end, let us compute the de Broglie wavelength of a classical object of mass 1 g, moving at a velocity of 1 cm/sec. The de Broglie wavelength is

$$\lambda = \frac{h}{mv} = \frac{6.62 \times 10^{-27} \text{ erg sec}}{1 \text{ g} \times 1 \text{ cm/sec}} = 6.62 \times 10^{-27} \text{ cm}$$

This is an undetectably small length. Compare it, for example, to the spacings of atoms in a crystal that are typically of the order of 10^{-8} cm.

It was well known, following the work of Max von Laue in 1912, that crystals can be used as diffraction gratings for x rays. However, it was not until 1927 that C. J. Davisson and L. H. Germer succeeded in observing the diffraction of electrons upon reflection by a nickel crystal. In the same year, G. P. Thomson reported his observation of the diffraction of electrons upon passing through thin films of platinum. As a historical sideline, it is interesting to note that Davisson began his study of the reflection of electrons by metals as early as 1919, and continued it because of the unexpected result that the reflected electrons had the same energy as the incident electrons.

Let us now outline the experiment of Davisson and Germer (see Figure 4.2). The interatomic distance in the nickel crystal was determined, from x-ray diffraction measurements, to be $d = 2.15 \times 10^{-8}$ cm. The incident electron beam had a kinetic energy of $K = 54$ eV. The de Broglie wavelength of the electrons, which is in terms of kinetic energy expressed by

$$\lambda = \sqrt{\frac{150.4 \text{ eV}}{K}} \times 10^{-8} \text{ cm} \tag{4.22}$$

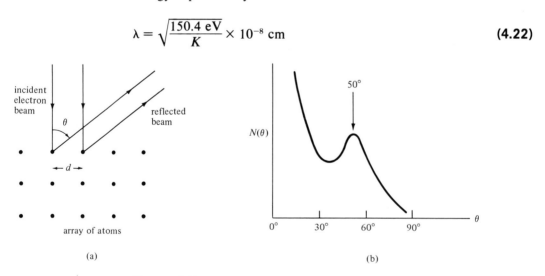

(a)　　　　　　　　　　　　　　(b)

Figure 4.2　(a) A schematic of the Davisson-Germer experiment: 54.0 eV electrons are diffracted by the atoms spaced $d = 2.15$ Å apart in a nickel crystal. (b) The observed distribution $N(\theta)$ of the reflected electrons as a function of the angle θ.

was computed to be

$$\lambda = 1.67 \times 10^{-8} \text{ cm}$$

Diffraction maxima would then occur at angles θ_n relative to the incident direction:

$$d \sin \theta_n = n\lambda \tag{4.23}$$

where n is an integer. The de Broglie hypothesis, accordingly, predicted that the first order, that is, $n = 1$, diffraction maximum would occur at the angle θ_1:

$$\sin \theta_1 = \frac{1.67 \times 10^{-8} \text{ cm}}{2.15 \times 10^{-8} \text{ cm}}$$

or

$$\theta_1 = 51°$$

Davisson and Germer observed a peak in the counting rate of the elastically scattered electrons at an angle of 50° from the incident direction. The slight difference between the predicted and the observed angles was accounted for by the greater speed of the electrons in the crystal than in vacuum. The increase in speed is a result of the crystal's attractive force on the electrons. The reality of the de Broglie wave of the electrons was thus established. It has since been observed that not only elementary particles such as electrons but composite systems such as atoms also exhibit wave properties precisely as predicted by the de Broglie hypothesis.

Example 4.1 Verify Eq. (4.22).

Solution. For a nonrelativistic electron, $p = \sqrt{2m_0 K}$, and thus the de Broglie wavelength is given by

$$
\begin{aligned}
\lambda &\approx \frac{h}{\sqrt{2m_0 K}} = \sqrt{\frac{h^2 c^2}{2m_0 c^2 K}} \\
&= \sqrt{\frac{(4.136 \times 10^{-15} \text{ eV sec})^2 (2.998 \times 10^{10} \text{ cm/sec})^2}{(2 \times 0.511 \times 10^6 \text{ eV})K}} \\
&= \sqrt{\frac{150.4 \text{ eV}}{K}} \times 10^{-8} \text{ cm}
\end{aligned}
$$

4.5 Relativistic Wave Equations

The photon is described by a wave equation. At the outset, the wave equation for a photon must be a linear differential equation inasmuch as photons obey the superposition principle, and must be consistent with the theory of relativity.

It is instructive to attempt the construction of a relativistic wave equation for the photon. To begin, recall that a photon of angular frequency ω and wave number k, traveling in the x direction, is described by a wave field of the form

$$\mathscr{E}(x, t) = \mathscr{E}_0 \cos (kx - \omega t) \tag{4.24}$$

with $\mathscr{E}_0$ denoting the amplitude of the wave field. Since the relativistic frequency-wave number invariant for the photon is zero,

$$\omega^2 = k^2 c^2 \tag{4.25}$$

we can write

$$\omega^2 \mathscr{E}(x, t) = c^2 k^2 \mathscr{E}(x, t) \tag{4.26}$$

The right-hand side of Eq. (4.26) is proportional to the second partial derivative with respect to x of the photon field $\mathscr{E}(x, t)$

$$c^2 k^2 \mathscr{E}(x, t) = -c^2 \frac{\partial^2 \mathscr{E}(x, t)}{\partial x^2} \tag{4.27}$$

The left-hand side of Eq. (4.26) is proportional to the second particle derivative with respect to time of the photon field $\mathscr{E}(x, t)$

$$\omega^2 \mathscr{E}(x, t) = -\frac{\partial^2 \mathscr{E}}{\partial t^2} \tag{4.28}$$

Then Eq. (4.26) can be written in a differential form

$$\frac{\partial^2 \mathscr{E}}{\partial t^2} = c^2 \frac{\partial^2 \mathscr{E}}{\partial x^2} \tag{4.29}$$

which is clearly satisfied by the photon field of Eq. (4.24). A wave equation has thus been constructed, and this wave equation turns out to describe all photons that propagate in free space.

A wave equation describing a material particle can be constructed in a similar manner. A free particle of rest mass m_0, moving in the x direction, is specified by a frequency ω and a wave number k. Assume, by analogy with the photon, that such a particle is described by a harmonic de Broglie wave field, which, in the complex representation, has the form

$$\psi(x, t) = A \exp (ikx - i\omega t) \tag{4.30}$$

where A is the amplitude of the particle field. Its relativistic frequency-wave number invariant is

$$\omega^2 - c^2 k^2 = \left(\frac{m_0 c^2}{\hbar}\right)^2 \tag{4.31}$$

so that the following equality holds

$$\omega^2 \, \psi(x, t) - c^2 k^2 \, \psi(x, t) = \left(\frac{m_0 c^2}{\hbar}\right)^2 \psi(x, t) \tag{4.32}$$

The first and the second members of Eq. (4.32) are, respectively, the second partial derivatives of $\psi(x, t)$

$$\omega^2 \, \psi(x, t) = -\frac{\partial^2 \psi}{\partial t^2} \tag{4.33}$$

$$-c^2 k^2 \, \psi(x, t) = c^2 \frac{\partial^2 \psi}{\partial x^2} \tag{4.34}$$

Hence Eq. (4.32) can be cast in a differential form

$$-\frac{\partial^2 \psi}{\partial t^2} + c^2 \frac{\partial^2 \psi}{\partial x^2} = \left(\frac{m_0 c^2}{\hbar}\right)^2 \psi$$

or

$$\frac{1}{c^2} \frac{\partial^2 \psi}{\partial t^2} - \frac{\partial^2 \psi}{\partial x^2} = -\left(\frac{m_0 c}{\hbar}\right)^2 \psi \tag{4.35}$$

which is a wave equation. Referred to as the **Klein-Gordon equation**, it describes the de Broglie wave field of a material particle in free space.

Let us now return to Eqs. (4.33) and (4.34) and make a very important observation. The first derivatives of the de Broglie wave field, $\psi = A \exp{(ikx - i\omega t)}$, are

$$-i\hbar \frac{\partial \psi}{\partial x} = \hbar k \psi \tag{4.36}$$

and

$$i\hbar \frac{\partial \psi}{\partial t} = \hbar \omega \psi \tag{4.37}$$

These equations are of the form

operator $\times$ function = number $\times$ function

where the function is the same on both sides of the equation. A differential equation of this type constitutes an important class of quantum mechanical equations, and is called an **eigenvalue equation.** The function and the "number" of the equation are referred to as an eigenfunction and an eigenvalue of the operator. Now the eigenvalue of Eq. (4.36) is the momentum, $p = \hbar k$, of the particle described by ψ, and hence the operator, $-i\hbar \; \partial/\partial x$, is identified as a momentum operator

$$p_{\mathrm{op}} = -i\hbar \; \frac{\partial}{\partial x} \tag{4.38}$$

Similarly, the observation that the eigenvalue of Eq. (4.37) is the energy $E = \hbar\omega$, of the particle leads to the identification of the operator, $i\hbar \; \partial/\partial t$, as an energy operator

$$E_{\mathrm{op}} = i\hbar \; \frac{\partial}{\partial t} \tag{4.39}$$

The de Broglie wave field ψ is then an eigenfunction of both the energy and momentum operators. It is a general feature of quantum mechanical eigenvalue equations that if a de Broglie field is an eigenfunction of an operator corresponding to a particular dynamical quantity, then its eigenvalues constitute the only values of the dynamical quantity that the particle is allowed to have.

Example 4.2 Construct an operator which, upon operating on $\psi = A \exp (ikx - i\omega t)$, yields $\hbar^2 k^2$ as its eigenvalue.

Solution. Since ψ is an eigenfunction of the momentum operator, its eigenvalue is the momentum of the particle described by ψ.

$$-i\hbar \; \frac{\partial \psi}{\partial x} = \hbar k \psi$$

A second operation by the momentum operator yields

$$-\hbar^2 \; \frac{\partial^2 \psi}{\partial x^2} = \hbar^2 k^2 \psi$$

Hence the desired quantum mechanical operator is

$$p^2_{\mathrm{op}} = -\hbar^2 \; \frac{\partial^2}{\partial x^2}$$

and its eigenvalue gives the square of momentum $p^2 = \hbar^2 k^2$. ●

The existence of quantum mechanical operators for the dynamical variables of classical mechanics provides a powerful method for mak-

Table 4.1   Quantum Mechanical Operators

Dynamical Variable or Observable	Symbols	Quantum Mechanical Operators
position	x y z	x y z
momentum	p_x p_y p_z	$-i\hbar \dfrac{\partial}{\partial x}$ $-i\hbar \dfrac{\partial}{\partial y}$ $-i\hbar \dfrac{\partial}{\partial z}$
energy	E	$i\hbar \dfrac{\partial}{\partial t}$
nonrelativistic kinetic energy	$\dfrac{p^2}{2m}$	$\dfrac{-\hbar^2}{2m} \nabla^2$
potential energy	$V(\mathbf{r})$	$V(\mathbf{r})$

ing a transition from classical mechanics to quantum mechanics. For example, the relativistic energy-momentum invariant of a material particle,

$$E^2 - c^2 p^2 = (m_0 c^2)^2$$

transforms into the quantum mechanical operator equation

$$(E_{\mathrm{op}})^2 - c^2 (p_{\mathrm{op}})^2 = (m_0 c^2)^2$$

which is simply the Klein-Gordon equation. This particular approach to quantum mechanics was first postulated by Erwin Schrödinger in 1926, and consists in transforming classical equations of motion to equivalent quantum mechanical operator equations.

4.6   The Schrödinger Equation

In 1926, Erwin Schrödinger discovered a nonrelativistic wave equation that successfully describes the de Broglie wave fields of material particles moving at nonrelativistic speeds. Called the Schrödinger equation, it can be constructed from the quantum mechanical operator interpretation of the dynamical variables of classical mechanics. The total energy E of a nonrelativistic particle of mass m in a potential $V(x)$ is

$$\frac{p^2}{2m} + V(x) = E \tag{4.40}$$

The corresponding quantum mechanical operator equation then governs the behavior of the de Broglie wave field $\psi(x, t)$ of such a particle

$$-\frac{\hbar^2}{2m}\frac{\partial^2\psi(x, t)}{\partial x^2} + V(x)\psi(x, t) = i\hbar\frac{\partial\psi(x, t)}{\partial t} \tag{4.41}$$

The three-dimensional version of Eq. (4.41) is

$$-\frac{\hbar^2}{2m}\nabla^2\psi(\mathbf{r}, t) + V(\mathbf{r})\psi(\mathbf{r}, t) = i\hbar\frac{\partial\psi(\mathbf{r}, t)}{\partial t} \tag{4.42}$$

where

$$\nabla^2 = \frac{\partial^2}{\partial x^2} + \frac{\partial^2}{\partial y^2} + \frac{\partial^2}{\partial z^2} \tag{4.43}$$

This is a wave equation and is referred to as the **time-dependent Schrödinger equation.** The solution $\psi(\mathbf{r}, t)$ of the Schrodinger equation is called a **wave function.**

The time-dependent Schrödinger equation is a partial differential equation involving both the position and time coordinates. It is thus more convenient to separate the equation into two parts, where one depends only on the position coordinate and the other only on the time coordinate. Such a separation of variables is achieved by writing the time-dependent wave function as a product of two single-variable functions

$$\psi(\mathbf{r}, t) = \psi(\mathbf{r})T(t) \tag{4.44}$$

Substitution in the time-dependent Schrödinger equation then yields

$$\left[\frac{-\hbar^2}{2m}\nabla^2\psi(\mathbf{r}) + V(\mathbf{r})\psi(\mathbf{r})\right]T(t) = i\hbar\psi(\mathbf{r})\frac{dT(t)}{dt} \tag{4.45}$$

which, upon dividing through by $\psi(\mathbf{r})\,T(t)$, reduces to

$$\frac{1}{\psi(\mathbf{r})}\left[-\frac{\hbar^2}{2m}\nabla^2\psi(\mathbf{r}) + V(\mathbf{r})\psi(\mathbf{r})\right] = \frac{i\hbar}{T(t)}\frac{dT}{dt} \tag{4.46}$$

The left-hand member of Eq. (4.46) is a function of $\mathbf{r}$ alone, and the right-hand member is a function of t alone. Each member of the equation may thus vary independently of the other, but both sides of the equation must maintain their equality. Such a situation obtains only if both sides of the equation are separately equal to the same constant, say E. That is,

$$\frac{-\hbar^2}{2m} \nabla^2 \psi(\mathbf{r}) + V(\mathbf{r})\psi(\mathbf{r}) = E\psi(\mathbf{r}) \tag{4.47}$$

and

$$\frac{dT(t)}{dt} = -i\left(\frac{E}{\hbar}\right)T(t) \tag{4.48}$$

Eq. (4.48) is easily integrated and yields

$$T(t) = e^{-itE/\hbar} \tag{4.49}$$

Hence the time dependence of the wave function is of the form

$$\psi(\mathbf{r}, t) = \psi(\mathbf{r})e^{-itE/\hbar} \tag{4.50}$$

Eq. (4.47) is independent of time and is therefore referred to as the **time-independent Schrödinger equation.** The time-independent Schrödinger equation is applicable to all physical systems whose potential energy is independent of time, that is, $\partial V/\partial t = 0$.

The time-independent Schrödinger equation can also be written

$$H\psi(\mathbf{r}) = E\psi(\mathbf{r}) \tag{4.51}$$

by defining a new quantum mechanical operator

$$H = \frac{-\hbar^2}{2m} \nabla^2 + V(\mathbf{r}) \tag{4.52}$$

H is called the Hamiltonian operator, and is the quantum mechanical equivalent of the classical, nonrelativistic total energy of a system. The solutions of the time-independent Schrödinger equation then constitute the eigenfunctions of the Hamiltonian operator, and the eigenvalues E of the Hamiltonian operator represent the only possible values of energy that a quantum system may assume.

Example 4.3 (a) Write the time-independent Schrödinger equation for a free, nonrelativistic particle moving in the x direction.

(b) Show that E represents the kinetic energy of the particle.

Solution. (a) Since the particle is free of all external forces, $V(x) = 0$. Hence the Schrödinger equation takes the form

$$\frac{-\hbar^2}{2m} \frac{d^2\psi(x)}{dx^2} = E\psi(x)$$

or

$$\frac{d^2\psi(x)}{dx^2} + \frac{2mE}{\hbar^2}\psi(x) = 0$$

Note that the Schrödinger equation for a free particle in one-dimensional motion is identical in structure to the Newtonian equation of motion for a simple harmonic oscillator, namely,

$$\frac{d^2x}{dt^2} + \omega^2 x = 0$$

whose solutions are familiar.

(b) The wave function of a free particle is of the form

$$\psi(x) = \exp\left(\frac{i\sqrt{2mE}}{\hbar}x\right)$$

The allowed values of the energy of the particle are given by the eigenvalues of the Hamiltonian operator

$$H = \frac{-\hbar^2}{2m}\frac{d^2}{dx^2}$$

or

$$\frac{-\hbar^2}{2m}\frac{d^2\psi(x)}{dx^2} = \frac{-\hbar^2}{2m}\left(\frac{-2mE}{\hbar^2}\right)\exp\left(\frac{i\sqrt{2mE}}{\hbar}x\right)$$

$$= E\exp\left(\frac{i\sqrt{2mE}}{\hbar}x\right)$$

E is thus an eigenvalue of the Hamiltonian operator and represents the energy a free particle is allowed to have. To relate E to the momentum of the particle, we calculate the eigenvalue of the momentum operator

$$-i\hbar\frac{d\psi(x)}{dx} = -i\hbar\left(i\frac{\sqrt{2mE}}{\hbar}\right)\psi(x)$$

and find that the allowed values of the momentum are

$$p = \sqrt{2mE}$$

so

$$E = \frac{p^2}{2m} = \tfrac{1}{2}mv^2$$

The total energy of a free, nonrelativistic particle is, therefore, equal to its kinetic energy, as, indeed, it should be.

4.7 The Probability Interpretation of the Wave Function

The interpretation of the Schrödinger wave function is similar to the interpretation of the photon field: The wave function is given a probabilistic interpretation. The probabilistic interpretation was proposed by Max Born in 1926. In this interpretation, the absolute square of a wave function gives the probability density of finding a particle described by the wave function at a particular point in space and time. Hence the probability density integrated over all space must be unity. That is,

$$\int |\psi(\mathbf{r}, t)|^2 \, dV = \int \psi^* \psi \, dV = 1 \qquad (4.53)$$

where dV is the volume element. When Eq. (4.53) holds for a wave function, the wave function is said to be **normalized.**

The normalization of a wave function is achieved by multiplying the wave function by a so-called normalization constant. As an example of the normalization procedure for a wave function, consider the wave function of a free nonrelativistic particle that is confined to one-dimensional motion along the x axis between two points $x = -L$ and $x = +L$. The wave function of such a particle is obtained from the time-independent Schrödinger equation

$$\frac{\hbar^2}{2m} \frac{d^2\psi}{dx^2} + E\psi = 0 \qquad (4.54)$$

and is of the form

$$\psi(x) = A e^{ikx} \qquad (4.55)$$

where A is the normalization constant, and $\hbar k$ is the momentum of the particle. Now the probability of finding the particle somewhere between the turning points is unity, that is,

$$\int_{-L}^{+L} \psi^* \psi \, dx = A^2 \int_{-L}^{+L} dx = 1 \qquad (4.56)$$

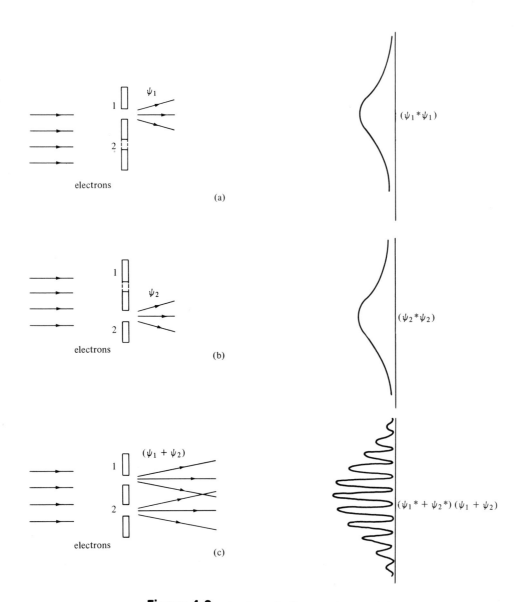

Figure 4.3 A schematic diagram of a two-slit interference experiment with electrons. (a) Electrons passing through slit 1 with slit 2 closed are described by ψ_1, and the distribution of the electrons at the "screen" is given by $\psi_1^* \psi_2$. (b) Electrons passing through slit 2 with slit 1 closed. (c) If electrons pass through both slits, an interference pattern in the distribution of the electrons is observed at the "screen."

whence the normalization constant A is

$$A = \frac{1}{\sqrt{2L}} \tag{4.57}$$

As a way of gaining further insight into the probabilistic interpretation of the wave function, consider the example of a "two-slit" experiment with electrons (Figure 4.3). The experiment begins by letting the electrons pass through slit 1 with slit 2 closed. If these electrons are described by a wave function ψ_1, their distribution on the "screen" is given by the probability density function $|\psi_1|^2 = \psi_1^*\psi_1$. Similarly, if the electrons are let through slit 2 with slit 1 closed, they are described by a wave function ψ_2 and their distribution on the screen is given by $|\psi_2|^2 = \psi_2^*\psi_2$. Now, if the electrons pass through both slits, they must be described by a new wave function ψ, which is simply the superposition of ψ_1 and ψ_2, namely,

$$\psi = \psi_1 + \psi_2 \tag{4.58}$$

The distribution of the electrons on the screen is then given by the probability density function of the total wave function

$$\psi^*\psi = \psi_1^*\psi_1 + \psi_2^*\psi_2 + \psi_1^*\psi_2 + \psi_2^*\psi_1 \tag{4.59}$$

The last two terms of Eq. (4.59) describe the interference between the two wave functions, and they give rise to the observed interference pattern of the electrons.

The probabilistic interpretation of the wave function is an essential part of the Schrödinger theory. The wave function itself is not observable. It is the probabilistic interpretation that connects the physical content of the wave function with measurement. And it is in this sense that quantum theory is a probabilistic theory.

4.8 Summary

Light exhibits both particle and wave properties. The photon is described by an electromagnetic field, and the probability of detecting the photon is given by the square of the amplitude of the wave field. In a beam of monochromatic radiation the square of the field amplitude is proportional to the photon flux.

Hypothesized by de Broglie in 1924, the wavelength of a material particle is determined by its momentum

$$\lambda = \frac{h}{mv}$$

The confirmation of the wave nature of matter came in 1927 in a series of experiments performed by Davisson and Germer, and independently by Thomson. The de Broglie waves are governed by quantum mechanical wave equations.

The transition from classical mechanics to quantum mechanics is made by postulating that for every dynamical variable of classical mechanics there exists a quantum mechanical operator. The momentum operator, for example, is

$$\mathbf{p}_{op} = - i\hbar \, \nabla$$

The time-independent Schrödinger equation is an eigenvalue equation

$$H\psi = E\psi$$

where

$$H = \frac{-\hbar^2}{2m} \, \nabla^2 + V(\mathbf{r})$$

The eigenvalues of the Hamiltonian operator correspond to the allowed energy states of a quantum system such as an atom. A particle, say an atomic electron, existing in a particular energy state is described by a corresponding eigenfunction, and the probability of finding the particle at any point in space is given by the square of the eigenfunction. The probability interpretation of quantum mechanics demands that eigenfunctions be normalized.

Problems

4.1 Calculate the photon flux of a beam of red light having an intensity of 1 W/cm². (Note that 1 W = 10^7 erg/sec and $\lambda_{red} = 7000$ Å.)

4.2 If a photon is emitted by an atom with no preferred direction in space, what is the probability of finding the photon per unit area at a distance r from the atom?

4.3 Deduce from the probabilistic interpretation of the radiation flux that the electric field amplitude of a photon emitted by an atom must vary inversely as the radial distance from the source.

4.4 (a) Justify the claim that light is a de Broglie wave.
(b) Comment on the statement made by Professor E. H. Wichmann: "The de Broglie wave and the particle are the *same thing:* there is nothing else."

4.5 (a) The phase velocity of a wave is defined by $v_\phi = \omega/k$. Show that the phase velocity of a photon is the velocity of light, and the phase velocity of an electron is greater than the velocity of light.
(b) The group velocity of a wave packet is given by $v_g = d\omega/dk$. Show that the group velocity of a de Broglie wave packet is equal to the velocity of the particle that it describes.
(c) Argue from (a) and (b) that a material particle cannot be described by a de Broglie wave of a precise wavelength.

4.6 Calculate the de Broglie wavelengths of (a) a 50-eV photon, (b) a 50-eV (kinetic energy) electron, and (c) a 50-eV neutron. (Take the rest energy of the neutron to be $E_0 = 940$ MeV.)

4.7 (a) Show that the velocity dependence of the de Broglie wavelength for a particle of rest mass m_0 is

$$\lambda = \frac{h}{m_0 c} \frac{\sqrt{1 - (v^2/c^2)}}{v/c}$$

(b) The quantity $(h/m_0 c)$ is called the Compton wavelength of the particle. Calculate the Compton wavelength of the electron.

4.8 (a) Express the de Broglie wavelength of a particle of rest mass m_0 as a function of its kinetic energy K.

(b) Show that your answer to (a) reduces to $\lambda = h/m_0 v$ for $(K/E_0) \ll 1$.

4.9 The average kinetic energy of a gas molecule at room temperature is 0.025 eV. Calculate the de Broglie wavelength of a hydrogen molecule at room temperature. Take the rest energy of a hydrogen molecule to be 1877 MeV.

4.10 (a) The resolving power of a microscope is a measure of its ability to yield the fine details of an object under study. The theoretical maximum resolving power of a microscope is equal to the wavelength of the illuminating "light," that is, a microscope cannot resolve details smaller than λ. A typical electron microscope is illuminated with electrons of an energy of 50 KeV. Calculate the theoretical maximum resolving power of such an electron microscope.

(b) Calculate the de Broglie wavelength of an electron that has been accelerated to a kinetic energy of 12.4 GeV (1 GeV $= 10^9$ eV). Would such an electron be useful in the study of any physical object?

4.11 A free particle of mass m confined to one-dimensional motion between two impenetrable walls is described by a wave function $\psi = \sin kx$. Calculate the nonrelativistic kinetic energy of the particle.

4.12 The potential energy of a simple harmonic oscillator of mass m, oscillating with angular frequency ω is

$$V(x) = \tfrac{1}{2} m \omega^2 x^2$$

(a) Write the time-independent Schrödinger equation for a simple harmonic oscillator.

(b) Given the eigenfunctions of the Hamiltonian operator

$$\psi_0 = e^{-(m\omega/2\hbar)x^2}$$

$$\psi_1 = 2\sqrt{\frac{m\omega}{\hbar}}\, x e^{-(m\omega/2\hbar)x^2}$$

$$\psi_2 = (2 - \frac{4m\omega}{\hbar} x^2) e^{-(m\omega/2\hbar)x^2}$$

calculate the energy eigenvalue for each eigenfunction.

4.13 (a) If a particle moves freely along the x axis and its wave function is an eigenfunction of the momentum operator, find the wave function.

(b) Show that such a particle can only have real and finite values for its momentum.

4.14 Show that the wave function of a free particle that is located at $x = x_0$ must vanish everywhere except at $x = x_0$.

4.15 (a) Find a wave function that is an eigenfunction of an angular momentum operator, $-i\hbar\, \partial/\partial\phi$.

(b) From the requirement that a physically acceptable wave function must be single-valued at every point in space, show that the eigenvalues of this angular momentum operator must be restricted to values $m\hbar$, where m is an integer.

(c) Why is it necessary to require that a wave function must be single-valued everywhere?

4.16 The total energy and hence the Hamiltonian of a falling body under gravity is

$$H = \frac{p_y^2}{2m} + mgy$$

Write the time-independent Schrödinger equation for the falling body.

4.17 Given a particle whose motion along the x axis is unbounded, calculate the probability of finding the particle at any point x.

4.18 A particle moving freely between two impenetrable walls at $x = 0$ and $x = L$ is described by a wave function of the form

$$\psi_n(x) = \sin\left(\frac{n\pi}{L}\right)x \qquad n = 1, 2, 3, \ldots$$

(a) Calculate the normalization constant of the wave function.

(b) Calculate the probability of finding the particle between $x = 0$ and $x = L/2$.

4.19 (a) Show that two wave functions which are of the same functional form but differ in their multiplicative constants are identical in their physical content.

(b) When is it necessary to normalize a wave function?

4.20 Prove that the probability density of a complex wave function must necessarily be a real function.

4.21 If the probability densities are added instead of the wave functions in the two-slit experiment with electrons, what would be the distribution of the electrons on the "screen" of observation?

Suggestion for Further Reading

Christy, R. W., and A. Pytte. *The Structure of Matter*. New York: W. A. Benjamin, Inc., 1965. Chapter 20.

Crawford, F. S., Jr. *Waves: Berkeley Physics Course*, vol. 3. New York: McGraw-Hill Book Company, 1965. Pp. 548–557.

Darrow, K. K. "Davisson and Germer." *Scientific American* (May 1948).

Ford, K. W. *Classical and Modern Physics*, vol. 3. Lexington, Mass.: Xerox College Publishing, 1974, Chapter 23.

Schrödinger, E. "What Is Matter?" *Scientific American* (September 1953).

Wichmann, E. H. *Quantum Physics: Berkeley Physics Course*, vol. 4. New York: McGraw-Hill Book Company, 1967. Chapter 5.

The hydrogen atom:
the time-independent theory

5.1　The Bohr Model

Upon excitation by a mechanism such as an electric discharge, a tube of hydrogen gas emits radiation. If the radiation is collimated by a narrow slit and then resolved by a device such as a prism or a diffraction grating, spectral components appear as a series of lines on a screen or a photographic plate. The spectrum of hydrogen is thus referred to as a **line spectrum.**

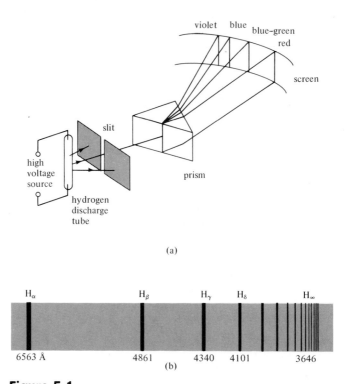

violet blue blue–green red

screen

slit

high
voltage
source

prism

hydrogen
discharge
tube

(a)

H_α H_β H_γ H_δ H_∞

6563 Å   4861   4340   4101   3646
(b)

Figure 5.1   (a) A schematic diagram of a prism spectrograph. (b) The Balmer series of hydrogen spectral lines. [From G. Herzberg, *Atomic Spectra and Atomic Structure.* (New York: Dover Publications, 1944). Used by permission of the publisher.]

Precision measurements of the visible lines of hydrogen were first made by A. J. Ångström during the latter half of the nineteenth century (see Figure 5.1). A pattern of regularities in the visible spectrum was discovered by J. J. Balmer in 1885. The Balmer pattern can be put in the form

$$\frac{1}{\lambda} = R\left(\frac{1}{2^2} - \frac{1}{n^2}\right) \qquad n = 3, 4, 5, 6, \ldots \tag{5.1}$$

where λ is the wavelength of the spectral line and R is the so-called Rydberg constant. Balmer's formula, however, yields no picture of the hydrogen atom. The red line of hydrogen, for example, is associated with integers 2 and 3, but the physical significance of such numbers is totally obscure.

Maxwell showed that electromagnetic waves are produced by accelerated charges. Hence the emission of electromagnetic radiation by hydrogen and other elements lent strong support to the idea of the electrical constitution of matter. But atoms are normally neutral in electric charge. Thus it was reasonable to view them as consisting of equal amounts of positive and negative charge. The question as to whether charge is associated with some fundamental constituents of matter was, however, still unresolved in the latter part of the nineteenth century.

In 1897, J. J. Thomson provided convincing evidence for the existence of an elementary particle with which negative charge was to be associated. The evidence emerged from his study of the so-called cathode rays that emanated from the negative terminal of a high-voltage source. Study of the properties of these rays identified them as streams of negatively charged particles to which the name "electron" was assigned. From Thomson's measurements of the ratio of the electron's charge to its mass, and a similar ratio for the hydrogen ion, he was led to conclude that the electron was some 2000 times less massive than the hydrogen ion.

His work on the electron led Thomson to propose a model for the atom (Figure 5.2). He pictured a neutral atom to consist of electrons embedded within a spherical distribution of positive charge. The mutual repulsion among the electrons would cause them to be uniformly distributed throughout the sphere of positive charge. This picture was often referred to as a "raisin-cake" model of the atom. The attractive feature of this model was that it provided a qualitative explanation for the emission of radiation from atoms. Thomson was, for example, able to show that in order for the vibrating electrons to produce visible light, the atom would have to have a radius of about 10^{-8} cm, in good agreement with the size of the atom deduced from the kinetic theory of gases. However, attempts to produce a theoretical spectrum in agreement with observation were not fruitful, and consequently the usefulness of the model diminished.

A direct test of the Thomson model of the atom was provided by E. Rutherford in 1911. The experimental setup consisted of an alpha-particle source, two diaphragms to collimate the alpha particles into a narrow parallel beam, a gold foil, and a detector of the alpha particles (Figure 5.3). The experiment was carried out by Rutherford's two able assistants, Geiger and Marsden.

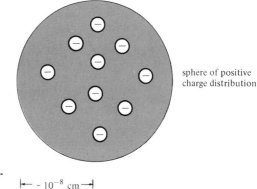

sphere of positive charge distribution

Figure 5.2 Thomson's "raisin-cake" model of the atom. $\vdash\!\!\!-\sim 10^{-8}\ \text{cm}\,\dashv$

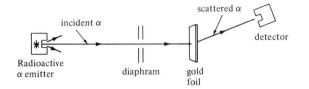

Figure 5.3 A schematic diagram of Rutherford's alpha-particle scattering experiment.

The "raisin-cake" model predicted that an alpha particle traversing a thin gold foil would suffer little deflection. The deflection of an alpha particle by a gold atom would be caused by the positive charge distribution as well as the embedded electrons. But the positive charge of gold was distributed in a sphere of radius which was some 10^5 times the size of an alpha particle. The electrons, on the other hand, were some 8000 times less massive than alpha particles.

The experimental result was a complete surprise. About one in every 20,000 of the incident alpha particles was caused to bounce back by a gold foil only 4×10^{-5} cm thick. This was a far cry from the prediction of the Thomson model that no alpha particles would deviate more than at most 1° from the incident path. So unexpected was the result that Rutherford is reported to have commented, "It was quite the most incredible event that has ever happened to me in my life. It was almost as incredible as if you had fired a 15-in. shell at a piece of tissue paper and it came back and hit you."

Rutherford established that the positive charge of an atom is concentrated within a region of radius 10^{-12} cm. He called this positive core the **nucleus** of an atom, and worked out a theory of alpha-particle scattering on the basis of his nuclear atom. In the nuclear atom, the alpha particles collide with the massive nuclei, and, therefore, scatter in all directions, including the backward direction. Furthermore, Rutherford succeeded in deriving a formula to describe the observed scattering results and thus put the hypothesis of the atomic nucleus on a sound basis (see Section 10.4 for details).

Armed with unquestionable experimental evidence for a nuclear atom, Rutherford proceeded to explore its further implications. First of all, he argued that the concentration of the atom's positive charge within the nucleus left the outer regions of the atom to carry the negative charge required to produce an electrically neutral atom. This outer region, then, would be occupied by electrons and would determine the overall size of the atom. Then there was the question of the force holding the electrons in this region. He assumed that the force was the well-known electrostatic force between opposite charges. Thus the atom was to be governed by the laws of electricity and magnetism. But an attractive force would pull the electrons toward the center of the force. This situation necessitated postulating a circular motion for

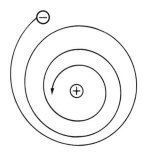

(a)

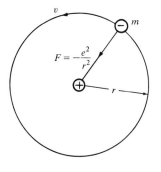

v

$F = -\dfrac{e^2}{r^2}$

m

r

(b)

Figure 5.4 (a) The planetary model of Rutherford's nuclear atom is classically unstable. (b) Atomic stability is achieved in Rutherford's planetary model by Bohr's quantum condition, $mvr = n\hbar$.

the electrons, and thus the nuclear atom emerged as a miniature solar system.

But a planetary atom, according to Maxwell's electromagnetic theory, would emit a continuous spectrum. An orbiting electron must emit radiation with a frequency determined by the frequency of revolution of the electron. But, as it radiates, the electron must give up some of its energy and hence must be pulled in closer to the nucleus. As it continues to emit radiation of continuously increasing frequency, the electron must finally collapse into the nucleus. Thus Maxwell's theory predicted for Rutherford's planetary atom a total collapse and a continuous spectrum, both of which were in violent contradiction with observation (Figure 5.4).

In 1913, Niels Bohr succeeded in applying Planck's quantum hypothesis to Rutherford's nuclear atom. He hypothesized that the circular motion of the electron in the hydrogen atom is governed by a quantum condition, namely,

$$mvr = n\hbar, \qquad n = 1, 2, 3, \ldots \tag{5.2}$$

where m and v are the mass and the velocity of the electron and r is the radius of the orbit. Note that Bohr's quantum condition is a condition on the angular momentum of the electron. Less obvious but perhaps more profound, the condition has in it a hint of an intimate relation between position and momentum.

Bohr, for the sake of simplicity, treated the hydrogen nucleus as a particle of infinite mass.[1] Then the circular motion of the electron about the nucleus is, in cgs units, described by

$$\frac{mv^2}{r} = \frac{e^2}{r^2} \tag{5.3}$$

where e is the magnitude of the charge of the electron as well as of the nucleus. If the Bohr condition, Eq. (5.2), is imposed on Eq. (5.3), the radius of the orbit becomes quantized according to

$$r = n^2 \frac{\hbar^2}{me^2}$$

or

$$r_n = n^2 a_0 \tag{5.4}$$

where a_0 is called the **Bohr radius** and has the numerical value

$$a_0 = \frac{\hbar^2}{me^2} = 0.53 \times 10^{-8} \text{ cm} \tag{5.5}$$

In Bohr's nonrelativistic approach to the hydrogen atom, the total energy of the electron, and thus of the atom, is simply the sum of the potential and kinetic energies of the electron:

$$E = \tfrac{1}{2}mv^2 + V(r) \tag{5.6}$$

where

$$V(r) = -\frac{e^2}{r} \tag{5.7}$$

Substitution of Eq. (5.3) into Eq. (5.6) simplifies the expression for the total energy to

$$E = -\frac{e^2}{2r} \tag{5.8}$$

which, upon substitution for r by Eq. (5.4), becomes quantized:

---

[1] The hydrogen nucleus is about 1836 times as massive as the electron.

$$E_n = -\frac{e^2}{2a_0 n^2} \tag{5.9}$$

Now, substituting for a_0 by Eq. (5.5), Eq. (5.9) can be cast in a more transparent and computationally easier form, which is

$$E_n = -\frac{mc^2\alpha^2}{2n^2} = -\frac{E_0}{n^2} \tag{5.10}$$

where

$$\alpha = \frac{e^2}{\hbar c} \approx \frac{1}{137} \tag{5.11}$$

$$E_0 = \frac{mc^2\alpha^2}{2} = 13.6 \text{ eV}$$

The constant α is a dimensionless number and is called the **fine-structure constant**. The rest energy of the electron has a numerical value of $mc^2 = 0.51$ MeV. Both of these numbers appear frequently and merit commitment to memory.

Bohr interpreted the quantized energies and quantized radii to be the only allowed ones for the hydrogen atom. He postulated that an electron in an allowed state does not radiate, and, in so doing, contradicted Maxwell's electromagnetic theory. To stress the nonradiating character of the electron orbiting in an allowed state, the allowed states were also referred to as stationary states.

Concerning the absorption and emission of radiation by atoms, Bohr postulated that these processes are associated with transitions between allowed states and that atomic transitions are governed by the principle of energy conservation. Consider, for example, an electron transition from the $n = 3$ state to the $n = 2$ state. Such a transition leads to the emission of a photon whose energy is

$$h\nu = E_3 - E_2$$

and whose wavelength is

$$\frac{1}{\lambda} = \frac{E_0}{hc}\left(\frac{1}{2^2} - \frac{1}{3^2}\right)$$

This result is identical to the Balmer formula for the red line of the hydrogen spectrum. In general, a downward transition from an n state to an m state leads to the emission of a photon whose wavelength is given by

$$\frac{1}{\lambda} = R\left(\frac{1}{m^2} - \frac{1}{n^2}\right) \qquad n > m \tag{5.12}$$

and the Rydberg constant R is determined by the Bohr theory to be

$$R = \frac{E_0}{hc} \qquad \text{(5.13)}$$

The successes of the Bohr theory were impressive. First of all, it provided a theoretical basis for atomic spectroscopy (Figure 5.5).

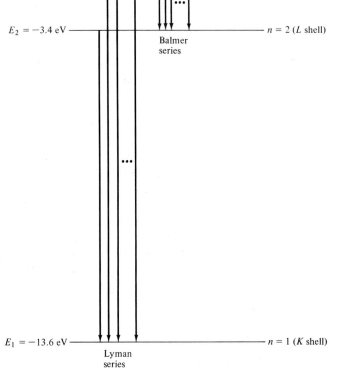

Figure 5.5 The quantized energy states of the hydrogen atom given by $E_n = -13.6$ eV/n^2.

The Balmer series and the Paschen series (transitions to $n = 3$), which were already known to exist, were immediately explained. Series resulting from downward transitions to the $n = 1$, $n = 4$, and $n = 5$ states were also predicted by the Bohr theory, and were observed, respectively, by Lyman in 1914, by Brackett in 1922, and by Pfund in 1924. The theory, furthermore, gave rise to the concept of atomic number through the experimental work of H. G. J. Moseley in 1913.

Example 5.1 (a) Calculate the frequency of the spectral line emitted when the hydrogen electron makes a transition from the $n = 2$ orbit to the $n = 1$ orbit.

(b) Calculate the frequencies of revolution of the electron in the two orbits.

Solution. (a) The energy of the emitted photon is

$$hv = E_2 - E_1 = -13.6 \text{ eV} \left(\frac{1}{2^2} - \frac{1}{1^2} \right)$$
$$= 12.2 \text{ eV}$$

The numerical value of Planck's constant is $h = 4.15 \times 10^{-15}$ eV sec, so

$$v = 2.94 \times 10^{15} \text{ Hz}$$

(b) The angular frequency of revolution of the electron in the nth orbit is given by

$$m\omega_n r^2 = n \frac{h}{2\pi}$$

Hence

$$v_n = \frac{\omega_n}{2\pi} = |E_n| \left(\frac{2}{nh} \right)$$

The numerical values are

$$v_1 = 6.55 \times 10^{15} \text{ rev/sec}$$
$$v_2 = 0.82 \times 10^{15} \text{ rev/sec}$$

Notice that the frequency of the emitted photon lies between the frequencies of revolution of the electron in the two orbits, that is, $v_1 > v > v_2$.

5.2 The Atomic Number

In 1913, Moseley made measurements of the wavelengths of photons that were emitted by metal targets upon bombardment by energetic electron beams (Figure 5.6). These photons were of short wavelengths, and as they were characteristic of the struck atoms, were referred to as the characteristic x rays of the atoms.

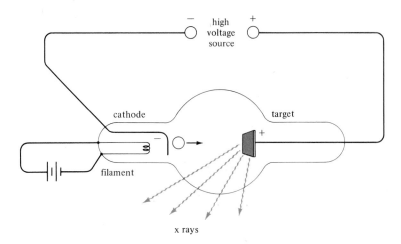

Figure 5.6 An x-ray tube.

Moseley noticed that the frequencies of the characteristic x rays of the elements increased with the atomic weights, paralleling the ordering of the elements in Mendeleyev's periodic table of the elements. He further noticed that the observed frequency of the most prominent x-ray spectral line of an element was expressible in the form:

$$\nu = cR(Z-1)^2\left(\frac{1}{1^2}-\frac{1}{2^2}\right) \tag{5.14}$$

where Z is an integer characteristic of the element. This fundamental number Z for aluminum, for instance, was found to be 13, and, since aluminum is the thirteenth element in the periodic table, Z was called the **atomic number.**

The physical significance of the atomic number is provided by the Bohr theory. A single-electron atom, whose electron is bound by a nucleus of charge Ze, would, upon a transition from the $n=2$ state to the $n=1$ state, emit a photon of frequency

$$\nu = cRZ^2\left(\frac{1}{1^2}-\frac{1}{2^2}\right) \tag{5.15}$$

The similarity between this result and the Moseley formula, Eq. (5.14), then suggests that the quantity $(Z-1)$ in the Moseley formula may be interpreted as the effective positive charge of the nucleus as "seen" by the jumping electron (Figure 5.7). The similarity further suggests that the electron transition leading to the emission of the most prominent x-ray line must be from $n=2$ to $n=1$, and that in the target atom the nucleus of charge Z must be shielded by an electron in the $n=1$ state. Moseley thus established that the periodicity of the elements

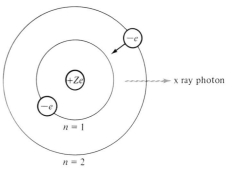

Figure 5.7 An effective nuclear charge of $(Z-1)e$ is "seen" by the electron making a transition from the $n = 2$ orbit to the $n = 1$ orbit. The result is the emission of the so-called K_α x-ray spectral line.

is a function of the atomic number rather than that of the atomic weights as postulated by Mendeleyev. Moreover, Moseley's work provided further credence to Bohr's picture of the atom.

The Bohr theory was the first quantum model of the atom. As its simplicity was its strength, its ad hoc nature and lack of coherence were its weaknesses. This was recognized from the beginning, as so well stated by Rutherford: "Your ideas . . . are very ingenious and seem to work out well: but the mixture of Planck's ideas with the old mechanics makes it very difficult to form a physical idea of what is the basis of it all." It is then not surprising that the Bohr theory failed to cope with the increasing sophistication of experimental spectroscopy. Nevertheless, it, more directly than any other theory, ushered in a quantum era in the domain of atoms, and was thus a precursor of all fundamental theories of the atom that were necessary and therefore had to be forthcoming.

5.3 Quantization of Energy

In the Bohr theory, the quantization of the orbit and the associated energy follow directly from the ad hoc quantum condition, $mvr = n\hbar$. In 1924, de Broglie suggested that the Bohr quantum condition might be rooted in the wave nature of the electron. An electron in the first Bohr orbit of hydrogen, for example, has a de Broglie wavelength

$$\lambda_1 = \frac{h}{mv_1} = 3.3 \times 10^{-8} \text{ cm}$$

Note that this de Broglie wavelength equals the circumference of the first orbit. In general, if a de Broglie wave is assumed to constitute a standing wave in an allowed Bohr orbit, then the circumference of an allowed orbit must be an integral multiple of the de Broglie wavelength:

$$2\pi r = n\lambda \qquad n = 1, 2, 3, \ldots \tag{5.16}$$

Now the equivalence of Eq. (5.16) to the Bohr quantum condition emerges when the de Broglie relation $\lambda = h/mv$ is applied:

$$2\pi r = n\frac{h}{mv}$$

or

$$mvr = n\hbar \tag{5.17}$$

Schrödinger was particularly impressed with the de Broglie hypothesis of matter waves. His excitement was perhaps heightened by his deep knowledge of wave phenomena in general and by his facility with the mathematical techniques dealing with them. When he succeeded in constructing a relativistic wave equation for matter waves and applied it to the hydrogen atom, he found that his results did not agree with experiment, and this led him to temporarily abandon the work. The event to rekindle his interest in the work was an invitation to give a colloquium on the de Broglie wave. In the course of reviewing the idea of matter waves Schrödinger modified his original wave equation so as to be nonrelativistic. When he applied this nonrelativistic wave equation to the hydrogen atom, the results were in complete agreement with experiment. He published his theory in 1926.

In the Schrödinger wave theory of the atom, the quantization of energy follows directly from the boundary conditions imposed on the wave function. The wave function is simply a solution of the Schrödinger wave equation. As an illustration, let us consider an electron constrained to one-dimensional motion in an infinite square well potential. The time-independent Schrödinger equation for such an electron is

$$-\frac{\hbar^2}{2m}\frac{d^2\psi(x)}{dx^2} = E\,\psi(x) \tag{5.18}$$

This can be cast in the form of the familiar simple harmonic oscillator equation

$$\frac{d^2\psi}{dx^2} + k^2\psi = 0 \tag{5.19}$$

where

$$k^2 = \frac{2mE}{\hbar^2} \tag{5.20}$$

The general solution of Eq. (5.19) is

$$\psi(x) = A\,\sin kx + B\cos kx \tag{5.21}$$

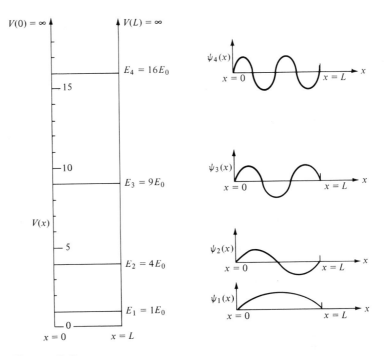

Figure 5.8 A particle in an infinite square well potential, in quantum physics, assumes a discrete set of energies, $E_n = n^2 E_0$, $n = 1, 2, 3, \ldots$, and is described by a wave function of the form $\psi_n(x) = A \sin(n\pi x/L)$.

where A and B are constants. Now, if we impose the boundary condition that the wave function vanish at the turning points of the electron, that is, $\psi(0) = \psi(L) = 0$, then the wave function takes the form (see Figure 5.8)

$$\psi(x) = A \sin kx \tag{5.22}$$

and the quantity k becomes quantized according to

$$\psi(L) = A \sin kL$$

or

$$kL = n\pi \qquad n = 1, 2, 3, \ldots \tag{5.23}$$

The quantization of k, in turn, implies that the energy of the electron must be quantized. The quantized energy is, from Eq. (5.20):

$$E_n = n^2 \frac{(\pi\hbar)^2}{2mL^2} \tag{5.24}$$

The quantized energy of the hydrogen atom is likewise a consequence of the boundary conditions imposed on the wave functions describing the hydrogen atom. In the hydrogen atom, the electron exists in the electrostatic potential of the proton, namely,

$$V(r) = \frac{-e^2}{r}$$

Hence the time-independent Schrödinger equation for hydrogen is

$$\frac{-\hbar^2}{2m} \nabla^2 \psi(\mathbf{r}) - \frac{e^2}{r} \psi(\mathbf{r}) = E\psi(\mathbf{r}) \tag{5.25}$$

This is an eigenvalue equation and will thus hold for only certain wave functions corresponding to certain energies. The solution of Eq. (5.25), which is outlined in Section 5.4, entails the requirements that the wave function must be single-valued everywhere and that the wave function must remain finite at $r = \infty$. The only acceptable solutions subject to such boundary conditions are those wave functions corresponding to the energy eigenvalues given by

$$E_n = -\frac{\alpha^2 mc^2}{2n^2} \qquad n = 1, 2, 3, \ldots \tag{5.26}$$

These energy eigenvalues of the Schrödinger equation are identical to the quantized energies given by the Bohr theory. In fact, it was this identity that constituted the first successful test of the Schrödinger theory.

5.4 The Solution of the Schrödinger Equation: An Outline

The spherical symmetry of the potential energy function of the hydrogen atom suggests the choice of a spherical coordinate system for the solution of the Schrödinger equation (Figure 5.9). The ∇^2 operator in spherical coordinates is

$$\frac{1}{r^2} \frac{\partial}{\partial r} \left(r^2 \frac{\partial}{\partial r} \right) + \frac{1}{r^2 \sin^2 \theta} \frac{\partial^2}{\partial \phi^2} + \frac{1}{r^2 \sin \theta} \frac{\partial}{\partial \theta} \left(\sin \theta \frac{\partial}{\partial \theta} \right) \tag{5.27}$$

in terms of which the time-independent Schrödinger equation for the hydrogen atom becomes

$$\frac{-\hbar^2}{2m} \nabla^2 \psi(r, \theta, \phi) - \frac{e^2}{r} \psi(r, \theta, \phi) = E\psi(r, \theta, \phi) \tag{5.28}$$

This is a partial differential equation of three variables, and the standard method of solution is to separate it into three ordinary differential

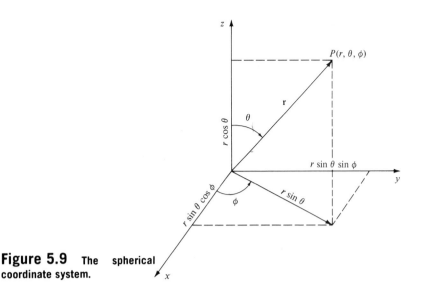

Figure 5.9 The spherical coordinate system.

equations. The separation of variables is achieved by assuming the solution to be in the form of a product of three single-variable functions

$$\psi(r,\ \theta,\ \phi) = R(r)\Theta(\theta)\Phi(\phi) \tag{5.29}$$

and by substituting it in the original equation, Eq. (5.28). The division of the resulting expression by $R\Theta\Phi$ yields a partial separation of variables

$$\frac{\sin^2\theta}{R}\frac{\partial}{\partial r}\left(r^2\frac{\partial R}{\partial r}\right) + \frac{\sin\theta}{\Theta}\frac{\partial}{\partial\theta}\left(\sin\theta\frac{\partial\Theta}{\partial\theta}\right) + \frac{2mr^2}{\hbar^2}\sin\theta\left(E + \frac{e^2}{r}\right)$$

$$= -\frac{1}{\Phi}\frac{\partial^2\Phi}{\partial\phi^2} \tag{5.30}$$

The left side of Eq. (5.30) is a function of r and θ and the right side is a function of ϕ, and thus each side can change independently of the other. Consequently, the equality will hold only if both sides of the equation are equal to the same constant. Denoting this constant by m_l^2, the right side of Eq. (5.30) becomes

$$\frac{d^2\Phi}{d\phi^2} + m_l^2\Phi = 0 \tag{5.31}$$

This is the first of the three ordinary differential equations, and is referred to as the **azimuthal wave equation.**

The left side of Eq. (5.30), upon dividing through by $\sin^2\theta$ and rearranging, takes the form

$$\frac{1}{R}\frac{\partial}{\partial r}\left(r^2\frac{\partial R}{\partial r}\right) + \frac{2mr^2}{\hbar^2}\left(E + \frac{e^2}{r}\right)$$

$$= \frac{m_l^2}{\sin^2\theta} - \frac{1}{\Theta\sin\theta}\frac{\partial}{\partial\theta}\left(\sin\frac{\partial\Theta}{\partial\theta}\right) \tag{5.32}$$

Again, the right and left sides of Eq. (5.32) are functions of different variables, and hence both sides of the equation must equal the same constant. This constant is most conveniently written in the form $l(l+1)$ where l is a parameter. With this choice of a constant Eq. (5.32) separates into two additional equations

$$\frac{d}{dr}\left(r^2\frac{dR}{dr}\right) + \frac{2mr^2}{\hbar^2}\left(E + \frac{e^2}{r}\right)R = l(l+1)R \tag{5.33}$$

and

$$-\frac{1}{\sin\theta}\frac{d}{d\theta}\left(\sin\theta\frac{d\Theta}{d\theta}\right) + \frac{m_l^2\Theta}{\sin^2\theta} = l(l+1)\Theta \tag{5.34}$$

Equations (5.33) and (5.34) are referred to as the **radial wave equation** and the **polar wave equation,** respectively. We have thus achieved the separation of the original Schrödinger equation into three ordinary differential equations.

The solution of the azimuthal wave equation, Eq. (5.31), is the simplest to obtain. Analogous to the classical equation of motion for a simple harmonic oscillator, it has a solution of the form

$$\begin{aligned}\Phi(\phi) &= e^{im_l\phi}\\ &= \cos m_l\phi + i\sin m_l\phi\end{aligned} \tag{5.35}$$

subject to the boundary condition that $\Phi(\phi)$ must be single-valued everywhere. Such a requirement implies

$$\Phi(\phi) = \Phi(\phi + 2\pi)$$

or

$$\Phi(0) = \Phi(2\pi) \tag{5.36}$$

More explicitly, the boundary condition implies

$$1 = \cos m_l 2\pi + i\sin m_l 2\pi$$

and is satisfied only by the integral values of m_l, that is,

$$m_l = 0, \pm 1, \pm 2, \pm 3, \ldots \tag{5.37}$$

Hence the acceptable solutions of the azimuthal wave equation exist only for integral values of m_l, and are accordingly labeled by m_l. The normalized azimuthal wave functions are

$$\Phi_{m_l}(\phi) = \frac{1}{\sqrt{2\pi}}\, e^{im_l\phi} \tag{5.38}$$

Example 5.2 Normalize the azimuthal wave function.

Solution. The normalization condition is

$$\int \psi^*\psi\, dV = 1$$

The volume element in spherical coordinates is

$$dV = r^2 \sin\theta\, dr\, d\theta\, d\phi$$

So

$$\int_0^\infty R^*R\, r^2\, dr \int_0^\pi \Theta^*\Theta \sin\theta\, d\theta \int_0^{2\pi} \Phi^*\Phi\, d\phi = 1$$

or

$$\int_0^{2\pi} \Phi_{m_l}^*\Phi_{m_l}\, d\phi = 1$$

Writing the azimuthal wave function

$$\Phi_{m_l}(\phi) = A\, e^{im_l\phi}$$

so that

$$\int_0^{2\pi} \Phi_{m_l}^*\Phi_{m_l}\, d\phi = A^2 \int_0^{2\pi} e^{i(m_l - m_l)\phi} d\phi = A^2 2\pi$$

Hence

$$A = \frac{1}{\sqrt{2\pi}} \quad \bullet$$

The solution of the polar equation, Eq. (5.34), is beyond the scope of our present purpose. We merely state that the polar equation, subject to the boundary condition that $\Theta(\theta)$ must be finite everywhere, has a class of solutions known as **associated Legendre polynomials**. These solutions exist only for those values of the parameter l that are equal to or greater than the absolute magnitude of m_l

$$l = |m_l|, |m_l| + 1, |m_l| + 2, \ldots \tag{5.39}$$

Thus l must be zero or a positive integer. For a given l, m_l assumes values

$$m_l = 0, \pm 1, \pm 2, \ldots, \pm l \tag{5.40}$$

A polar wave function is labeled by both l and m_l

$$\Theta(\theta) = A_{lm_l} \Theta_{lm_l}(\theta) \tag{5.41}$$

The normalization constant A_{lm_l} is also labeled by l and m_l.

Example 5.3 Normalize the polar wave function for $l = 1$, $m_l = 0$.

Solution. The normalization condition for $\Theta(\theta)$ is

$$\int_0^\pi \Theta^2 \sin\theta \, d\theta = 1$$

Writing $\Theta_{10} = A_{10} \cos\theta$,

$$\int_0^\pi A_{10}^2 \cos^2\theta \sin\theta \, d\theta = -A_{10}^2 \left.\frac{\cos^3\theta}{3}\right]_0^\pi = \frac{2A_{10}^2}{3} = 1$$

Hence

$$A_{10} = \sqrt{\frac{3}{2}} \quad \bullet$$

Table 5.1   Polar Wave Functions

l	m_l	Θ_{lm_l}	Normalization Constant A_{lm_l}	Functional Form
0	0	Θ_{00}	$\dfrac{1}{\sqrt{2}}$	1
1	0	Θ_{10}	$\sqrt{\dfrac{3}{2}}$	$\cos\theta$
1	± 1	$\Theta_{1\pm1}$	$\sqrt{\dfrac{3}{4}}$	$\sin\theta$
2	0	Θ_{20}	$\sqrt{\dfrac{5}{8}}$	$3\cos^2\theta - 1$
2	± 1	$\Theta_{2\pm1}$	$\sqrt{\dfrac{15}{4}}$	$\sin\theta\cos\theta$
2	± 2	$\Theta_{2\pm2}$	$\sqrt{\dfrac{15}{16}}$	$\sin^2\theta$

In solving the radial wave equation, Eq. (5.33), it turns out to be convenient to rewrite the equation by defining

$$n^2 = \frac{-\alpha^2 \, mc^2}{2E} \qquad (5.42)$$

The solutions of the equation, subject to the boundary condition that $R(r)$ must not diverge as r approaches infinity, are then found to exist only for those values of n such that

$$n = l + 1, \, l + 2, \, l + 3, \, \ldots$$

or

$$n = 1, \, 2, \, 3, \, \ldots$$
$$l = 0, \, 1, \, 2, \, 3, \, \ldots, \, n - 1 \qquad (5.43)$$

This integral condition on n imposes a condition on the energy of the hydrogen atom, namely,

$$E = \frac{-\alpha^2 mc^2}{2n^2} = \frac{-13.6 \text{ eV}}{n^2} \qquad (5.44)$$

This is a crucial result: The energy of the hydrogen atom is quantized in the Schrödinger theory precisely as in the Bohr theory, although the

Table 5.2   Normalized Radial Wave Functions $R_{n,l}(r)$

n	l	Normalization Constant	e^{-r/na_0}	$\left(\dfrac{r}{a_0}\right)^l$	$L_{n,l}(r)$
1	0	$\dfrac{2}{a_0^{3/2}}$	e^{-r/a_0}	1	1
2	0	$\dfrac{1}{2\sqrt{2}\,a_0^{3/2}}$	$e^{-r/2a_0}$	1	$\left(2 - \dfrac{r}{a_0}\right)$
2	1	$\dfrac{1}{2\sqrt{6}\,a_0^{3/2}}$	$e^{-r/2a_0}$	$\left(\dfrac{r}{a_0}\right)$	1
3	0	$\dfrac{2}{81\sqrt{3}\,a_0^{3/2}}$	$e^{-r/3a_0}$	1	$\left(27 - 18\dfrac{r}{a_0} + \dfrac{2r^2}{a_0^2}\right)$
3	1	$\dfrac{4}{81\sqrt{6}\,a_0^{3/2}}$	$e^{-r/3a_0}$	$\left(\dfrac{r}{a_0}\right)$	$\left(6 - \dfrac{r}{a_0}\right)$
3	2	$\dfrac{4}{81\sqrt{30}\,a_0^{3/2}}$	$e^{-r/3a_0}$	$\left(\dfrac{r}{a_0}\right)^2$	1

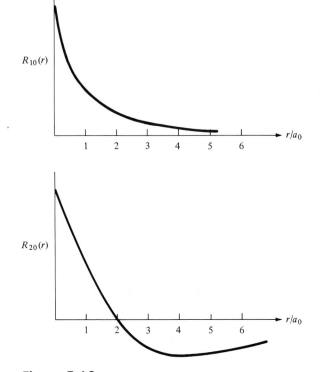

Figure 5.10 The first two radial wave functions.

interpretation of the number n is different in the two theories. The radial wave functions are labeled by n and l, and are of the form

$$R_{nl}(r) = e^{-r/na_0} \left(\frac{r}{a_0}\right)^l L_{n,l}(r) \tag{5.45}$$

where $L_{n,l}(r)$ is a class of functions called **associated Laguerre polynomials** and a_0 is the Bohr radius (see Figure 5.10).

Example 5.4 Normalize the radial wave function for $n = 2$, $l = 1$.

Solution. With the radial wave function written in the form

$$R(r) = A_{n,l} R_{nl}(r)$$

the radial probability integral becomes

$$\int_0^\infty |R(r)|^2 \, r^2 \, dr = A_{n,l} \int_0^\infty |R_{nl}(r)|^2 \, r^2 \, dr = 1$$

This integral for $n = 2$, $l = 1$ is

$$\frac{A_{21}^2}{a_0^2} \int_0^\infty e^{-r/a_0} r^4 \, dr$$

which is, by use of the table of integrals, evaluated

$$\frac{A_{21}^2}{a_0^2} \frac{4!}{\left(\frac{1}{a_0}\right)^5} = A_{21}^2 \, 24 \, a_0^3$$

Setting it equal to unity,

$$A_{21} = \frac{1}{2\sqrt{6} \, a_0^{3/2}} \quad \bullet$$

The numerical value of a_0 as well as that of E is computed from the radial wave function by substituting it in the radial wave equation. Consider, for example, the ground state radial wave function

$$R_{10}(r) = \frac{2}{a_0^{3/2}} e^{-r/a_0} \tag{5.46}$$

Substitution in the wave equation, Eq. (5.33), yields

$$\left(\frac{r^2}{a_0^2} - \frac{2r}{a_0}\right) R_{10} + \left(\frac{2 \, mEr^2}{\hbar^2} + \frac{2 \, me^2 \, r}{\hbar^2}\right) R_{10} = 0 \tag{5.47}$$

which, upon rearrangement, reduces to

$$\left(\frac{2 \, me^2}{\hbar^2} - \frac{2}{a_0}\right) r + \left(\frac{1}{a_0^2} + \frac{2mE}{\hbar^2}\right) r^2 = 0 \tag{5.48}$$

Clearly, both terms of Eq. (5.48) must separately equal zero, that is,

$$\frac{2me^2}{\hbar^2} - \frac{2}{a_0} = 0$$

yielding

$$a_0 = \frac{\hbar^2}{me^2} = 0.529 \text{ Å} \tag{5.49}$$

and

$$\frac{1}{a_0^2} + \frac{2mE}{\hbar^2} = 0$$

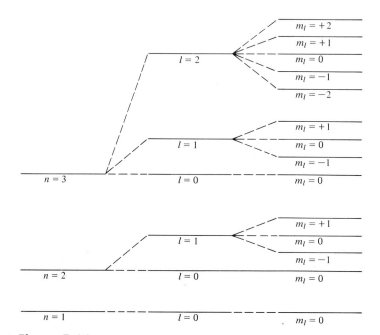

Figure 5.11 A diagramatic display of the rule governing the spatial quantum numbers *n, l, m₁*.

yielding

$$E = -\frac{\alpha^2 mc^2}{2} \cong -13.6 \text{ eV} \tag{5.50}$$

We have now completed our attempt to outline the solution of the time-independent Schrödinger equation for the hydrogen atom. The physically acceptable solutions are the eigenfunctions

$$\psi_{nlm_l}(r, \theta, \phi) = R_{nl}(r)\Theta_{lm_l}(\theta)\Phi_{m_l}(\phi) \tag{5.51}$$

and they exist only for integral values of the constants *n, l, m₁*, namely,

$$
\begin{aligned}
&n = 1, 2, 3, 4, \ldots \\
&0 \le l \le n - 1 \\
&-l \le m_l \le l
\end{aligned}
\tag{5.52}
$$

These constants are related to the three dimensions of space, thus being referred to as **spatial quantum numbers,** and the rule governing them is displayed in Figure 5.11. Wave functions differing in their quantum numbers are linearly independent, and the total number of linearly independent solutions for a given n is n^2:

$$\sum_{l=0}^{n-1} (2l+1) = 1 + 3 + \ldots + (2n-1)$$

$$= \left[\frac{1 + (2n-1)}{2}\right] n = n^2 \tag{5.53}$$

Table 5.3 Wave Functions of Hydrogen Atoms and Hydrogenlike Ions of Nuclear Charge Ze.

$$\psi_{100} = \frac{1}{\sqrt{\pi}} \left(\frac{Z}{a_0}\right)^{3/2} \cdot e^{-Zr/a_0}$$

$$\psi_{200} = \frac{1}{4\sqrt{2\pi}} \left(\frac{Z}{a_0}\right)^{3/2} \cdot \left(2 - \frac{Zr}{a_0}\right) \cdot e^{-Zr/2a_0}$$

$$\psi_{210} = \frac{1}{4\sqrt{2\pi}} \left(\frac{Z}{a_0}\right)^{3/2} \cdot \frac{Zr}{a_0} \cdot e^{-Zr/2a_0} \cos\theta$$

$$\psi_{21\pm1} = \frac{1}{8\sqrt{\pi}} \left(\frac{Z}{a_0}\right)^{3/2} \cdot \frac{Zr}{a_0} \cdot e^{-Zr/2a_0} \sin\theta \cdot e^{\pm i\phi}$$

$$\psi_{300} = \frac{1}{81\sqrt{3\pi}} \left(\frac{Z}{a_0}\right)^{3/2} \cdot \left(27 - 18\frac{Zr}{a_0} + \frac{2Z^2r^2}{a_0^2}\right) e^{-Zr/3a_0}$$

$$\psi_{310} = \frac{1}{81}\sqrt{\frac{2}{\pi}} \left(\frac{Z}{a_0}\right)^{3/2} \left(6 - \frac{Zr}{a_0}\right) \cdot \frac{Zr}{a_0} e^{-Zr/3a_0} \cos\theta$$

$$\psi_{31\pm1} = \frac{1}{81\sqrt{\pi}} \left(\frac{Z}{a_0}\right)^{3/2} \left(6 - \frac{Zr}{a_0}\right) \cdot \frac{Zr}{a_0} \cdot e^{-Zr/3a_0} \sin\theta \cdot e^{\pm i\phi}$$

$$\psi_{320} = \frac{1}{81\sqrt{6\pi}} \left(\frac{Z}{a_0}\right)^{3/2} \cdot \frac{Z^2r^2}{a_0^2} \cdot e^{-Zr/3a_0} \cdot (3\cos^2\theta - 1)$$

$$\psi_{32\pm1} = \frac{1}{81\sqrt{\pi}} \left(\frac{Z}{a_0}\right)^{3/2} \cdot \frac{Z^2r^2}{a_0^2} \cdot e^{-Zr/3a_0} \cdot \sin\theta \cdot \cos\theta \cdot e^{\pm i\phi}$$

$$\psi_{32\pm2} = \frac{1}{162\sqrt{\pi}} \left(\frac{Z}{a_0}\right)^{3/2} \cdot \frac{Z^2r^2}{a_0^2} \cdot e^{-Zr/3a_0} \cdot \sin^2\theta \cdot e^{\pm 2i\phi}$$

The wave functions of the hydrogen atom for the first three values of n are listed in Table 5.3, and a photographic representation of the electron probability densities of hydrogen for several low-lying energy states is given in Figure 5.12.

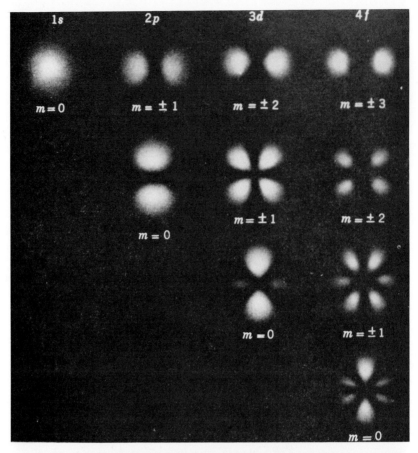

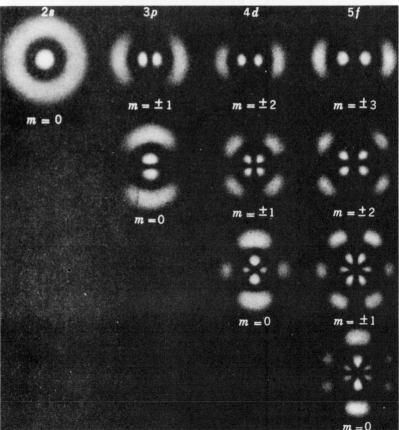

5.5 The Interpretation of the Spatial Quantum Numbers

The time-independent Schrödinger equation is an energy eigenvalue equation

$$H\psi = E\psi \tag{5.54}$$

with the total energy of the system represented by the Hamiltonian operator, $-(\hbar^2/2m)\nabla^2 + V(r)$. The eigenfunctions of the Hamiltonian operator are specified by all three spatial quantum numbers n, l, m_l, but the energy eigenvalues depend only on the so-called **principal quantum number** n. The energy of the hydrogen atom is thus degenerate with respect to the quantum numbers l and m_l, and there exists for a given n, an n^2 number of linearly independent eigenfunctions sharing the same energy eigenvalue. The physical basis for this fact is the strict r^{-1} dependence of the potential field of the hydrogen atom. Deviations from the pure r^{-1} dependence of the potential function occur in multielectron atoms and are caused by the shielding of the nucleus by electrons (see Section 7.6 for details). The energy of such atoms is not degenerate with respect to l. The energy of the lithium atom, for example, depends on n as well as on l. The degeneracy with respect to m_l arises from the central character of the force field, and its removal requires the presence of external fields such as a magnetic field.

For the physical interpretation of l, let us rewrite the polar equation, Eq. (5.34), to include the coordinate ϕ

$$-\frac{1}{\sin\theta}\frac{\partial}{\partial\theta}\left(\sin\theta\,\frac{\partial\Theta\Phi}{\partial\theta}\right) - \frac{\partial^2\Theta\Phi}{\partial\phi^2} = l(l+1)\Theta\Phi \tag{5.55}$$

Defining a new angular function

$$Y(\theta,\phi) = \Theta(\theta)\,\Phi(\phi) \tag{5.56}$$

which is called a **spherical harmonic,** we can cast Eq. (5.55) in the form of an eigenvalue equation

$$-\left[\frac{1}{\sin\theta}\frac{\partial}{\partial\theta}\left(\sin\theta\,\frac{\partial}{\partial\theta}\right) + \frac{\partial^2}{\partial\phi^2}\right]Y(\theta,\phi) = l(l+1)Y(\theta,\phi) \tag{5.57}$$

Figure 5.12 Photographic representation of the electron probability density for low-lying energy eigenstates of the hydrogen atom. [Reproduced from R. B. Leighton, *Principles of Modern Physics.* (New York: McGraw-Hill Book Company, 1959). Used by permission of the publisher.]

Multiplied by $\hbar^2$, the operator of Eq. (5.57) is precisely the quantum mechanical operator representing the square of the orbital angular momentum vector, $\mathbf{L} = \mathbf{r} \times \mathbf{p}$.

$$\mathbf{L}_{op}^2 = -\hbar^2 \left[\frac{1}{\sin\theta} \frac{\partial}{\partial\theta} \left(\sin\theta \frac{\partial}{\partial\theta} \right) + \frac{\partial^2}{\partial\phi^2} \right] \tag{5.58}$$

and, operating on the total eigenfunctions of the hydrogen atom, yields

$$\mathbf{L}_{op}^2 \, \psi_{nlm_l}(r, \theta, \phi) = l(l+1)\hbar^2 \, \psi_{nlm_l}(r, \theta, \phi) \tag{5.58'}$$

Now it is a general feature of quantum mechanics that the eigenvalues of a quantum mechanical operator are the only possible results of a precise measurement of the physical quantity represented by the operator. Consequently, the quantity $l(l + 1)\hbar^2$, as an eigenvalue of $\mathbf{L}_{op}^2$, gives the square of the orbital angular momentum vector of the atom in the state described by $\psi_{nlm_l}(r, \theta, \phi)$

$$\mathbf{L}^2 = l(l+1)\hbar^2 \tag{5.59}$$

or

$$L = \sqrt{l(l+1)} \; \hbar \tag{5.60}$$

The orbital angular momentum of a quantum state thus depends only on the quantum number l, which is aptly called the **orbital angular momentum quantum number.** Moreover, the natural unit for angular momentum is $\hbar$.

The energy eigenfunction of the hydrogen atom is also an eigenfunction of a quantum mechanical operator representing the z component of the orbital angular momentum

$$L_{z_{op}} = -i\hbar \frac{\partial}{\partial\phi} \tag{5.61}$$

which yields

$$L_{z_{op}} \, \psi_{nlm_l}(r, \theta, \phi) = m_l \hbar \psi_{nlm_l}(r, \theta, \phi) \tag{5.62}$$

The eigenvalue $m_l \hbar$ gives the z component of the orbital angular momentum vector $\mathbf{L}(l)$

$$L_z(l) = m_l \hbar \tag{5.63}$$

The possible values of m_l, for a given l, are

$$m_l = 0, \pm 1, \pm 2, \pm 3, \ldots, \pm l$$

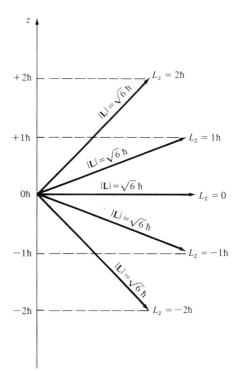

Figure 5.13 The space quantization of the orbital angular momentum vector L for $I = 2$.

implying that $L(l)$ can assume only $(2l + 1)$ possible orientations relative to a given, say, z, direction in space (Figure 5.13). Such a direction in space may be established by an external magnetic field, and the energy of an atom in a magnetic field is not degenerate with respect to m_l. Thus the quantum number m_l is called the **magnetic quantum number.**

The orientational quantization of the orbital angular momentum vector is referred to as space quantization. In terms of the polar angle θ, space quantization is described by

$$\cos \theta = \frac{m_l}{\sqrt{l(l+1)}} \tag{5.64}$$

5.6   The Zeeman Effect

The energy of an isolated atom is degenerate with respect to the magnetic quantum number because its force field as a central force does not prefer any particular direction in space. If a direction is established by an external magnetic field, the orbital angular momentum of an atom undergoes space quantization as a result of an interaction

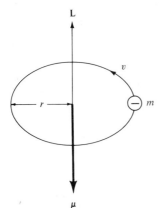

Figure 5.14 The orbital magnetic moment vector associated with a circular orbit of an atom.

between the magnetic moment of the atom and the external magnetic field.

The orbital magnetic moment of an atom has its origin in the orbital angular momentum of the atom. In quantum mechanics as well as in classical electromagnetism the orbital magnetic moment is proportional to the orbital angular momentum with the proportionality constant called the **gyromagnetic ratio.** Defined as $\mu = i\pi r^2/c$, the magnetic moment of a Bohr orbit, for example, is (see Figure 5.14)

$$\mu = \frac{-e}{2mc}\,\mathbf{L} \qquad (5.65)$$

since the current of the orbiting electron is

$$i = \frac{-ev}{2\pi r} \qquad (5.66)$$

with $-e$ and m denoting the charge and mass of the electron.

The quantum mechanical gyromagnetic ratio for the atomic magnetic moment is written as

$$\frac{\boldsymbol{\mu}_l}{\mathbf{L}} = -\frac{g_l e}{2mc} \qquad (5.67)$$

although g_l, called the orbital g factor, has the value

$$g_l = 1 \qquad (5.68)$$

The magnitude of the atomic magnetic moment is

$$\mu_l = g_l \frac{e\hbar}{2mc}\,\sqrt{l(l+1)} \qquad (5.69)$$

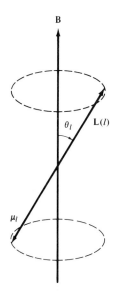

Figure 5.15 In a given angular momentum state, μ_l and thus L precess about the magnetic field B so that the magnetic energy and θ_l remain constant.

and gives rise to a natural unit for atomic magnetic moments. Called the **Bohr magneton,** its numerical value is

$$\mu_B = \frac{e\hbar}{2mc} = 9.27 \times 10^{-21} \text{ erg/gauss} \tag{5.70}$$

In terms of the Bohr magneton the atomic magnetic moment is

$$\boldsymbol{\mu}_l = -\frac{g_l \mu_B}{\hbar} \, \mathbf{L}$$

or

$$|\boldsymbol{\mu}_l| = g_l \mu_B \, \sqrt{l(l+1)} \tag{5.71}$$

The classical expression for the potential energy ΔE_m of a magnetic moment $\boldsymbol{\mu}$ in a magnetic field $\mathbf{B}$ is

$$\Delta E_m = -\boldsymbol{\mu} \cdot \mathbf{B} \tag{5.72}$$

It turns out that the same expression can be used for the potential energy of an atomic magnetic moment in an external magnetic field, provided the associated angular momentum vector is treated quantum mechanically. This means that the potential energy of an atomic magnetic moment will assume $(2l + 1)$ values, that is,

$$\begin{aligned}
\Delta E_{m_l} &= \left(\frac{\mu_B}{\hbar}\right) \mathbf{L} \cdot \mathbf{B} \\
&= m_l \mu_B \, |\mathbf{B}| \tag{5.73}
\end{aligned}$$

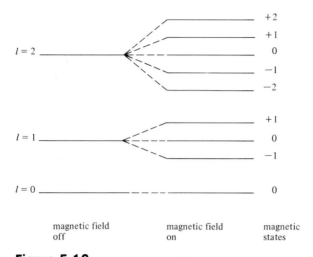

magnetic field off   magnetic field on   magnetic states

Figure 5.16   Energy level splittings of angular momentum states by an external magnetic field.

corresponding to $(2l + 1)$ possible orientations of the magnetic moment vector with respect to the magnetic field direction. Hence an atom in an angular momentum state l, whose energy, in the absence of an external magnetic field, is E_l, can now exist in any one of the $(2l + 1)$ possible magnetic substates whose energy E_{m_l} is (see Figure 5.16)

$$E_{m_l} = E_l + \Delta E_{m_l} \tag{5.74}$$

The spectral lines emitted from atoms in an external magnetic field are governed by a selection rule on the magnetic quantum number (see Chapter 6 for its derivation)

$$\Delta m_l = 0, \pm 1 \tag{5.75}$$

and thus consist of three components. The central component is identical to the spectral line without the magnetic field and arises from a transition that involves no change in the magnetic quantum number, that is, $\Delta m_l = 0$. Transitions for $\Delta m_l = \pm 1$ give rise to the outer components whose frequencies are, respectively, higher and lower than the frequency of the central component. The frequencies of the three spectral lines are

$$\nu_+ = \nu_0 + \frac{\mu_B B}{h} \qquad \Delta m_l = +1$$

$$\nu_0 = \nu_0 \qquad\qquad \Delta m_l = 0 \tag{5.76}$$

$$\nu_- = \nu_0 - \frac{\mu_B B}{h} \qquad \Delta m_l = -1$$

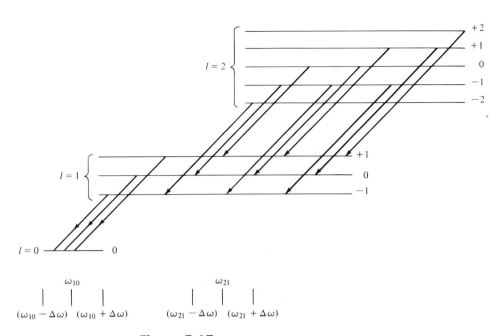

Figure 5.17 The Zeeman splitting of spectral lines: $\Delta\omega = \mu_B B/\hbar$.

This phenomenon, in which the spectral line of an atom splits into three components in the presence of a magnetic field, is referred to as the normal Zeeman effect (Figure 5.17). The Zeeman effect was first observed by Pieter Zeeman in 1892.

5.7 Space Quantization and Electron Spin

The concept of space quantization is implicit in the analysis of the Zeeman effect. The $(2l + 1)$ multiplet structure of an energy level, however, is not directly observed in the Zeeman effect.

In 1921, O. Stern and W. Gerlach proposed, and subsequently performed, an experiment that would explicitly display the space quantization of atoms. Their experiment was based on a well-known fact in electrodynamics that a magnetic moment $\boldsymbol{\mu}$ in an inhomogeneous magnetic field experiences a translational force. If the magnetic field is inhomogeneous in the z direction, the translational force is also in the z direction and is given by

$$F_z = -\frac{\partial(-\boldsymbol{\mu} \cdot \mathbf{B})}{\partial z} = \mu_z \frac{\partial B}{\partial z} \tag{5.77}$$

Now an atomic magnetic moment has the form

$$\boldsymbol{\mu} = -\frac{g_l \mu_B}{\hbar} \mathbf{L}$$

or

$$\mu_z = -m_l g_l \mu_B \tag{5.78}$$

Hence the translational force that an atomic magnetic moment experiences in an inhomogeneous magnetic field is

$$F_z = -m_l g_l \mu_B \frac{\partial B}{\partial z} \tag{5.79}$$

which is clearly quantized. Thus, upon passing through a region of an inhomogeneous magnetic field, atoms in an orbital angular momentum state l get separated into a $(2l + 1)$ number of spatial components.

Stern and Gerlach performed their initial experiment with neutral silver atoms (Figure 5.18). A beam of silver atoms was caused to traverse a nonuniform magnetic field at right angles to the field, whereupon the atoms emerged separated into two beams, depositing at two different spots on the photographic plate. Their observation thus succeeded in establishing a sort of space quantization. However, their result could not be interpreted in terms of the magnetic quantum number m_l, because an atomic moment associated with the electron's orbital motion must be space-quantized into an odd [i.e., $(2l + 1)$] number of components.

A result similar to the Stern-Gerlach observation was obtained using hydrogen atoms. Note, however, that a hydrogen beam cannot be space-quantized by an inhomogeneous magnetic field into two beams by the orbital magnetic moment of the atom because the angular moment of a ground-state hydrogen atom is zero, that is, $l = 0$. The observed space quantization of hydrogen atoms, therefore, must be traced to something other than the orbital motion of the electron. The possibility thus emerges that the observed space quantization may be associated with the electron itself, that is, the electron's intrinsic magnetic moment. Now a magnetic moment, in quantum theory, is always

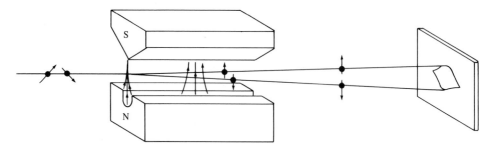

Figure 5.18   A schematic diagram of the Stern-Gerlach apparatus.

associated with an angular momentum. Hence, if the existence of an intrinsic angular momentum is postulated for the electron, the intrinsic magnetic momentum of the electron must be expressed in a form analogous to the orbital magnetic moment, namely,

$$\boldsymbol{\mu} = -\frac{g_s \mu_B}{\hbar} \mathbf{S}$$

or

$$\mu_z = -m_s g_s \mu_B \tag{5.80}$$

Here $\mathbf{S}$ denotes the intrinsic, or spin, angular momentum vector of the electron; g_s is called the spin g factor; and $m_s \hbar$ represents the z components of $\mathbf{S}$.

Now the number of possible values of m_s can be deduced from the observed multiplicity of the space-quantized components of the hydrogen beam, and this number is 2. Furthermore, the observed multiplicity determines the spin quantum number s according to

$$\text{multiplicity} = 2s + 1 \tag{5.81}$$

which yields

$$s = \tfrac{1}{2} \tag{5.82}$$

Hence the magnitude of the spin vector $\mathbf{S}$, by analogy with that of the orbital angular momentum vector, is

$$|\mathbf{S}| = \sqrt{s(s + 1)}\,\hbar = \frac{\sqrt{3}}{2}\,\hbar \tag{5.83}$$

and the z components of the spin vector $\mathbf{S}$ are (see Figure 5.19)

$$S_z = m_s \hbar$$

where

$$m_s = -\tfrac{1}{2}, +\tfrac{1}{2} \tag{5.84}$$

The spin g factor, or the gyromagnetic ratio, of the electron may be measured in the following way. Equation (5.79) can, for a hydrogen atom, be recast in the form

$$\frac{F_z}{\mu_B (\partial B / \partial z)} = -m_s g_s \tag{5.85}$$

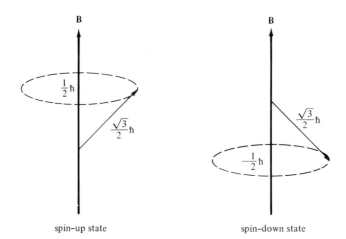

spin–up state   spin–down state

Figure 5.19   The spin angular momentum vector **S** precesses about a magnetic field **B**. Its projection on the axis of precession is $+\frac{1}{2}\hbar$ or $-\frac{1}{2}\hbar$.

Now the translational force F_z, which causes the atoms to deviate from the initial path, can be evaluated from the measurement of the splitting of the beam. The nonuniform magnetic field can be measured directly. It is found experimentally that

$$\frac{F_z}{\mu_B(\partial B/\partial z)} \approx \pm 1$$

whence the spin g factor is

$$g_s \approx 2 \tag{5.86}$$

In the relativistic quantum theory of the electron, which was developed by P. A. M. Dirac in 1928, the spin g factor is predicted to be precisely 2. The currently accepted value is $g_s = 2.003192$, and the slight deviation from $g_s = 2$ is fully accounted for by quantum electrodynamics, which was developed independently by R. P. Feynman, J. S. Schwinger, and S. I. Tomonaga in the 1940s.

Electron spin is real. Hence the spatial quantum numbers (n, l, m_l) alone do not completely specify the quantum state of an electron. The spin quantum number s must now be added.

5.8   The Spin-Orbit Interaction

The question now arises: Does electron spin affect atomic structure? That is, is there an internal magnetic field in the atom? To explore this question, begin with the Bohr model of the hydrogen atom. An orbiting electron, according to classical electrodynamics, gives rise to a magnetic field, which, at the position of the nucleus, is simple to calculate. However, the question of importance here is whether there is a magnetic field at the position of the electron. Let us therefore view the atom in the rest frame of the electron. In this frame the nucleus orbits the electron at velocity $-\mathbf{v}$ and produces a magnetic field according to Ampère's law

$$\mathbf{B} = \frac{-\mathbf{v} \times \mathbf{E}}{c} \tag{5.87}$$

The electric field $\mathbf{E}$ produced by the nucleus is

$$\mathbf{E} = \frac{e\mathbf{r}}{r^3} \tag{5.88}$$

Consequently, the magnetic field at the position of the electron is

$$\mathbf{B} = \frac{-e\mathbf{v} \times \mathbf{r}}{cr^3}$$
$$= \frac{e\mathbf{r} \times (m\mathbf{v})}{cr^3 m}$$
$$= \left(\frac{e}{mcr^3}\right)\mathbf{L} \tag{5.89}$$

Here $\mathbf{L}$ is the orbital angular momentum of the electron.

The existence of an atomic magnetic field suggests an interaction between the magnetic field and the spin magnetic moment of the electron. The classical expression for the orientational potential energy of a magnetic moment $\boldsymbol{\mu}$ in a magnetic field $\mathbf{B}$ is $\Delta E = -\boldsymbol{\mu} \cdot \mathbf{B}$, and would be valid in the rest frame of the electron. However, because atoms are observed in the laboratory frame of reference, a Lorentz transformation from the rest frame of the electron to the laboratory frame is necessary. In the laboratory frame the orientational energy of the spin magnetic moment of an electron in its atomic magnetic field is[2]

$$\Delta E = -\frac{1}{2}\boldsymbol{\mu} \cdot \mathbf{B} \tag{5.90}$$

The only difference in the two frames is thus a factor of 1/2. This result along with the substitution of Eqs. (5.80) and (5.89) yields

[2] See, for example, R. M. Eisberg, *Fundamentals of Modern Physics* (New York: Wiley, 1961), pp. 339–344.

$$\Delta E = \frac{e g_s \mu_B}{2mc\hbar r^3} \mathbf{S} \cdot \mathbf{L} \tag{5.91}$$

which, because of the term $\mathbf{S} \cdot \mathbf{L}$, is referred to as the spin-orbit interaction.

The transition from the classical to the quantum mechanical treatment of the spin-orbit interaction is made by "averaging" the classical interaction energy

$$\langle \Delta E \rangle = \int_0^\infty \psi_{nlm_l}^* \, \Delta E \, \psi_{nlm_l} \, dV \tag{5.92}$$

Of the two variables in Eq. (5.92), the averaging of r^{-3} is accomplished by evaluating the integral

$$\left\langle \frac{1}{r^3} \right\rangle = \int_0^\infty \psi_{nlm_l}^* \left(\frac{1}{r^3} \right) \psi_{nlm_l} \, dV$$

$$= \frac{1}{n^3 l(l + \frac{1}{2})(l + 1) a_0^3} \tag{5.93}$$

whereupon Eq. (5.92) takes the compact form

$$\langle \Delta E \rangle = \frac{\alpha^2 |E_n|}{nl(l + \frac{1}{2})(l + 1)} \langle \mathbf{S} \cdot \mathbf{L} \rangle \tag{5.94}$$

with

$$\alpha = \frac{e^2}{\hbar c} \approx \frac{1}{137}$$

Example 5.5   Calculate $\langle r^{-3} \rangle$ for the $n = 2$, $l = 1$, $m_l = 0$ state of the hydrogen atom.

Solution.

$$\left\langle \frac{1}{r^3} \right\rangle = \int_0^\infty \psi_{210}^* \left(\frac{1}{r^3} \right) \psi_{210} \, dV$$

$$= \frac{1}{32\pi a_0^3} \int_0^\infty \int_0^\pi \int_0^{2\pi} \frac{r^2}{a_0^2} e^{-r/a_0} \left(\frac{1}{r^3} \right) \cos^2\theta r^2 \sin\theta \, dr \, d\theta \, d\phi$$

$$= \frac{2\pi}{32\pi a_0^3} \frac{\cos^3\theta}{3} \Big]_\pi^0 \int_0^\infty \frac{r}{a_0^2} e^{-r/a_0} \, dr$$

From the table of integrals,

$$\int_0^\infty \frac{r}{a_0^2} e^{-r/a_0} \, dr = 1$$

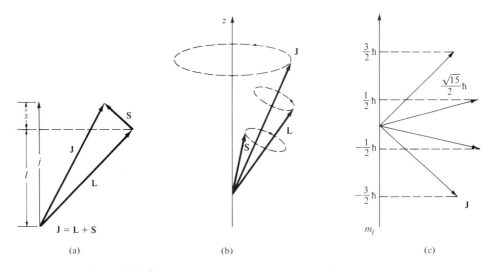

(a)   (b)   (c)

Figure 5.20 (a) When L and S are coupled "parallel," the total angular momentum quantum number is given by $j = l + s$. (b) L and S precess fast about J, and J in turn precesses slowly about a given axis such as provided by a magnetic field. (c) The possible z-components of the total angular momentum vector J for $j = \frac{3}{2}$.

Hence

$$\left\langle \frac{1}{r^3} \right\rangle = \frac{1}{24a_0^3} \quad \bullet$$

The quantum mechanical treatment of the term $\mathbf{S} \cdot \mathbf{L}$ proceeds as follows. Begin with the observation that the spin vector $\mathbf{S}$ and the orbital angular momentum vector $\mathbf{L}$ are now coupled, thus giving rise to a total angular momentum vector $\mathbf{J}$

$$\mathbf{J} = \mathbf{L} + \mathbf{S} \tag{5.95}$$

If $\mathbf{S}$ and $\mathbf{L}$ are coupled "parallel," the total angular momentum quantum number j is given by (see Figure 5.20)

$$j = l + s = l + \tfrac{1}{2} \tag{5.96}$$

and the coupled state is called the "spin-up" state of the electron. If $\mathbf{S}$ and $\mathbf{L}$ are coupled "antiparallel," the total angular momentum vector is given by (see Figure 5.21)

$$j = l - s = l - \tfrac{1}{2} \tag{5.97}$$

and the coupled state is called the "spin-down" state of the electron. The total angular momentum of an atomic state with a spin-orbit interaction is thus

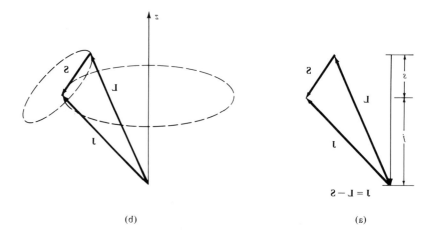

Figure 5.21 (a) When *L* and *S* are coupled "antiparallel," the total angular momentum quantum number is given by *j = l − s*. (b) *L* and *S* precess fast about the vector *J*, and *J* in turn precesses slowly about the *z*-axis established by a magnetic field.

$$\mathbf{J} = \mathbf{L} + \mathbf{S}$$

with

$$
\begin{aligned}
j &= l + \tfrac{1}{2} \quad &&\text{spin-up} \\
j &= l - \tfrac{1}{2} \quad &&\text{spin-down}
\end{aligned}
\tag{5.98}
$$

The quantum mechanics of the total angular momentum is analogous to that of **L** and **S**, so that its magnitude is given by

$$J = \sqrt{j(j + 1)}\,\hbar \tag{5.99}$$

We are now ready to calculate **S · L**. Squaring **J**

$$\mathbf{J}^2 = \mathbf{L}^2 + \mathbf{S}^2 + 2\,\mathbf{S} \cdot \mathbf{L}$$

we obtain

$$\mathbf{S} \cdot \mathbf{L} = \tfrac{1}{2}[\,j(j + 1) - l(l + 1) - s(s + 1)\,] \tag{5.100}$$

Let us now calculate the spin-orbit interaction energy for the $n = 2$, $l = 1$ state, or the $2p$ state in the spectroscopic notation, of the hydrogen atom. The calculation of **S · L** for the $2p$ state is straightforward

$$
\begin{aligned}
\mathbf{S} \cdot \mathbf{L} &= \tfrac{1}{2}\hbar^2 \quad &&\text{spin-up} \\
\mathbf{S} \cdot \mathbf{L} &= -\hbar^2 \quad &&\text{spin-down}
\end{aligned}
$$

The average value of r^{-3} for the $2p$ state is

$$\left\langle \frac{1}{r^3} \right\rangle = \frac{1}{24 a_0^3}$$

The spin-orbit energy for the $2p$ state is thus

$$\Delta E = \frac{\alpha^2 |E_2|}{6 \hbar^2} \mathbf{S} \cdot \mathbf{L} \qquad \qquad \textbf{(5.101)}$$

More specifically, the spin-orbit energy of the spin-up state is

$$\Delta E^+ = \frac{\alpha^2 |E_2|}{6 \hbar^2} (\tfrac{1}{2} \hbar^2)$$

and the spin-orbit energy of the spin-down state is

$$\Delta E^- = \frac{\alpha^2 |E_2|}{6 \hbar^2} (-\hbar^2)$$

Consequently, the total spin-orbit splitting of the energy level E_2 is (see Figure 5.22)

$$\Delta E_2 = \frac{\alpha^2 |E_2|}{6 \hbar^2} \left[\tfrac{1}{2} \hbar^2 - (-\hbar^2) \right]$$
$$= 4.52 \times 10^{-5} \text{ eV}$$

or

$$\frac{\Delta E_2}{|E_2|} = 1.3 \times 10^{-5} \qquad \qquad \textbf{(5.102)}$$
$$\approx \alpha^2$$

The spin-orbit splitting of an energy level is also referred to as the fine structure of the level.

All energy levels, except the $l = 0$ state, possess fine structure, and the multiplicity of the fine structure is $2s + 1 = 2$. Historically, it was this doublet structure of energy levels that led S. Goudsmit and G. E. Uhlenbeck to postulate electron spin in 1925.

The spectroscopic notation for an atomic state must now include

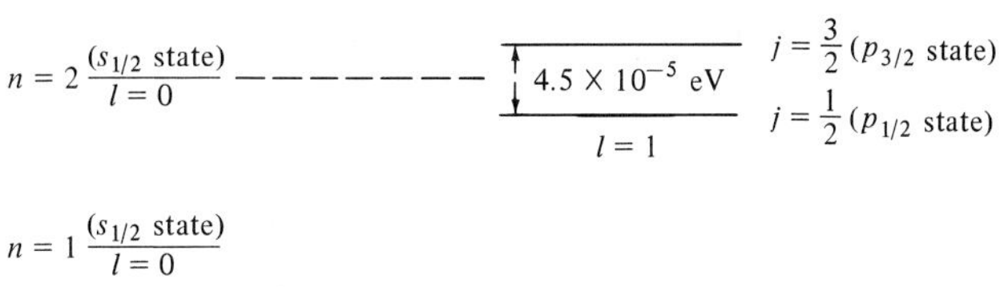

$$n = 2 \frac{(s_{1/2} \text{ state})}{l = 0} \; - - - - - - \; \overline{} \updownarrow \; 4.5 \times 10^{-5} \text{ eV} \quad \frac{}{} \; j = \frac{3}{2} \; (p_{3/2} \text{ state})$$
$$l = 1 \qquad \qquad j = \frac{1}{2} \; (p_{1/2} \text{ state})$$

$$n = 1 \frac{(s_{1/2} \text{ state})}{l = 0}$$

Figure 5.22 The fine structure splitting of the $2p$ state of the hydrogen atom.

Table 5.4   Spectroscopic Notation

l :	0	1	2	3	4	5
symbol :	s	p	d	f	g	h

the fine structure splitting. Each member of a doublet is represented by its total angular momentum quantum number j. The orbital angular momentum state is denoted by a letter symbol, as in Table 5.4. The two states of E_2, for example, are designated by $2\,^2P_{3/2}$ and $2\,^2P_{1/2}$; the first numeral is the principal quantum number, the capital letter symbol gives the orbital angular momentum state of the atom, the superscript indicates the doublet structure of the state and is read "doublet," and the subscript denotes the total angular momentum quantum number of the state.

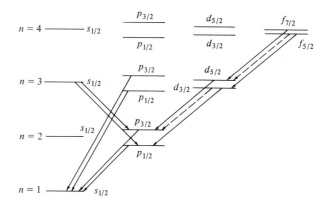

Figure 5.23   Origins of the spectral doublets and triplets with the selection rule $\Delta j = 0, \pm 1$. The dotted line is very much weaker than the other two lines of a triplet.

It is now clear that the spin quantum number must be added to the three spatial quantum numbers to specify a quantum state. If the spin-orbit interaction is ignored, an eigenfunction is specified by the quantum numbers (n, l, m_l, m_s). On the other hand, if the spin-orbit coupling is included, the quantum numbers l and m_l must be replaced by j and m_j, because j and m_j and not l and m_l have definite values by virtue of the conservation of the total angular momentum.

5.9   Summary

The Bohr theory of the hydrogen atom is a quantum version of Rutherford's classical planetary model, and is based on the ad hoc quantum condition $mvr = nh/2\pi$, which yields the quantization of energy and the existence of allowed orbits. The atom is rendered stable by the hypothesis of the nonradiating character of the electron in allowed orbits, and the emission and the absorption of spectral lines are accounted for by the hypothesis of quantum transitions of the electron between allowed orbits.

Moseley's observations on the characteristic x-ray spectra of the elements led to the concept of the atomic number, and the fundamental significance of the atomic number was provided by the Bohr theory.

Insight into the possible origin of Bohr's ad hoc quantum condition was first achieved by de Broglie, viewing the atomic electron as a standing wave.

The wave theory of the atom was proposed by Schrödinger. The time-independent Schrödinger equation is an energy eigenvalue equation

$$H\psi = E\psi$$

where H is the Hamiltonian operator, $-(\hbar^2/2m)\,\nabla^2 + V(\mathbf{r})$, representing the total energy of the system. The physically acceptable solutions of the equation exist only for a discrete set of energies, namely, the energy eigenvalues, and thus the quantization of energy emerges in a natural way. The energy eigenfunctions of the hydrogen atom are simultaneously eigenfunctions of the angular momentum operators L^2_{op} and $L_{z_{op}}$, and this fact implies that all three quantum numbers n, l, m_l have definite values. The relations among the spatial quantum numbers are as follows: For a given n, $0 \leq l \leq n - 1$, and for a given l, $-l \leq m \leq l$, with all quantum numbers assuming integer values.

The splitting of a spectral line into three components by an external magnetic field is called the Zeeman effect and is a result of quantum transitions subject to the selection rule $\Delta m_l = 0, \pm 1$. Although implicit in the Zeeman effect, a direct confirmation of the space quantization of angular momentum is provided by the Stern-Gerlach experiment.

All atoms have an internal magnetic field. The internal magnetic field is associated with the orbital angular momentum of the atom and space-quantizes the spin angular momentum of the electron. The resulting interaction is called the spin-orbit coupling and is responsible for the doublet structure, or the fine structure, of all but the $l = 0$ energy levels. The fine structure splitting of an energy level E_n is of the order of $\alpha^2 |E_n|$ where α is the fine structure constant.

Problems

5.1 The readily observable components of the Balmer series of the hydrogen atom are the red, blue, blue-violet, and violet lines. Calculate their energies and wavelengths.

5.2 Will the hydrogen atom in the ground state absorb visible light? Explain.

5.3 (a) Calculate the numerical value of the Rydberg constant for the case where the mass of the proton is treated as infinite compared to the mass of the electron.

(b) Since the mass of the proton is 1836 times the mass of the electron, the Rydberg constant must be corrected for the finite mass of the proton. This can be done by replacing the mass of the electron by its reduced mass defined by

$$\mu = \frac{m}{1 + (m/M)}$$

where m and M are the masses of the electron and the proton, respectively. Calculate the Rydberg constant corrected for the finite mass of the hydrogen nucleus.

(c) The mass of the deuteron, which is the nucleus of the deuterium atom, is 3670 times the mass of the electron. If the wavelength of the red line of the Balmer series of the hydrogen atom is 6562.80 Å, calculate the wavelength of the corresponding line in the deuterium spectrum.

(d) A positronium is an atom consisting of an electron and a positron. Calculate its ground state energy.

5.4 (a) Calculate the velocity of the electron in the $n = 1$ orbit of the hydrogen atom.

(b) Calculate the de Broglie wavelength of the electron in the first Bohr orbit, and compare it to the radius of the orbit.

(c) Calculate the de Broglie wavelength of an electron in the nth Bohr orbit, and show that it is expressible as $\lambda_n = n\lambda_1$, where λ_1 is the de Broglie wavelength of the electron in the ground state.

5.5 The Compton wavelength and the "classical radius" of the electron are defined as $\lambda_C = \hbar/mc$ and $r_c = e^2/mc^2$, respectively, where m and e are the rest mass and the charge of the electron.

(a) Calculate the ratio of these quantities to the Bohr radius of the hydrogen atom, expressing the ratios in terms of the fundamental constants $\hbar$, c, and e.

(b) Give the numerical values of λ_C and r_c, taking the fine structure constant to be $1/137$.

(c) Calculate the Compton wavelength of the pion, or pi-meson, defined as $\lambda_\pi = \hbar/m_\pi c$, where the pion mass is $m_\pi = 273.15\, m$, and compare it with the "classical radius" of the electron.

5.6 The following are the wavelengths in Ångströms of the most prominent x-ray spectral lines of some of the elements that Moseley measured: 3.357, 2.766, 2.521, 2.295, 1.945, 1.548. Identify the elements.

5.7 Calculate the energy of the most intense x-ray spectral line that an aluminum target would emit upon bombardment with energetic electrons.

5.8 If an electron is confined to one-dimensional motion between two infinite potential walls which are separated by a distance equal to the Bohr radius, calculate the energies of the three lowest states of motion.

5.9 If a particle is trapped between two infinite potential walls separated by a distance L, its nth state of motion is described by a wave function $\psi_n = A_n \sin (n\pi x/L)$. If $\int_0^L |\psi_n|^2 dx = 1$, then ψ_n is said to be normalized and $\int_0^L |\psi_n|^2 dx$ gives the probability of finding the particle in the nth state between $x = 0$ and $x = L$.

(a) What must A_1 be in order for the ground state wave function to be normalized?

(b) What is the probability of finding the particle in the ground state between $x = 0$ and $x = L/2$?

5.10 Consider a particle of mass m under a simple harmonic type of restoring force. If its classical frequency of oscillation is ω, then its potential energy is $V(x) = \frac{1}{2}m\omega^2 x^2$. The time-independent Schrödinger equation for such a particle is

$$-\frac{\hbar^2}{2m}\frac{d^2\psi}{dx^2} + \frac{1}{2}m\omega^2 x^2\psi = E\psi$$

(a) Show that a wave function $\psi_0 = A_0 e^{-bx^2}$ is a solution of the Schrödinger equation for $b = m\omega/2\hbar$ and $E = \frac{1}{2}\hbar\omega$.

(b) The energy levels are given by $E_n = (n + \frac{1}{2})\hbar\omega$, $n = 0, 1, 2, 3, \dots$. Show that a particle in the $n = 1$ state is described by a wave function $\psi_1 = A_1 x e^{-bx^2}$.

(c) Show that if $x = 0$ is the position of equilibrium, then the classical turning points of the particle in the ground state, that is, $E = \frac{1}{2}\hbar\omega$, are $x = \pm x_0 = \pm \sqrt{\hbar/m\omega}$.

(d) Calculate the momentum p of the particle in the ground state and show that $x_0 p = \hbar$.

(e) Argue why a particle in a simple harmonic potential can never be at rest.

5.11 (a) Show that for the electron in the ground state of hydrogen the product of its momentum and orbit radius is equal to Planck's constant, that is, $pa_0 = \hbar$.

(b) Argue why a hydrogen atom cannot be a point object.

(c) Calculate the magnitude of the angular momentum vector of the hydrogen atom in the ground state.

(d) Bohr's assumed value of the ground state angular momentum of hydrogen is different from that given by quantum mechanics. Argue why Bohr's incorrect value of the ground state angular momentum still led to a correct value of the ground state energy.

5.12 (a) Compute the magnitude of the angular momentum vector of hydrogen in the $l = 3$ state.

(b) If such an atom is oriented relative to an external magnetic field in the z direction, what are the possible z components of the atom's angular momentum vector?

5.13 The angular momentum quantum number l can be interpreted in terms of the shape of the electron orbit. If the semimajor and the semiminor axes of an elliptical orbit are denoted by a and b, respectively, then the following relationship holds:

$$\frac{b}{a} = \frac{l+1}{n}$$

(a) Sketch the allowed orbits for $n = 3$.

(b) The outermost electron of the Na atom ($Z = 11$) exists normally in the $n = 3$ state. The remaining 10 electrons fill up the $n = 1$ and $n = 2$ states and shield the nucleus. Show that the energy of the outermost electron in the $n = 3$, $l = 2$ state is -1.5 eV.

(c) The energies of the outermost electron in the Na atom in the $n = 3$, $l = 0$ and the $n = 3$, $l = 1$ states are -5.1 eV and -3.0 eV, respectively. Give a qualitative explanation for the l dependence of the energy levels of atomic sodium.

5.14 The emission of a photon by an excited atom conserves energy and momentum. Show that the energy difference between two stationary states is greater than the energy of the emitted photon by $\hbar^2\omega^2/2Mc^2$, where ω is the frequency of the photon and M is the mass of the atom.

5.15 Calculate the energy eigenvalues of the bound state eigenfunctions ψ_{211} and ψ_{31-1} of hydrogen.

5.16 Calculate the quantum mechanical average of the potential energy of the hydrogen atom, $V(r) = -e^2/r$, in the $n = 1$ state.

5.17 Normalize the polar wave function Θ_{21}.

5.18 Normalize the radial wave function $R_{30}(r)$.

5.19 Calculate the orbital magnetic moments of the hydrogen atom in the $1S$ and $2P$ states.

5.20 Calculate the magnitude of a magnetic field required to split a spectral line of 4500 Å into three Zeeman components with a separation of 0.0283 Å between the adjacent components.

5.21 Calculate the magnitude of the internal magnetic field of the hydrogen atom that a $2p$ electron would experience.

5.22 Show that the angle θ between $\mathbf{J}$ and $\mathbf{L}$ is given by

$$\cos \theta = \frac{j(j+1) + l(l+1) - s(s+1)}{2\sqrt{j(j+1)l(l+1)}}$$

where $\mathbf{J}$ and $\mathbf{L}$ are the total and the orbital angular momentum vectors.

5.23 (a) If the total angular momentum vector $\mathbf{J}$ and the associated magnetic moment vector $\boldsymbol{\mu}$ are defined by

$$\mathbf{J} = \mathbf{L} + \mathbf{S}$$

and

$$\boldsymbol{\mu} = \boldsymbol{\mu}_l + \boldsymbol{\mu}_s$$

show that $\mathbf{J}$ and $\boldsymbol{\mu}$ do not lie along the same line.

(b) Show that the component of $\boldsymbol{\mu}$ along the direction of $\mathbf{J}$ is given by

$$\mu_j = g_j \left(\frac{e\hbar}{2mc}\right) \sqrt{j(j+1)}$$

where the so-called Lande g factor is

$$g_j = 1 + \frac{j(j+1) + s(s+1) - l(l+1)}{2\,j(j+1)}$$

$$\left[\text{By definition, } \boldsymbol{\mu}_j = -g_j \left(\frac{e}{2mc}\right) \mathbf{J}.\right]$$

(c) If an atom in an angular momentum state j is subjected to an external magnetic field B, show that the magnetic energy for the Zeeman effect is given by

$$\Delta E_{m_j} = g_j \left(\frac{e\hbar}{2mc}\right) B m_j$$

5.24 If an atom is subjected to a weak magnetic field, the spin-orbit interaction remains intact and consequently the total angular momentum vector **J** is coupled to the magnetic field. Therefore the quantum states of an atom in a weak field are specified by the quantum numbers (n, l, j, j_j). In a strong magnetic field, on the other hand, the spin vector **S** and the orbital angular momentum vector **L** are separately coupled to the magnetic field, and both **S** and **L** are constants of the motion. A quantum state of the atom in a strong field is then specified by (n, l, m_l, m_s).

(a) Show that the magnetic energy of an atomic electron in an angular momentum state l in a strong magnetic field B is given by

$$\Delta E_{m_l} = \frac{e\hbar}{2mc} B \,(2m_s + m_l)$$

(b) Show that the $2p$ state of the hydrogen atom splits into five components under a strong magnetic field. (This is called a Paschen-Back effect.)

(c) If the energy difference between two adjacent components is to be 10 times the spin-orbit splitting in the $2p$ state of hydrogen, what must be the magnitude of the external magnetic field?

Suggestions for Further Reading

Christy, R. W., and A. Pytte. *The Structure of Matter*. New York: W. A. Benjamin, Inc., 1965. Chapters 20 and 21.

Ford, K. W. *Classical and Modern Physics*, vol. 3. Lexington, Mass.: Xerox College Publishing, 1974. Chapter 23.

Eisberg, R. M. *Fundamentals of Modern Physics*. New York: John Wiley and Sons, Inc., 1961. Chapter 10.

McGervey, J. D. *Introduction to Modern Physics*. New York: Academic Press, 1971. Chapters 6 and 7.

Weidner, R. T., and R. L. Sells *Elementary Modern Physics.* Boston: Allyn and Bacon, Inc., 1960. Chapter 5.

Wichmann, E. H. *Quantum Physics: Berkeley Physics Course,* vol. 4. New York: McGraw-Hill Book Company, 1967. Chapter 8.

Ziock, K. *Basic Quantum Mechanics.* New York: John Wiley & Sons, Inc., 1969. Chapter 8.

The hydrogen atom:
the time-dependent theory

6.1 The Stationary State

The Bohr theory of the hydrogen atom postulates that an electron in allowed orbits does not radiate. Such a postulate was necessitated by the observation that normal atoms are stable and excited atoms emit spectral lines. The postulate, however, was conspicuously inconsistent with Maxwell's electrodynamics. According to Maxwell, an electron in a circular orbit must radiate a wave of frequency equal to the frequency of revolution of the electron. Clearly, a reconciliation between Maxwell and Bohr was desirable, and a physical basis rather than an ad hoc hypothesis was needed for the nonradiating character of the allowed state.

In the Schrödinger theory, the allowed, or stationary, states of the hydrogen atom are described by the solutions of the time-dependent Schrödinger equation

$$-\frac{\hbar^2}{2m} \nabla^2 \psi(r, \theta, \phi, t) - \frac{e^2}{r} \psi(r, \theta, \phi, t) = i\hbar \frac{\partial \psi(r, \theta, \phi, t)}{\partial t} \tag{6.1}$$

Because the potential energy of the hydrogen atom is independent of time, the solutions of the equation are of the form

$$\psi_{nlm_l}(r, \theta, \phi, t) = \psi_{nlm_l}(r, \theta, \phi) \, e^{-itE_n/\hbar} \tag{6.2}$$

where the functions $\psi_{nlm_l}(r, \theta, \phi)$ are the eigenfunctions of the time-independent Schrödinger equation. The wave function of a stationary state is thus an oscillatory function with the angular frequency of oscillation determined by $\omega_n = |E_n|/\hbar$. The probability density of a stationary state wave function, however, is not oscillatory in time; in fact, it is independent of time

$$\begin{aligned} \rho_{nlm}(r, \theta, \phi) &= \psi^*_{nlm_l}(r, \theta, \phi, t) \, \psi_{nlm_l}(r, \theta, \phi, t) \\ &= \psi^*_{nlm_l}(r, \theta, \phi) \, \psi_{nlm_l}(r, \theta, \phi) \end{aligned} \tag{6.3}$$

The solutions of the time-dependent Schrödinger equation are, for this reason, said to describe the stationary states of an atom.

The charge density of an atom in an eigenstate is also independent of time. The charge density of an atom is determined by its probability density inasmuch as the charge and the electron are inseparable. Now, according to Maxwell, a static charge density does not radiate, and so the physical basis for the nonradiating character of a stationary state is its static charge density.

The allowed orbits of the Bohr theory are given a probability interpretation in the Schrödinger theory. Let $P_{nl}(r)$ denote the radial probability of a stationary state (n, l, m_l), namely, the probability of finding an electron of energy E_n on a spherical shell about the nucleus between the radial distances r and $r + dr$. (See Figure 6.1.) Then

$$\int_0^\infty P_{nl}(r) \, dr = 1$$

or

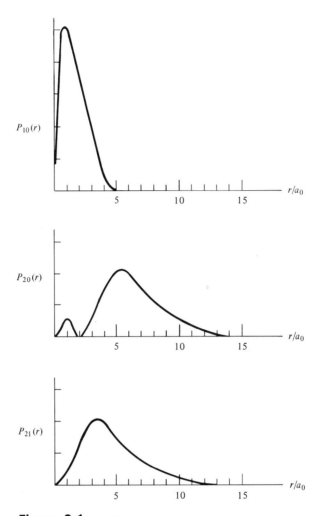

Figure 6.1 Radial probability densities.

$$\int_0^\infty \int_0^\pi \int_0^{2\pi} |\psi|^2 r^2 \sin\theta \, dr \, d\theta \, d\phi$$

$$= \int_0^\infty r^2 |R_{nl}|^2 dr \int_0^\pi \int_0^{2\pi} |\Theta_{lm_l}|^2 |\Phi_{m_l}|^2 \sin\theta \, d\theta \, d\phi = 1 \qquad \textbf{(6.4)}$$

so that

$$P_{nl}(r) \, dr = r^2 \, |R_{nl}(r)|^2 \, dr \qquad\qquad\qquad\qquad \textbf{(6.5)}$$

The dependence of the radial probability density on n and l comes as no surprise inasmuch as the radial wave function itself does.

The radii of the Bohr orbits resemble the most probable radial distances of the electron. For the ground state of the hydrogen atom, the radial wave function is

$$R_{10}(r) = \frac{2}{\sqrt{a_0^3}} e^{-r/a_0}$$

and the radial probability density is

$$P_{10}(r) = \frac{4}{a_0^3} r^2 e^{-2r/a_0} \qquad \text{(6.6)}$$

The maximum of $P_{10}(r)$ occurs at

$$r_m = a_0 \qquad \text{(6.7)}$$

which is precisely the radius of the first Bohr orbit.

The radial probability density of the 2s electron is

$$P_{20}(r) \propto \left(2 - \frac{r}{a_0}\right)^2 r^2 e^{-r/a_0} \qquad \text{(6.8)}$$

which has two maxima, namely, at $r = 0.764\, a_0$ and $r = 5.236\, a_0$. Since

$$\frac{P_{20}(5.236a_0)}{P_{20}(0.764a_0)} = 3.678 \qquad \text{(6.9)}$$

the 2s electron is about four times more likely to be found at $r = 5.236a_0$ than at $r = 0.764a_0$. Neither of these two values corresponds to the radius of the second Bohr orbit.

The radial probability density for the 2p electron is

$$P_{21}(r) \propto \left(\frac{r}{a_0}\right)^2 r^2 e^{-r/a_0} \qquad \text{(6.10)}$$

Its maximum occurs at $r = 4a_0$, which we recognize as the radius of the second Bohr orbit. In general, the maximum of the probability density of the $n, l = n - 1$ state occurs at $r = n^2 a_0$ in precise agreement with the radius of the nth Bohr orbit.

6.2   The Mixing of Stationary States

Consider the superposition, or the mixing, of two stationary wave functions $\psi_{nlm_l}(\mathbf{r}, t)$ and $\psi_{n'l'm_{l'}}(\mathbf{r}, t)$. Since their energies E_n and $E_{n'}$ depend only on the principal quantum numbers and not on the angular momentum and magnetic quantum numbers, we can suppress the latter numbers and write

$$\psi_{n+n'}(\mathbf{r},\, t) = \psi_n(\mathbf{r})e^{-itE_n/\hbar} + \psi_{n'}(\mathbf{r})e^{-itE_{n'}/\hbar} \tag{6.11}$$

The superposed wave function $\psi_{n+n'}(\mathbf{r},\, t)$ is clearly a solution of the time-dependent Schrödinger equation and must describe a state as physically realizable as the stationary wave functions do. For a physical picture of the new state we resort to the probability interpretation, and, calculating the absolute square of the superposed wave function

$$\begin{aligned} P_{n+n'}(\mathbf{r},\, t) = \psi_n^*(\mathbf{r})\psi_n(\mathbf{r}) + \psi_{n'}^*(\mathbf{r})\psi_{n'}(\mathbf{r}) + \psi_n^*(\mathbf{r})\psi_{n'}(\mathbf{r})e^{-it(E_{n'}-E_n)/\hbar} \\ + \psi_n(\mathbf{r})\psi_{n'}^*(\mathbf{r})e^{it(E_{n'}-E_n)/\hbar} \end{aligned} \tag{6.12}$$

we observe that, whereas the first two terms are independent of time, the last two terms are oscillatory functions of time. Their angular frequency of oscillation is determined by the energies of the stationary states giving rise to the mixed state, and is identical to the Bohr relation

$$\omega_{n'n} = \frac{E_{n'} - E_n}{\hbar} \tag{6.13}$$

The probability density, moreover, gives the charge density of the new state, which is clearly oscillatory and therefore must radiate. We are tnus led to the important conclusion that an atom can exist not only in a nonradiating stationary state but also in a radiating nonstationary, mixed state.

The question now arises: How is the mixing of two stationary states actually achieved in an atom? An atom possesses a discrete set of natural, or resonant, frequencies given by the Bohr relation of Eq. (6.13). Suppose an atom, initially in state n, is exposed to electromagnetic radiation from an external source. If a photon of angular frequency equal to one of the resonant frequencies of the atom, say, $\omega_{n'n}$, is present in the spectrum of the incident radiation, such a photon can cause the mixing of the states n and n' in the atom, and, in so doing, transfer its energy to the atom. The mixed state has an oscillatory charge density, and thus describes an electron oscillating between the two stationary states. The oscillatory motion of the electron persists so as to complete the transfer of radiation energy, and the atom, upon absorption of the radiant energy, assumes the final state n'.

If an atom, initially in the higher state n', is exposed to radiation of angular frequency $\omega_{n'n}$, the states n' and n can also mix, and the oscillatory motion of the electron in the mixed state leads eventually to the transition of the atom to the lower state n. Such a transition is accompanied by the emission of a photon of energy $\hbar\omega_{n'n}$ and is referred to as induced, or stimulated, emission of radiation. A remarkable feature of induced emission is that the induced radiation is in phase with the inducing radiation, and they have the same propagation direction and the same polarization in addition to having the same energy. It is this feature of induced emission that underlies the phenomenon of the laser.

The description of resonant absorption and induced emission in terms of oscillating charge densities is quantum mechanical in origin but classical in flavor. Atoms, unlike classical systems, absorb and radiate energy discontinuously. That is, atoms absorb and emit photons. Therefore, the proper interpretation of the oscillatory charge density of an atom is that it gives the probability of the absorption and the emission of a photon by the atom.

It is an experimental fact that an excited atom radiates spontaneously. Unlike the process of induced emission, an excited atom undergoing spontaneous emission does not require the presence of an external electromagnetic field. Spontaneous emission is rather a result of the interaction between the electron and the internal radiation field of the atom. The Coulomb field of the nucleus is static and consequently does not cause spontaneous emission. A proper treatment of spontaneous emission, therefore, requires a quantized radiation field. Since the quantized radiation field belongs in a quantum field theory called quantum electrodynamics, the process of spontaneous radiation in particular, and the interaction of light with the atom in general, cannot be treated rigorously within the framework of the Schrödinger theory. However, results in good agreement with observation can be obtained by judicially combining the Schrödinger theory with classical electrodynamics.

Example 6.1   Show explicitly that the probability density of a particle in an infinite square potential well of width a in a mixed state is an oscillatory function.

Solution.   The stationary state wave functions are

$$\psi_n(x,\ t) = \psi_n(x)e^{-itE_n/\hbar}$$

where

$$\psi_n(x) = \sqrt{\frac{2}{a}} \sin \frac{n\pi x}{a}$$

and

$$E_n = n^2 \frac{\pi^2 \hbar^2}{2ma^2}$$

The probability density for the mixed state of, say, $\psi_1(x,\ t)$ and $\psi_2(x,\ t)$ is

$$P_{1+2}(x,\ t) = [\psi_1^*(x,\ t) + \psi_2^*(x,\ t)][\psi_1(x,\ t) + \psi_2(x,\ t)]$$

$$\propto \sin^2\frac{\pi x}{a} + \sin^2\frac{2\pi x}{a} + \sin\frac{\pi x}{a}\sin\frac{2\pi x}{a}\left[e^{-(it/\hbar)(E_2-E_1)}\right]$$

$$+ \sin\frac{\pi x}{a}\sin\frac{2\pi x}{a}\left[e^{(it/\hbar)(E_2-E_1)}\right]$$

$$= \sin^2\frac{\pi x}{a} + \sin^2\frac{2\pi x}{a}$$

$$+ 2\sin\frac{\pi x}{a}\sin\frac{2\pi x}{a}\left[\frac{e^{(it/\hbar)(E_2-E_1)} + e^{-(it/\hbar)(E_2-E_1)}}{2}\right]$$

$$= \sin^2\frac{\pi x}{a} + \sin^2\frac{2\pi x}{a} + 2\sin\frac{\pi x}{a}\sin\frac{2\pi x}{a}\cos\omega_{21}t$$

where

$$\omega_{21} = \frac{E_2 - E_1}{\hbar}$$

6.3   The Atom as a Damped Oscillator

The oscillator behavior of an atom in a mixed state implies that an atom may be treated as an oscillator insofar as its radiative processes are concerned. In this so-called oscillator model of the atom, an electron of mass m is viewed as bound to the nucleus by a restoring force, $-m\omega_0^2 x$, and its oscillation as damped by a damping force, $-m\Gamma\dot{x}$, so as to produce radiation. Moreover, the motion of the atomic oscillator is described by Newton's equation

$$m\ddot{x} = -m\omega_0^2 x - m\Gamma\dot{x}$$

or

$$\ddot{x} + \Gamma\dot{x} + \omega_0^2 x = 0 \tag{6.14}$$

whose general solution is

$$x = Ae^{-\Gamma t/2}\cos(\omega_1 t + \theta) \tag{6.15}$$

with

$$\omega_1^2 = \omega_0^2 - \frac{\Gamma^2}{4} \tag{6.16}$$

Here ω_0 is the natural frequency of the oscillator, that is, the angular frequency of the atom in a mixed state, and Γ is the damping constant

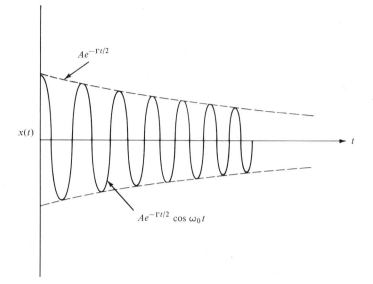

Figure 6.2 The amplitude of a damped oscillator as a function of time.

due to radiation. A and θ are arbitrary constants required in the general solution and are determined by the initial conditions. (See Figure 6.2.)

The total energy of the oscillator is

$$E(t) = \tfrac{1}{2}m\dot{x}^2 + \tfrac{1}{2}m\omega_0^2 x^2 \tag{6.17}$$

In the important case of weak damping, $\Gamma \ll \omega_0$, we can set $\omega_1 \quad \omega_0$ and treat the exponential factor, $\exp{(-\Gamma t/2)}$, as being constant during any one cycle of oscillation. In this approximation,

$$E(t) = E(0)\, e^{-\Gamma t} \tag{6.18}$$

implying that the energy of a weakly damped oscillator decays exponentially. This is precisely how a system of excited atoms emits its radiant energy, and thus a radiating atom is quite adequately described by a weakly damped oscillator.

The damping constant of an oscillator is related to its mean-life or lifetime. The energy of an oscillator after a time lapse of

$$\tau = \frac{1}{\Gamma} \tag{6.19}$$

is

$$E(\tau) = E\left(\frac{1}{\Gamma}\right) = E(0)\, e^{-1} \tag{6.20}$$

The quantity τ then represents the duration of the oscillatory motion in which the initial energy of the oscillator decreases by the factor $1/e$. But such a duration of time defines what is called the mean-life or lifetime of an oscillator. Equation (6.19) thus establishes an inseparable linkage between the lifetime and the damping constant of an oscillator.

The lifetime of an excited atom can be calculated by analogy with classical electrodynamics. A simple radiating device in classical electrodynamics is the oscillating electric dipole. A classical dipole oscillating at a frequency ω_0 emits radiant energy at the rate

$$P = \frac{\omega_0^4}{3c^3}(ed)^2 \tag{6.21}$$

where the quantity (ed) is the magnitude of the dipole and c is the speed of light.

The application of this classical expression to a quantum system requires a decisive assumption: An atom oscillating at ω_0 for a duration τ must emit a photon of energy $\hbar\omega_0$. Then the power output of an atomic oscillator is

$$P = \frac{\hbar\omega_0}{\tau} \tag{6.22}$$

which can be equated to the power output of an equivalent classical dipole

$$\frac{\hbar\omega_0}{\tau} = \frac{\omega_0^4}{3c^3}(ed)^2$$

or

$$\frac{1}{\tau} = \frac{\omega_0^3}{3\hbar c^3}(ed)^2 \tag{6.23}$$

The lifetime of an excited atom is thus inversely proportional to the square of the magnitude of its electric dipole moment and the cube of the angular frequency of its oscillation.

As an example, let us estimate the lifetime of the $n = 3$ excited state of the hydrogen atom leading to the emission of a red spectral line of the Balmer series. The energy of the emitted photon is

$$\hbar\omega_0 = E_3 - E_2 = \frac{5}{72}\frac{\alpha^2 mc^2}{}$$

The magnitude of the oscillating electric dipole in the hydrogen atom can reasonably be taken as the product of the electron charge and the Bohr radius, that is,

$$d = \frac{\hbar}{\alpha mc}$$

Substituting these values in Eq. (6.23), we obtain

$$\tau = 5.62 \times 10^{-7} \text{ sec}$$

This order of magnitude calculation yields a result well within the range of the observed lifetimes of excited atoms emitting visible light, which are typically of the order 10^{-7} sec to 10^{-9} sec.

6.4 Dipole Radiation and the Selection Rules

The most prominent mode of atomic radiation is the so-called electric dipole radiation. The atomic electric dipole is calculated by quantum mechanically averaging its classical expression $e\mathbf{r}$, that is, by evaluating the so-called expectation value of the quantum mechanical operator representing the classical electric dipole

$$\langle e\mathbf{r} \rangle = \int \psi^* \, e\mathbf{r} \, \psi \, dV \tag{6.24}$$

Our primary concern here is not with the actual evaluation of the expectation values of the electric dipole operator but rather with the conditions leading to nonvanishing electric dipole moments. The electric dipole of an atom in a stationary state vanishes. We have already established that the stationary state does not radiate, but, even if it could radiate, the stationary state would still not be able to radiate electromagnetic waves of the electric dipole type because its dipole moment is zero.

Nonstationary states, on the other hand, can have electric dipole moments. The atomic dipole of a nonstationary state, which is given by

$$\langle e\mathbf{r} \rangle_{n'n} = \int \psi^*_{n'l'm'_l}(\mathbf{r}, t) \, e\mathbf{r} \, \psi_{nlm_l}(\mathbf{r}, t) \, dV$$

$$= e^{it(E_{n'} - E_n)/\hbar} \int \psi^*_{n'l'm'_l}(\mathbf{r}) \, e\mathbf{r} \, \psi_{nlm_l}(\mathbf{r}) \, dV \tag{6.25}$$

oscillates with an angular frequency, $\omega_{n'n} = (E_{n'} - E_n)/\hbar$, and radiates a so-called electric dipole photon whose energy is $\hbar\omega_{n'n}$. The emission rate of electric dipole radiation is proportional to the square of the dipole integral

$$\langle e\mathbf{r} \rangle_{l'l} = \int \psi_{n'l'm'_l}(\mathbf{r}) \, e\mathbf{r} \, \psi_{nlm_l}(\mathbf{r}) \, dV \tag{6.26}$$

A salient feature of the electric dipole integral of Eq. (6.26) is that the electric dipole integral has a nonvanishing value only if the angular momentum quantum numbers differ by unity

$$l' - l = \pm 1$$

or

$$\Delta l = \pm 1 \tag{6.27}$$

Called a selection rule, such a condition on the angular momentum quantum number implies that an atom radiating and absorbing photons of the electric dipole type can only exist in those nonstationary states consistent with the selection rule. Atomic transitions obeying the selection rule are referred to as **allowed** transitions (Figure 6.3). A transition from a d state to an s state, for example, is not allowed, and

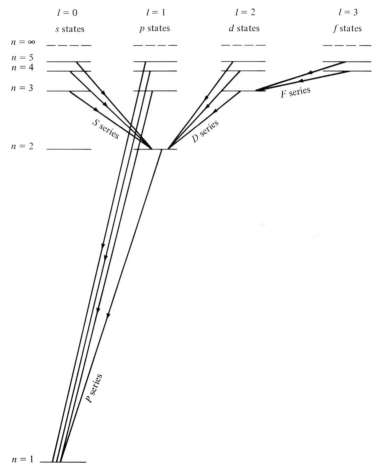

Figure 6.3　Electric dipole transitions of the hydrogen atom.

does not emit or absorb radiation of the electric dipole type. Transitions violating the selection rule can, nevertheless, emit and absorb radiation of other types, namely, of the electric quadrupole and the magnetic dipole type, and these are referred to as **forbidden** transitions. The probability of emission and absorption of photons by atoms is typically 10^3 or higher for the allowed transition than for the forbidden transition.

The physical basis for the selection rule on the angular momentum quantum number is the conservation of angular momentum. An isolated system, such as a hydrogen atom, conserves angular momentum, and thus exists in a definite angular momentum state. A change by unity in the angular momentum of the atom must, therefore, result in the emission of a photon carrying one unit of angular momentum. The angular momentum of the electric dipole photon is, indeed, $1\hbar$.

Example 6.2 Show that the electric dipole moment of a nonstationary state mixing ψ_{100} and ψ_{200} which violates the selection rule on the angular momentum quantum number, vanishes.

Solution. The electric dipole vector $e\mathbf{r}$, in the spherical coordinate system, has components

$ex = er \sin \theta \cos \phi$
$ey = er \sin \theta \sin \phi$
$ez = er \cos \theta$

The expectation value of ez is

$$\langle ez \rangle = e \int \psi_{100}^*(r, \theta, \phi) \, r \cos \theta \, \psi_{200}(r, \theta, \phi) \, r^2 \, \sin \theta \, dr \, d\theta \, d\phi$$

The polar part of the integral is

$$\int_0^\pi \Theta_{00} \, \Theta_{00} \cos \theta \sin \theta \, d\theta = \int_0^\pi \cos \theta \sin \theta \, d\theta = 0$$

Thus

$$\langle ez \rangle_{l'l} = \langle ez \rangle_{00} = 0$$

For the evaluation of $\langle ex \rangle$, the azimuthal part of the integral yields

$$\int_0^{2\pi} \Phi_{00} \, \Phi_{00} \, d\phi \propto \int_0^{2\pi} \cos \phi \, d\phi = 0$$

So

$$\langle ex \rangle = 0$$

Similarly,

$$\langle ey \rangle = 0 \qquad \bullet$$

An atomic electric dipole in an external magnetic field can oscillate either parallel or perpendicular to the field. If it oscillates parallel to the magnetic field, that is, in the z direction, then its expectation value is

$$\langle ez \rangle_{l'l} = e \int \psi_{n'l'm_l'}^* z \psi_{nlm_l} \, dV \qquad \textbf{(6.28)}$$

which, in polar coordinates, takes the form

$$\langle er \cos \theta \rangle_{l'l} \propto e \int_0^\infty R_{n'l'}^* R_{nl} r^3 dr \int_0^\pi \Theta_{l'm_l'}^* \Theta_{lm_l} \cos \theta \sin \theta \, d\theta$$

$$\times \int_0^{2\pi} \Phi_{m_l'}^* \Phi_{m_l} \, d\phi \qquad \textbf{(6.29)}$$

The azimuthal part of the integral is explicitly

$$\int_0^{2\pi} e^{-im_l'\phi} e^{im_l\phi} \, d\phi = \int_0^{2\pi} e^{i(m_l - m_l')\phi} \, d\phi \qquad \textbf{(6.30)}$$

and has a nonvanishing value only if

$$m_l - m_l' = 0$$

or

$$\Delta m_l = 0 \qquad \textbf{(6.31)}$$

This condition on m_l constitutes a selection rule on the magnetic quantum number, and a transition consistent with this selection rule emits a photon which, viewed at right angles to the magnetic field, is polarized in the z direction.

The expectation value of an electric dipole oscillating at right angles to the magnetic field, namely, in the xy plane is

$$\langle e(x \pm iy) \rangle_{l'l} = e \int \psi_{n'l'm_l'}^* (x \pm iy) \psi_{nlm_l} \, dV \qquad \textbf{(6.32)}$$

In polar coordinates,

$$x \pm iy = r \sin \theta (\cos \phi \pm i \sin \phi) = r \sin \theta \, e^{\pm i\phi} \qquad \textbf{(6.33)}$$

Substitution of Eq. (6.33) into Eq. (6.32) and evaluation by analogy with Eq. (6.29) yields

$$\langle er\sin\theta\,e^{\pm i\phi}\rangle = \int(\cdots)dr\int(\cdots)d\theta\int_0^{2\pi}e^{i(m_l-m_l'\pm1)\phi}\,d\phi \qquad \textbf{(6.34)}$$

which is nonvanishing only if

$$m_l - m_l' \pm 1 = 0$$

or

$$\Delta m_l = \pm 1 \qquad \textbf{(6.35)}$$

This is another selection rule on the magnetic quantum number. Now the x and y components of the electric dipole are out of phase by $\pi/2$, because $i = e^{i\pi/2}$. As a result, transitions subject to the selection rule $\Delta m_l = \pm 1$ emit photons that, viewed parallel to the magnetic field, are circularly polarized.

We have already seen that the selection rule on the magnetic quantum number accounts for the Zeeman splitting of atomic energy levels in an external magnetic field. Our investigation of the origin of the selection rule has now produced information about the polarization of the spectral lines in the Zeeman effect (Figure 6.4). If the spectral lines are viewed perpendicular to the magnetic field **B**, all three Zeeman components are linearly polarized; the electric field vector **E** of the central line, which is subject to $\Delta m_l = 0$, is polarized parallel to **B**, and the outer spectral lines are polarized perpendicular to **B**. If the spectral

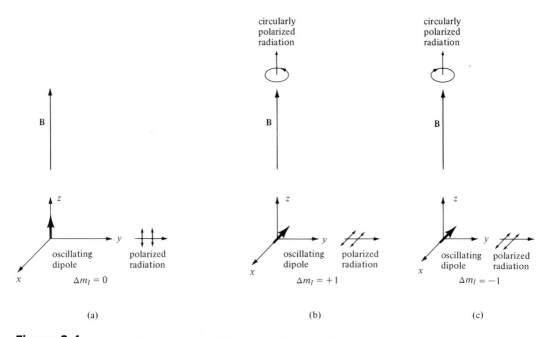

Figure 6.4   Polarization of the spectral lines in the Zeeman effect.

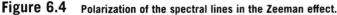

lines are viewed along **B,** the central component is absent because the photon is a transverse wave, and the outer components are circularly polarized.

6.5   Spectral Line Widths

The absorption rate of radiant energy of frequency ω by a damped classical oscillator of a natural frequency ω_0 is proportional to

$$A(\omega) \propto \frac{1}{(\omega - \omega_0)^2 + (\Gamma/2)^2} \qquad (6.36)$$

The absorption of radiant energy by an oscillator thus constitutes a resonance phenomenon, and the plot of the absorption rate $A(\omega)$ as a function of the incident frequency ω is called a resonance curve (Figure 6.5). The maximum value of the absorption rate obtains at $\omega = \omega_0$, and the half maximum value obtains when

$$(\omega - \omega_0)^2 = \left(\frac{\Gamma}{2}\right)^2$$

or

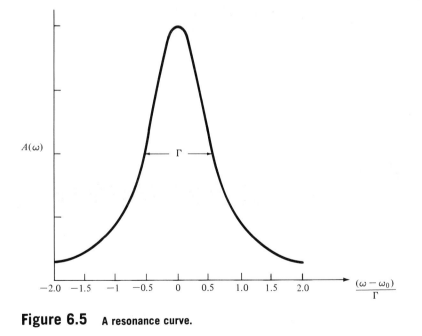

Figure 6.5   A resonance curve.

$$\omega - \omega_0 = \pm \frac{\Gamma}{2} \qquad (6.37)$$

If $\Delta\omega$ denotes the full width, or simply the width, of the resonance curve at the half maximum value of $A(\omega)$, then

$$\Delta\omega = \Gamma \qquad (6.38)$$

so that the damping constant Γ can also be called the natural line width. But the damping constant of an oscillator determines the lifetime of the oscillator according to Eq. (6.19)

$$\Gamma = \frac{1}{\tau}$$

and thus

$$\Delta\omega = \frac{1}{\tau} \qquad (6.39)$$

implying that the width and the lifetime of an oscillator have a reciprocal relation.

The process of spontaneous emission of energy by an atomic oscillator is similar to the process of absorption. Since an excited atom has a finite lifetime, the emitted spectral line must, according to Eq. (6.39), have an intrinsic line width. A typical lifetime of an excited atom emitting visible light is $\tau = 10^{-8}$ sec. Consequently, the intrinsic line width $\Delta\lambda$ of a spectral line of $\lambda = 5000$ Å, for example, is of the order

$$\frac{\Delta\lambda}{\lambda} = \frac{\Delta\omega}{\omega} = \frac{10^8 \text{ sec}^{-1}}{3.77 \times 10^{15} \text{ sec}^{-1}} = 2.6 \times 10^{-8}$$

or

$$\Delta\lambda = 1.3 \times 10^{-4} \text{ Å}$$

The natural line widths of visible light are clearly very small. The measurement of intrinsic spectral line widths are further complicated by the so-called Doppler broadening, which arises from the thermal motions of the source atoms, and by the so-called collision broadening, which arises from the rapid collisions of atoms at high pressures. The intrinsic spectral line width can nevertheless be measured by reducing the temperature and pressure of the source.

6.6 Level Widths and the Uncertainty Principle

The intrinsic widths of spectral lines imply intrinsic widths in the energy levels of the atom (Figure 6.6). Consider, for example, the transition of a hydrogen atom from the $n = 2$ state to the ground state. The energy of the emitted photon is

$$\hbar\omega_{21} = E_2 - E_1$$

with an intrinsic line width defined by

$$\Delta\omega_{21} = \frac{|\Delta E_2|}{\hbar} \equiv \frac{\Delta E_2}{\hbar} \tag{6.40}$$

According to Eq. (6.39), the line width is the reciprocal of the lifetime of the oscillator, which, in the present example, is the lifetime of the excited state E_2, and so

$$\frac{1}{\tau_2} = \frac{\Delta E_2}{\hbar}$$

or

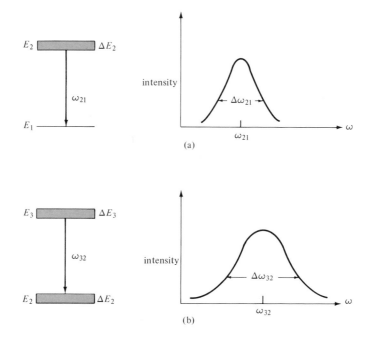

(a)

(b)

Figure 6.6 Energy level widths are related to spectral line widths: (a) $\hbar\,\Delta\omega_{21} = \Delta E_2$, (b) $\hbar\,\Delta\omega_{32} = \Delta E_3 + \Delta E_2$.

$$\tau_2 \, \Delta E_2 = \hbar \tag{6.41}$$

Thus there exists an inseparable linkage between the lifetime and the level width of the $n = 2$ excited state. This result is not unique to the $n = 2$ state but applies to all energy levels: The product of the lifetime and width of an energy level equals Planck's constant

$$\tau \Delta E = \hbar \tag{6.42}$$

An excited atom emits its photon sometime during its lifetime. The lifetime of an excited state therefore constitutes an uncertainty in the time at which the emission of a photon takes place. With this uncertainty denoted, for emphasis, by Δt, Eq. (6.42) may be rewritten

$$\Delta t \, \Delta E = \hbar \tag{6.43}$$

This is an **uncertainty relation** enunciated by Werner Heisenberg in 1927, and reveals a fundamental limit to the precision with which time and energy can be specified simultaneously. The energy of a quantum state cannot be determined precisely within a finite time, and the energy width is an inherent property of all systems having a finite lifetime.

There is a classical analog of the time-energy uncertainty relation. A damped oscillator must radiate a finite wave train, and a finite wave train must be observed within a finite time Δt that must be at least as great as the lifetime of the oscillator. The width in the frequency of the wave train is given by $\Delta \omega = \Delta E / \hbar$. The product of the observation time and the frequency width of the wave train is then obtained from Eq. (6.42)

$$\Delta t \, \Delta \omega \geqslant 1 \tag{6.44}$$

This is a well-known time-frequency uncertainty relation in wave theory, and states that only if the frequency of a wave can be measured over a long time can the uncertainty in the frequency be made small.

If a wave train can be observed for a time Δt, it must have a spatial extension Δx given by

$$\Delta x = c \, \Delta t \tag{6.45}$$

with c denoting the propagation speed of the wave. Substitution of Eq. (6.45) into Eq. (6.44) yields

$$\Delta x \, \Delta \omega \geqslant c$$

which, by use of the dispersion relation, $\omega = ck$ or $\Delta \omega = c \, \Delta k$, takes the form

$$\Delta x \, \Delta k \geqslant 1 \qquad\qquad\qquad (6.46)$$

with Δk denoting the width in the wave number. It follows from Eq. (6.46) that if the wavelength is to be well defined, the wave must be infinitely extended in space.

A similar uncertainty relation holds for a material particle. By use of the de Broglie relation $p = \hbar k$, Eq. (6.46) can be cast in the form

$$\Delta x \, \Delta p \geqslant \hbar \qquad\qquad\qquad (6.47)$$

This is the **Heisenberg position-momentum uncertainty relation** formulated in 1927, and implies that the position and the momentum of a particle cannot be measured with unlimited precision at the same time. But, of course, it is possible to determine the position of a particle precisely if the momentum of the particle need not be specified at the same time.

The Heisenberg uncertainty relation underlies the stability of the atom. Consider the hydrogen atom in the ground state. If we denote the uncertainty in the position of the electron by r and the uncertainty in the momentum by p, then

$$pr = \hbar$$

or

$$p = \frac{\hbar}{r} \qquad\qquad\qquad (6.48)$$

The total energy of the electron is

$$E = \frac{p^2}{2m} - \frac{e^2}{r} = \frac{\hbar^2}{2mr^2} - \frac{e^2}{r} \qquad\qquad\qquad (6.49)$$

The minimum value of the energy function occurs at $r = r_0$, which is obtained by setting the derivative of E with respect to r equal to zero

$$\frac{dE}{dr} = -\frac{\hbar^2}{mr^3} + \frac{e^2}{r^2} = 0$$

or

$$r_0 = \frac{\hbar^2}{me^2}$$

which is precisely the Bohr radius. Although the Bohr radius determines the size of the hydrogen atom, it is clear that the uncertainty principle renders meaningless the notion of well-defined orbits. The uncertainty in the position of the electron is an inherent property of the atom.

The Heisenberg uncertainty principle holds profound implications for quantum theory. The principle, first of all, recognizes the existence of a minimal disturbance upon any phenomenon observed, which cannot be eliminated by increased refinements of the measuring device. An interaction between the observer and the observed phenomenon is simply unavoidable, and the observer must be regarded as an integral part of the phenomenon. In fact, it is meaningless, even in principle, to speak of the existence of a quantum phenomenon per se.

The uncertainty principle, secondly, implies that a physical theory must be founded ultimately on quantities that can be measured. The notions of a purely classical particle and a purely classical wave, for example, do not correspond to reality in the microscopic realm. What is observed is the duality of matter and radiation, and the uncertainty principle is a natural consequence of such dualism. A correct quantum theory, therefore, must be founded upon the uncertainty principle.

6.7   Lifetimes and the Transition Probability

The decay of an excited atom is governed by an exponential law. If λ denotes the probability that an atom emits a photon per unit time, the number of atoms remaining in the excited state at any time t is given by

$$N(t) = N(0) \, e^{-\lambda t} \tag{6.50}$$

where $N(0)$ represents the initial number of atoms in the excited state. The lifetime τ of the excited state is then defined by

$$\tau = \frac{\displaystyle\int_0^\infty t N(t) \, dt}{\displaystyle\int_0^\infty N(t) \, dt} = \frac{\displaystyle\int_0^\infty t e^{-\lambda t} \, dt}{\displaystyle\int_0^\infty e^{-\lambda t} \, dt} \tag{6.51}$$

which, upon integration by parts, reduces to

$$\tau = \frac{1}{\lambda} \tag{6.52}$$

Now the lifetime of an excited state really means the duration of an atomic oscillation that results in the emission of a photon. We have already seen in Eq. (6.23) that

$$\frac{1}{\tau} \propto \omega_{ij}^3 \, (ed)_{ij}^2 \tag{6.53}$$

where $\omega_{ij} = (E_i - E_j)/\hbar$ and $(ed)_{ij}$ is the electric dipole moment that arises from the superposition of eigenstates i and j. Thus the transition probability per unit time, or the transition rate, from state i to state j is proportional to the cube of the frequency of the emitted photon and the square of the electric dipole moment.

Transition rates yield the relative intensities of spectral lines. The longer the lifetime of an excited state is, the fainter the emitted spectral line. The calculation of spectral intensities entails the calculation of atomic dipole moments, which is, more often than not, tedious. A correct description of the interaction between atoms and radiation is thus possible in the Schrödinger theory, but is, nevertheless, incomplete inasmuch as radiation is not quantized. A quantum theory of radiation has been developed and is called quantum electrodynamics, but the quantum electrodynamic treatment of radiation is well beyond the scope of our present work.

6.8   Summary

The solutions of the time-dependent Schrödinger equation describe the so-called stationary states of the hydrogen atom. Although the wave function of a stationary state has an oscillatory time dependence, its probability density and, equivalently, its charge density does not change with time. Consequently, the atom in a stationary state does not radiate.

The radiation processes of the hydrogen atom are described by nonstationary wave functions that are constructed by mixing two stationary wave functions. The probability density of a nonstationary state is an oscillatory function with its angular frequency determined by the energy difference of the stationary states. The oscillatory nature of the nonstationary state gives rise to an oscillator model of the atom, in which the atom is viewed as a weakly damped oscillator with radiation arising from the damping. The interaction between electromagnetic radiation and the atom is described by the oscillator model of the atom by treating radiation classically and the quantum system with which the radiation interacts quantum mechanically. The radiation of light by an atom, for example, is described by analogy with the radiation of the classical electric dipole with the atomic dipole moment calculated quantum mechanically.

Electric dipole radiation is the most prominent mode of atomic radiation, and is subject to selection rules on the angular momentum and magnetic quantum numbers. Transitions consistent with the selection rules are called allowed transitions, and are some 10^3 times more probable than the forbidden transitions that radiate photons of the electric quadrupole and the magnetic dipole types.

A fundamental feature of wave theory is the uncertainty relations. The quantum mechanical Heisenberg uncertainty principle formalizes the existence of a fundamental limit in the precision with which certain pairs of dynamical variables can be defined at the same time. The stability of the atom, for example, is a manifestation of the position-momentum uncertainty relation and the inherent level width of an excited state of the time-energy uncertainty relation. Moreover, the act of observation is an integral part of the observed phenomenon, and the notions of precise path and precise energy are not operationally defined for a quantum system. The knowledge of a quantum system is limited to that allowed by the uncertainty principle.

Problems

6.1 Verify that Eq. (6.2) is a solution of the time-dependent Schrödinger equation for the hydrogen atom.

6.2 The radial probability density $P_{nl}(r)$ of Eq. (6.5) is appreciable within a shell about an average radius given by

$$\langle r_{nl} \rangle \equiv \int_0^\infty r P_{nl}(r) \, dr$$

$$= n^2 a_0 \left\{ 1 + \tfrac{1}{2} \left[1 - \frac{l(l+1)}{n^2} \right] \right\}$$

Verify the preceding expression for the average radius for the $1s$ and $2p$ states by evaluating the expectation values.

6.3 Calculate the radius of the Bohr orbit for the $n = 3$ state of the hydrogen atom using the radial probability density. (Radial wave functions are given in Table 5.2.)

6.4 Calculate the quantum mechanical average, or the expectation value, of the potential energy of the hydrogen atom in the ground state.

6.5 Show that Eq. (6.15) is a solution of the damped harmonic oscillator equation given by Eq. (6.14).

6.6 Show that the energy of a weakly damped oscillator decays according to $E(t) = E(0) \exp(-\Gamma t)$.

6.7 Obtain a rough estimate of the lifetime of an excited atom that emits a visible photon of wavelength 5000 Å by treating the atom as an electric dipole oscillator of size $d = 10^{-8}$ cm.

6.8 (a) Taking the lifetime of the first excited state of atomic hydrogen to be of the order 10^{-8} sec, calculate the damping constant due to the radiation of a photon.

(b) Is the treatment of an atom as a weakly damped oscillator justified?

6.9 Transitions between energy states can be either upward or downward. Argue on the basis of the oscillator model of the atom that the probability for a downward transition must be proportional to the probability for an upward transition.

6.10 In the processes of absorption and emission of radiation by an atom, it is observed that the emitted wave has a definite phase relation to the incident wave. Argue why this experimental fact cannot satisfactorily be explained by Bohr's "jump-picture" of atomic transitions. Compare, for example, the transit time of a photon across an atom to the lifetime of an excited state.

6.11 (a) Calculate the z component of the electric dipole moment of the hydrogen atom for the transition from the $2p$ ($l = 1$) state to the $1s$ ($l = 0$) state.
(b) Calculate the lifetime of the $2p$ state.
(c) Calculate the intrinsic level width of the $2p$ state.
(d) Calculate the line width of the emitted spectral line.

6.12 In the ground state of atomic hydrogen the electron can be pictured as existing somewhere within a distance Δx from the position of the nucleus with Δx interpreted as the uncertainty in the position of the electron.
(a) If Δx is taken to be the Bohr radius, calculate the uncertainty in the momentum of the electron.
(b) If the uncertainty relation is written as $rp = \hbar$, the energy of the hydrogen atom can be expressed by

$$E(p) = \frac{p^2}{2m} - \frac{e^2 p}{\hbar}$$

Calculate the minimum energy of the hydrogen atom.

6.13 (a) If a proton is confined within a spherical nucleus whose radius is equal to the Compton wavelength of the pion, that is, $r = \hbar/m_\pi c = 1.4 \times 10^{-13}$ cm, calculate the inherent kinetic energy of the proton. (Take $m_p/m_\pi = 6.7$, where m_p and m_π are the masses of the proton and the pion. The rest energy of the pion is 139.58 MeV.)
(b) If the proton exists in an attractive nuclear potential, what is the maximum potential energy of the proton in the nucleus?

6.14 Consider a nonrelativistic quantum particle of mass m attached to a spring of a spring constant $k = m\omega^2$.
(a) If its motion is described by $x = A_0 \cos \omega t$, what is the maximum value p_0 of its momentum?
(b) If the amplitude A_0 is interpreted as the uncertainty in the position of the particle, show that $p_0 = \sqrt{m\hbar\omega}$.
(c) If p_0 is interpreted as the uncertainty in momentum, show that the particle must have an inherent energy $E_0 = \frac{1}{2}\hbar\omega$. This energy is called the zero-point energy of a harmonic oscillator.
(d) Argue why such a particle can never be at rest.
(e) Show that the distance between the equilibrium position and the "classical turning point" for the zero-point vibration is $\sqrt{\hbar/m\omega}$.

6.15 Carry out the integration of Eq. (6.51).

6.16 Consider a gas of 1000 hydrogen atoms all existing initially in the $n = 3$ energy state.

(a) If all downward transitions are assumed to be equally probable, what is the total number of photons emitted when all the atoms have returned to the ground state?

(b) If the dipole moments associated with the downward transitions, $n = 3 \rightarrow n = 2$ and $n = 3 \rightarrow n = 1$, are assumed to be equal, calculate the relative intensities of the emitted spectral lines.

Suggestions for Further Reading

Crawford, F. S., Jr. *Waves: Berkeley Physics Course*, vol. 3. New York: McGraw-Hill Book Company, 1965, Sections 3.2 and 7.5.

McGervey, J. D. *Introduction to Modern Physics.* New York: Academic Press, 1971. Chapter 9.

Wichmann, E. H. *Quantum Physics: Berkeley Physics Course*, vol. 4. New York: McGraw-Hill Book Company, 1967. Chapters 3 and 6.

Ziock, K. *Basic Quantum Mechanics.* New York: John Wiley & Sons, Inc., 1969. Chapter 8.

Electrons and the exclusion principle

7.1 Indistinguishable Particles

In classical physics, identical particles are distinguishable. Their distinguishability has its origin in the classical equation of motion, which yields a completely traceable path for a particle. Thus if they are initially given a distinguishable identity, two identical particles maintain their distinguishable identity at all instants of time.

Identical quantum particles, on the other hand, are not distinguishable. Their indistinguishability follows from the uncertainty principle, namely, that the position and momentum of a particle cannot simultaneously be determined with unlimited precision. As a result, the path of a quantum particle cannot be traced precisely, and the distinguishable nature of identical classical particles becomes meaningless for quantum particles.

In quantum theory the indistinguishable nature of identical particles is given a probability interpretation. Consider a system of two identical, indistinguishable particles. Such a system is described by a wave function, which is a function of the dynamical variables of both particles, that is, $\psi(\mathbf{r}_1, \mathbf{r}_2)$. If the particles are exchanged, the system is described by a wave function $\psi(\mathbf{r}_2, \mathbf{r}_1)$. The indistinguishability of the particles, then, implies that the probability of finding one particle at $\mathbf{r}_1$ and another at $\mathbf{r}_2$ must remain the same under the interchange of particles. That is,

$$|\psi(\mathbf{r}_1, \mathbf{r}_2)|^2 = |\psi(\mathbf{r}_2, \mathbf{r}_1)|^2 \tag{7.1}$$

or

$$\psi(\mathbf{r}_1, \mathbf{r}_2) = \pm\psi(\mathbf{r}_2, \mathbf{r}_1) \tag{7.2}$$

If the sign of the wave function remains unchanged under a particle interchange, the wave function is called **symmetric;** if the sign of the wave function is reversed under a particle interchange, the wave function is called **antisymmetric.** It then follows from Eq. (7.2) that a wave function describing a system of indistinguishable, identical particles must be either symmetric or antisymmetric.

The construction of a symmetric or an antisymmetric wave function describing a system of two particles is straightforward for the case of zero interaction between the particles. In such an approximation each particle exists independent of the other, and the Schrödinger equation for the system separates into two independent Schrödinger equations, one for each particle. As a result, the total wave function of the system is expressible as a product of the wave functions describing the individual particles,

$$\psi(\mathbf{r}_1, \mathbf{r}_2) = \psi_\alpha(\mathbf{r}_1)\,\psi_\beta(\mathbf{r}_2) \tag{7.3}$$

where $\alpha = (n, l, m_l, m_s)_1$ and $\beta = (n, l, m_l, m_s)_2$ specify the quantum states of particle 1 and particle 2, respectively. Equation (7.3), however, does not satisfy the condition that particles 1 and 2 are indistinguishable, because, under an exchange of particles, it transforms to

$$\psi(\mathbf{r}_2, \mathbf{r}_1) = \psi_\alpha(\mathbf{r}_2)\,\psi_\beta(\mathbf{r}_1) \tag{7.4}$$

with the result that the probability density of the system does not remain invariant under the operation of particle exchange. That is,

$$|\psi(\mathbf{r}_1, \mathbf{r}_2)|^2 \neq |\psi(\mathbf{r}_2, \mathbf{r}_1)|^2 \qquad (7.5)$$

Therefore a wave function constructed as a simple product of the individual wave functions fails to describe a system of identical particles that are also indistinguishable. The sum and the difference of the wave functions defined by Eqs. (7.3) and (7.4), however, yield a symmetric and an antisymmetric wave function for a system of two identical, indistinguishable particles, namely,

$$\psi_S(\mathbf{r}_1, \mathbf{r}_2) = \psi_\alpha(\mathbf{r}_1)\, \psi_\beta(\mathbf{r}_2) + \psi_\alpha(\mathbf{r}_2)\, \psi_\beta(\mathbf{r}_1) \qquad (7.6)$$

$$\psi_A(\mathbf{r}_1, \mathbf{r}_2) = \psi_\alpha(\mathbf{r}_1)\, \psi_\beta(\mathbf{r}_2) - \psi_\alpha(\mathbf{r}_2)\, \psi_\beta(\mathbf{r}_1) \qquad (7.7)$$

The subscripts, S and A, refer to the symmetric and the antisymmetric character of the total wave function under the operation of particle exchange

$$\psi_S(\mathbf{r}_1, \mathbf{r}_2) = \psi_S(\mathbf{r}_2, \mathbf{r}_1) \qquad (7.8)$$
$$\psi_A(\mathbf{r}_1, \mathbf{r}_2) = -\psi_A(\mathbf{r}_2, \mathbf{r}_1) \qquad (7.9)$$

Their probability densities are clearly invariant under the operation of particle exchange

$$|\psi_S(\mathbf{r}_1, \mathbf{r}_2)|^2 = |\psi_S(\mathbf{r}_2, \mathbf{r}_1)|^2 \qquad (7.10)$$
$$|\psi_A(\mathbf{r}_1, \mathbf{r}_2)|^2 = |\psi_A(\mathbf{r}_2, \mathbf{r}_1)|^2 \qquad (7.11)$$

Hence both the symmetric and the antisymmetric wave functions satisfy the condition of indistinguishability of the particles. However, a system of identical particles is described by either a symmetric wave function or an antisymmetric wave function, and the symmetry character of the wave function is an intrinsic property of the particles. A system of particles with spin $\frac{1}{2}$, such as electrons, protons, and neutrons, is described by an antisymmetric wave function. A symmetric wave function, on the other hand, describes a system of particles with integral values of spin such as photons and mesons.

Example 7.1 Write out the Schrödinger equation for a system of two noninteracting electrons.

Solution. The Hamiltonian operator for the system is the sum of the Hamiltonian operators for the individual electrons

$$H = H_1 + H_2$$

where

$$H_1 = -\frac{\hbar^2}{2m}\nabla_1^2$$

$$H_2 = -\frac{\hbar^2}{2m}\nabla_2^2$$

The Schrödinger equation for the system is

$$H\,\psi(\mathbf{r}_1, \mathbf{r}_2) = E\,\psi(\mathbf{r}_1, \mathbf{r}_2)$$

where the total energy of the system is the sum of the energies of the individual electrons

$$E = E_1 + E_2$$

7.2   The Exclusion Principle

In 1925, Wolfgang Pauli discovered, from his analysis of helium spectra, an important property of electrons, namely, that no two electrons in an atom exist in the same quantum state. This property of electrons arises in consequence of the antisymmetric character of the wave functions by which a system of electrons is described. Consider, for example, the antisymmetric wave function of Eq. (7.7): It vanishes for $\alpha = \beta$

$$\psi_A = \psi_\alpha(\mathbf{r}_1)\,\psi_\alpha(\mathbf{r}_2) - \psi_\alpha(\mathbf{r}_2)\,\psi_\alpha(\mathbf{r}_1) = 0 \qquad (7.12)$$

so that the probability density for the two electrons sharing the same set of quantum numbers is zero. Such a result is referred to as the Pauli exclusion principle and applies not only to electrons but to all spin-$\frac{1}{2}$ particles.

Suppose, contrary to the prediction of the exclusion principle, that all three electrons of the lithium atom exist in the $n = 1$ shell. The theoretical expression for the ionization energy of an atom can be written in the form

$$E_i = \frac{Z_{\text{eff}}^2\,13.6}{n^2}\text{ eV} \qquad (7.13)$$

which, for our hypothetical lithium atom, becomes

$$E_i = Z_{\text{eff}}^2 \, 13.6 \text{ eV} \qquad (7.14)$$

Here Z_{eff} represents the effective nuclear charge that the third, or the least lightly bound, electron experiences as a result of the partial shielding of the nucleus by the other two electrons, and it is estimated by measuring the ionization energy of an atom. The experimental ionization energy of lithium is $E_i = 5.36$ eV, thus yielding for its effective nuclear charge

$$Z_{\text{eff}} = \sqrt{\frac{5.36}{13.6}} = 0.63 \qquad (7.15)$$

This value of Z_{eff}, however, is physically unrealizable because the nuclear charge, $Z = 3$, of the lithium nucleus cannot be shielded to an effective charge of $Z_{\text{eff}} = 0.63$ by two electrons; rather Z_{eff} must be at least equal to unity. The third electron of the lithium atom, therefore, cannot exist in the $n = 1$ shell.

The electron configuration of the lithium atom consistent with the exclusion principle, then, places the third electron in the $n = 2$ shell. The theoretical expression for the ionization energy of such an electron is

$$E_i = \frac{Z_{\text{eff}}^2 \, 13.6 \text{ eV}}{2^2} \qquad (7.16)$$

which yields a physically reasonable value for Z_{eff}, namely,

$$Z_{\text{eff}} = \sqrt{\frac{5.36}{3.4}} = 1.26 \qquad (7.17)$$

A value of Z_{eff} greater than unity implies that the filled $n = 1$ shell does not completely shield the nucleus and the outermost electron partially penetrates the filled core of the atom (Figures 7.1 and 7.2).

The exclusion principle gives rise to the shell structure of the atom (Figure 7.3). The shells of an atom are specified by the principle quantum number n, and are designated by the letter symbols $K, L, M,$

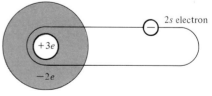

Figure 7.1 The 2s electron of the Li atom penetrates the 1S closed shell.

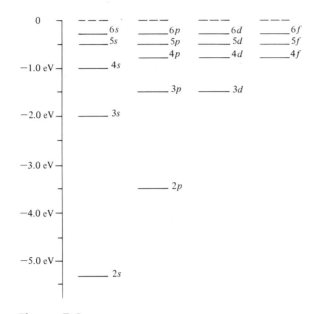

Figure 7.2 An energy level diagram of the neutral Li atom.

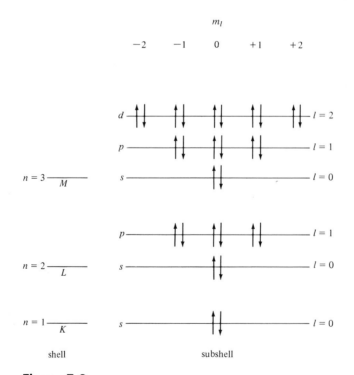

Figure 7.3 Schematic representation of the shell structure of the atom. The arrows indicate electrons occupying available states.

N for $n = 1, 2, 3, 4$. The subshells, which are specified by the angular momentum quantum number l, are given the letter symbols s, p, d, f for $l = 0, 1, 2, 3$. In this notation, the electron configurations for helium and lithium in the ground state are $1s^2$ and $1s^2 2s^1$, respectively. The superscript simply indicates the number of electrons sharing common n and l values. Electron configurations are designated in this way because electrons in the same subshell have the same energy, aside from the small fine structure effect. The energies of the subshells belonging to the same shell, however, are not the same; for a given n, the $l = 0$ subshell has the lowest energy and the $l = n - 1$ subshell has the highest energy.

Table 7.1   Electron Configurations for the Ground States of Some Elements

Element	Subshells				
	$n = 1, l = 0$	$n = 2, l = 0$	$n = 2, l = 1$	$n = 3, l = 0$	$n = 3, l = 1$
H	$1s^1$				
He	$1s^2$				
Li	$1s^2$	$2s^1$			
Be	$1s^2$	$2s^2$			
B	$1s^2$	$2s^2$	$2p^1$		
C	$1s^2$	$2s^2$	$2p^2$		
N	$1s^2$	$2s^2$	$2p^3$		
O	$1s^2$	$2s^2$	$2p^4$		
F	$1s^2$	$2s^2$	$2p^5$		
Ne	$1s^2$	$2s^2$	$2p^6$		
Na	$1s^2$	$2s^2$	$2p^6$	$3s^1$	
S	$1s^2$	$2s^2$	$2p^6$	$3s^2$	$3p^4$

The physical basis for the periodic nature of the elements is the exclusion principle. The columns of the periodic table consist of elements that have the same number of electrons in the outermost subshell. Oxygen and sulfur, for example, both have four electrons in the outermost p subshell, and are thus members of the same column. Since the chemical properties of the same column are similar, the chemical properties must clearly depend on how the subshells are filled. In fact, the s and p subshells play a major role in determining the chemical properties of atoms, and thus give rise to the periodicity of eight in the main body of the periodic table (see Figure 7.4).

Periodic Table

Period	I A										

Key

Carbon ← Name
Atomic number → 6 **C** ← Symbol
12.011 ← Atomic mass (weight)*
2.4 ← Electron configuration

Metalloids   Nonmetals

Period 1

Hydrogen
1 **H**
1.0080
1

II A

Period 2

Lithium	Beryllium
3 **Li**	4 **Be**
6.941	9.01218
2,1	2,2

Period 3

Sodium	Magnesium
11 **Na**	12 **Mg**
22.9898	24.305
2,8,1	2,8,2

Transition elements

III B	IV B	V B	VI B	VII B	VIII B

Period 4

Potassium	Calcium	Scandium	Titanium	Vanadium	Chromium	Manganese	Iron	Cobalt
19 **K**	20 **Ca**	21 **Sc**	22 **Ti**	23 **V**	24 **Cr**	25 **Mn**	26 **Fe**	27 **Co**
39.102	40.08	44.9559	47.90	50.9414	51.996	54.9380	55.847	58.9332
2,8,8,1	2,8,8,2	2,8,9,2	2,8,10,2	2,8,11,2	2,8,13,1	2,8,13,2	2,8,14,2	2,8,15,2

Period 5

Rubidium	Strontium	Yttrium	Zirconium	Niobium	Molybdenum	Technetium	Ruthenium	Rhodium
37 **Rb**	38 **Sr**	39 **Y**	40 **Zr**	41 **Nb**	42 **Mo**	43 **Tc**	44 **Ru**	45 **Rh**
85.4678	87.62	88.9059	91.22	92.9064	95.94	98.9062	101.07	102.9055
2,8,18,8,1	2,8,18,8,2	2,8,18,9,2	2,8,18,10,2	2,8,18,12,1	2,8,18,13,1	2,8,18,14,1	2,8,18,15,1	2,8,18,16,1

Period 6

Cesium	Barium	Lanthanum	Hafnium	Tantalum	Tungsten	Rhenium	Osmium	Iridium
55 **Cs**	56 **Ba**	57 **La**	72 **Hf**	73 **Ta**	74 **W**	75 **Re**	76 **Os**	77 **Ir**
132.9055	137.34	138.9055	178.49	180.9479	183.85	186.2	190.2	192.22
-18,18,8,1	-18,18,8,2	-18,18,9,2	-18,32,10,2	-18,32,11,2	-18,32,12,2	-18,32,13,2	-18,32,14,2	-18,32,15,2

Period 7

Francium	Radium	Actinium		
87 **Fr**	88 **Ra**	89 **Ac**	104	105
(223)	226.0254	(227)		
-18,32,18,8,1	-18,32,18,8,2	-18,32,18,9,2		

Lanthanide series

Cerium	Praseodymium	Neodymium	Promethium	Samarium	Europium
58 **Ce**	59 **Pr**	60 **Nd**	61 **Pm**	62 **Sm**	63 **Eu**
140.12	140.9077	144.24	(145)	150.4	151.96
-18,20,8,2	-18,21,8,2	-18,22,8,2	-18,23,8,2	-18,24,8,2	-18,25,8,2

Actinide series

Thorium	Protactinium	Uranium	Neptunium	Plutonium	Americium
90 **Th**	91 **Pa**	92 **U**	93 **Np**	94 **Pu**	95 **Am**
232.0381	231.0359	238.029	237.0482	(242)	(243)
-18,32,18,10,2	-18,32,20,9,2	-18,32,21,9,2	-18,32,22,9,2	-18,32,24,8,2	-18,32,25,8,2

*Numbers in parentheses are mass numbers of most stable isotopes.

Figure 7.4   The periodic table of the elements.

of the Elements

		VII A	VIII A
		Hydrogen	Helium
		$_1$H	$_2$He
		1.0080 1	4.00260 2

III A	IV A	V A	VI A	VII A	VIII A
Boron	Carbon	Nitrogen	Oxygen	Fluorine	Neon
$_5$B	$_6$C	$_7$N	$_8$O	$_9$F	$_{10}$Ne
10.81 2,3	12.011 2,4	14.0067 2,5	15.9994 2,6	18.9984 2,7	20.179 2,8
Aluminum	Silicon	Phosphorus	Sulfur	Chlorine	Argon
$_{13}$Al	$_{14}$Si	$_{15}$P	$_{16}$S	$_{17}$Cl	$_{18}$Ar
26.9815 2,8,3	28.086 2,8,4	30.9738 2,8,5	32.06 2,8,6	35.453 2,8,7	39.948 2,8,8

	I B	II B					
Nickel	Copper	Zinc	Gallium	Germanium	Arsenic	Selenium	Bromine	Krypton
$_{28}$Ni	$_{29}$Cu	$_{30}$Zn	$_{31}$Ga	$_{32}$Ge	$_{33}$As	$_{34}$Se	$_{35}$Br	$_{36}$Kr
58.71 2,8,16,2	63.546 2,8,18,1	65.37 2,8,18,2	69.72 2,8,18,3	72.59 2,8,18,4	74.9216 2,8,18,5	78.96 2,8,18,6	79.904 2,8,18,7	83.80 2,8,18,8
Palladium	Silver	Cadmium	Indium	Tin	Antimony	Tellurium	Iodine	Xenon
$_{46}$Pd	$_{47}$Ag	$_{48}$Cd	$_{49}$In	$_{50}$Sn	$_{51}$Sb	$_{52}$Te	$_{53}$I	$_{54}$Xe
106.4 2,8,18,18	107.868 2,8,18,18,1	112.40 2,8,18,18,2	114.82 2,8,18,18,3	118.69 2,8,18,18,4	121.75 2,8,18,18,5	127.60 2,8,18,18,6	126.9045 2,8,18,18,7	131.30 2,8,18,18,8
Platinum	Gold	Mercury	Thallium	Lead	Bismuth	Polonium	Astatine	Radon
$_{78}$Pt	$_{79}$Au	$_{80}$Hg	$_{81}$Tl	$_{82}$Pb	$_{83}$Bi	$_{84}$Po	$_{85}$At	$_{86}$Rn
195.09 -18,32,17,1	196.9665 -18,32,18,1	200.59 -18,32,18,2	204.37 -18,32,18,3	207.2 -18,32,18,4	208.9806 -18,32,18,5	(210) -18,32,18,6	(210) -18,32,18,7	(222) -18,32,18,8

Gadolinium	Terbium	Dysprosium	Holmium	Erbium	Thulium	Ytterbium	Lutetium
$_{64}$Gd	$_{65}$Tb	$_{66}$Dy	$_{67}$Ho	$_{68}$Er	$_{69}$Tm	$_{70}$Yb	$_{71}$Lu
157.25 -18,25,9,2	158.9254 -18,27,8,2	162.50 -18,28,8,2	164.9303 -18,29,8,2	167.26 -18,30,8,2	168.9342 -18,31,8,2	173.04 -18,32,8,2	174.97 -18,32,9,2
Curium	Berkelium	Californium	Einsteinium	Fermium	Mendelevium	Nobelium	Lawrencium
$_{96}$Cm	$_{97}$Bk	$_{98}$Cf	$_{99}$Es	$_{100}$Fm	$_{101}$Md	$_{102}$No	$_{103}$Lr
(247) -18,32,25,9,2	(249) -18,32,26,9,2	(251) -18,32,28,8,2	(254) -18,32,29,8,2	(253) -18,32,30,8,2	(256) -18,32,31,8,2	(254) -18,32,32,8,2	(257) -18,32,32,9,2

7.3 The Exchange Effect

Consider, again, a system of two electrons. The spin vector S of the system is the vector sum of the spin vectors S_1 and S_2 of the individual electrons:

$$S = S_1 + S_2 \tag{7.18}$$

The total spin quantum number S is given by

$$S = s_1 + s_2 = \tfrac{1}{2} + \tfrac{1}{2} = 1 \tag{7.19}$$

if the individual spin vectors are "parallel." The system is then said to be in a **triplet** state, because the total spin vector S can assume three possible z components, namely, $m_S = -1, 0, +1$. On the other hand, if the individual spin vectors are "antiparallel," the total spin quantum number S is given by

$$S = s_1 - s_2 = \tfrac{1}{2} - \tfrac{1}{2} = 0 \tag{7.20}$$

and the system is said to be in a **singlet** state. In short, the total spin quantum number S determines the multiplicity of a quantum state of the system according to multiplicity $= 2S + 1$ (see Figure 7.5).

The spin state is described by a spin wave function. It turns out

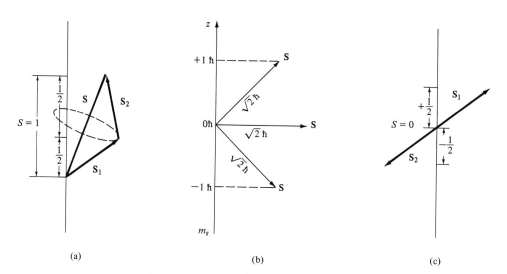

Figure 7.5   (a) When two spin vectors S_1 and S_2 are "parallel," the total spin quantum number is $S = s_1 + s_2 = 1$. (b) A total spin quantum number of $S = 1$ gives rise to a triplet state. (c) Two "antiparallel" spin vectors cancel and give rise to a singlet state, that is, $S = 0$.

that a singlet state is described by an antisymmetric spin wave function σ_- and a triplet state is described by a symmetric spin wave function σ_+. Now, a system of electrons must be described by a total wave function $\Psi(\mathbf{r}, \mathbf{S})$ which is antisymmetric. The total wave function $\Psi(\mathbf{r}, \mathbf{S})$ is most simply constructed as the product of the spatial wave function $\psi(\mathbf{r})$ and the spin wave function σ. If the system is in a singlet state, then the total wave function is (see Example 7.2)

$$\Psi_-(\mathbf{r}, S = 0) = \psi_+(\mathbf{r})\,\sigma_- \tag{7.21}$$

which implies that its spatial wave function must be symmetric, and hence $\psi(\mathbf{r}) = \psi_+(\mathbf{r})$. Note that the product of a symmetric function and an antisymmetric function yields an antisymmetric function. On the other hand, if the system is in a triplet state, its spatial wave function must be antisymmetric. This follows from the fact that the total wave function is antisymmetric and the spin function is symmetric (see Example 7.2):

$$\Psi_-(\mathbf{r}, S = 1) = \psi_-(\mathbf{r})\,\sigma_+ \tag{7.22}$$

The symmetry property of the spatial wave function is related to the important quantum mechanical notion that identical particles are indistinguishable, and gives rise to a purely quantum mechanical effect referred to as the **exchange effect.**

Example 7.2 Construct the symmetric and the antisymmetric spin wave functions for a system of two noninteracting electrons.

Solution. Let α and β be the spin wave functions of an electron for the spin states $m_s = +\frac{1}{2}$ and $m_s = -\frac{1}{2}$, respectively. Then the total spin wave function of a two-electron system can, in the approximation of no interactions involving the spin, be constructed as a product of the individual spin wave functions, namely,

$$
\begin{array}{cccc}
\alpha_1\alpha_2 & \alpha_1\beta_2 & \beta_1\alpha_2 & \beta_1\beta_2 \\
m_s. \ (+\tfrac{1}{2})(+\tfrac{1}{2}) & (+\tfrac{1}{2})(-\tfrac{1}{2}) & (-\tfrac{1}{2})(+\tfrac{1}{2}) & (-\tfrac{1}{2})(-\tfrac{1}{2})
\end{array}
$$

where the subscripts label the electrons. The products $\alpha_1\alpha_2$ and $\beta_1\beta_2$ remain the same, that is, symmetric, when the electrons are interchanged. But the other two products $\alpha_1\beta_2$ and $\beta_1\alpha_2$ are neither symmetric nor antisymmetric with respect to a particle exchange, implying that they cannot serve as the total spin wave functions of a two-electron system. However, a linear combination of them leads to a symmetric spin wave function

$$\alpha_1\beta_2 + \beta_1\alpha_2$$

and an antisymmetric spin wave function

$$\alpha_1\beta_2 - \beta_1\alpha_2$$

Hence there are three different spin wave functions that are symmetric

$$\sigma_+ = \begin{cases} \alpha_1\alpha_2 & m_S = +1 \\ \alpha_1\beta_2 + \beta_1\alpha_2 & m_S = 0 \\ \beta_1\beta_2 & m_S = -1 \end{cases}$$

and they correspond to the three possible values of m_S of the triplet state. The triplet state of a two-electron system is therefore described by a symmetric spin wave function. The singlet state of a two-electron spin system, on the other hand, is described by an antisymmetric spin wave function

$$\sigma_- = \alpha_1\beta_2 - \beta_1\alpha_2 \qquad \bullet$$

   We now explore the implications of the symmetry property of the spatial wave function. To begin, consider the singlet state of a two-electron system. Its spatial wave function is symmetric, and is expressible in the form

$$\begin{aligned} \psi_+(\mathbf{r}) &= \psi_a(\mathbf{r}_1)\,\psi_b(\mathbf{r}_2) + \psi_b(\mathbf{r}_1)\,\psi_a(\mathbf{r}_2) \\ &\equiv \psi_{ab} + \psi_{ba} \end{aligned} \qquad (7.23)$$

where a and b denote the spatial quantum numbers (n, l, m_l). The probability density of the system is

$$\psi_+^2 = \psi_{ab}^2 + \psi_{ba}^2 + 2\psi_{ab}\psi_{ba} \qquad (7.24)$$

whose physical meaning surfaces when each term is interpreted. ψ_{ab}^2 gives the probability of finding one electron in state a at position $\mathbf{r}_1$ and the other electron in state b at position $\mathbf{r}_2$. ψ_{ba}^2 may be similarly interpreted. In order to interpret the cross term $\psi_{ab}\psi_{ba}$, consider the situation where $\mathbf{r}_1$ and $\mathbf{r}_2$ are nearly equal, that is, $\mathbf{r}_1 \approx \mathbf{r}_2$. Then we have

$$\begin{aligned} \psi_{ab} &\approx \psi_{ba} \\ \psi_{ab}\,\psi_{ba} &\approx \psi_{ab}^2 \approx \psi_{ba}^2 \end{aligned} \qquad (7.25)$$

which leads to a total probability density

$$\psi_+^2 \approx 2(\psi_{ab}^2 + \psi_{ba}^2) \qquad (7.26)$$

The probability density is effectively doubled by the cross term when the electrons are in close proximity. Consequently, the electrons in a singlet state are found close together. Referred to as an exchange effect, this phenomenon is purely quantum mechanical in origin and has no classical analog.

The spatial wave function of a triplet state is antisymmetric. Its square, which gives the probability distribution of the electrons, is

$$\psi_-^2 = \psi_{ab}^2 + \psi_{ba}^2 - 2\psi_{ab}\psi_{ba} \tag{7.27}$$

The probability density approaches zero when the electrons are brought in close proximity, that is, $\mathbf{r}_1 \approx \mathbf{r}_2$. Consequently, the electrons are not found close together; rather, they tend to stay apart. This fact has nothing to do with the electrostatic repulsion between electrons but is a purely quantum mechanical exchange phenomenon.

7.4 Ground State of Helium

Helium has two electrons. The ground state of the helium atom is a singlet state, that is, the electron spin vectors are antiparallel. That the helium ground state cannot be a triplet state follows from the symmetry property of the spatial wave function. A triplet state is described by an antisymmetric spatial wave function of the form

$$\psi_- = \psi_a(\mathbf{r}_1)\,\psi_b(\mathbf{r}_2) - \psi_b(\mathbf{r}_1)\,\psi_a(\mathbf{r}_2) \tag{7.28}$$

The helium electrons share the same set of spatial quantum numbers (n, l, m_l) in the ground state, that is, $a = b$. As a result, the spatial wave function of the triplet state vanishes:

$$\psi_- = \psi_a(\mathbf{r}_1)\,\psi_a(\mathbf{r}_2) - \psi_a(\mathbf{r}_1)\,\psi_a(\mathbf{r}_2) = 0 \tag{7.29}$$

This is, of course, consistent with the exclusion principle.

The spatial wave function of a singlet state is symmetric. This implies that the exchange effect causes the electrons in the ground state of helium to be found close together. In fact, the wave functions associated with the individual electrons overlap completely.

The electrons of the helium atom experience an electrostatic repulsion. As a first approximation to the helium atom, however, let us ignore all interactions between the electrons and picture helium as consisting of a nucleus of charge $Z = 2$ and two electrons moving about the nucleus independently of each other. The Schrödinger equation for the helium atom, then, separates into two hydrogenlike equations, each with a quantized set of energies

$$E_n = -\frac{Z^2 13.6}{n^2} \text{eV} = -\frac{54.4}{n^2} \text{eV} \tag{7.30}$$

Note that Eq. (7.30) gives the energy of a singly ionized helium atom. The ground state energy E_g of the normal atom is the sum of the ground state energies of the ionized atoms and is thus twice the value given by Eq. (7.30):

$$E_g(\text{theo}) = 2E_1 = -108.8 \text{ eV} \tag{7.31}$$

By the same token, the ionization energy, which is the minimum energy required to remove one of the electrons in the ground state, is E_i (theo) $= 54.4 \text{ eV}$.

Now the observed value of the ionization energy is $E_i(\text{obs}) = 24.8 \text{ eV}$, and the energy required to remove the second electron is 54.4 eV. Hence the observed ground state energy of helium is

$$E_g(\text{obs}) = -(54.4 + 24.8) \text{ eV} = -79.2 \text{ eV} \tag{7.32}$$

The difference in the theoretical and the observed values is of the order

$$\Delta E = (108.8 - 79.2) \text{ eV} = 29.6 \text{ eV} \tag{7.33}$$

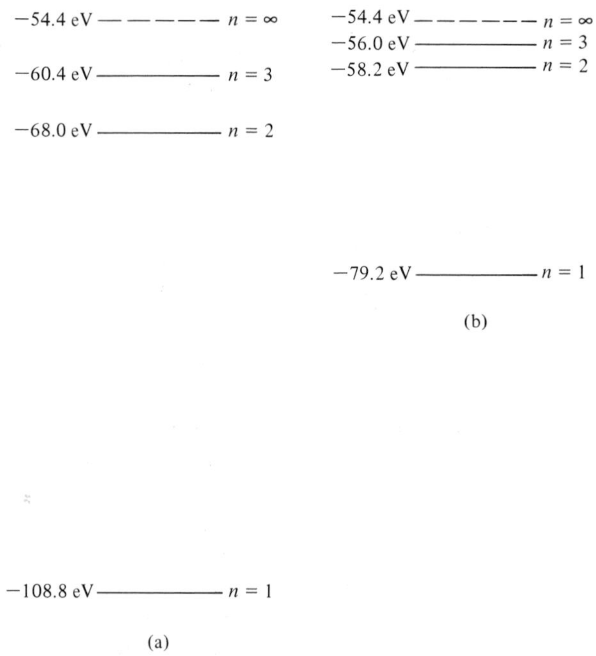

$-54.4 \text{ eV} \text{ —} \text{—} \text{—} \text{—} \text{—} \; n = \infty$

$-60.4 \text{ eV} \text{———} \; n = 3$

$-68.0 \text{ eV} \text{———} \; n = 2$

$-54.4 \text{ eV} \text{—} \text{—} \text{—} \text{—} \text{—} \; n = \infty$
$-56.0 \text{ eV} \text{———} \; n = 3$
$-58.2 \text{ eV} \text{———} \; n = 2$

$-79.2 \text{ eV} \text{———} \; n = 1$

(b)

$-108.8 \text{ eV} \text{———} \; n = 1$

(a)

Figure 7.6 Excited states of the neutral helium atom with one electron remaining in the $n = 1$ state. (a) The excited electron interacts only with the nucleus and not with the electron in the $n = 1$ state. (b) The excited electron interacts with the nucleus as well as with the other electron in the $n = 1$ state.

and must be attributed to the electrostatic and exchange effects (Figure 7.6).

Suppose we view one of the electrons as existing in the Coulomb field of the nucleus shielded by the other. Then the hydrogenlike expression for the ionization energy becomes valid, and hence

$$E_i = Z_{\text{eff}}^2 13.6\,\text{eV} \tag{7.34}$$

where Z_{eff} is the effective charge of the nucleus. The effective charge now emerges from the observed value of the ionization energy, that is,

$$Z_{\text{eff}} = \sqrt{\frac{24.8}{13.6}} = 1.35 \tag{7.35}$$

and is greater than unity. This implies that one of the electrons partially penetrates the screening of the nucleus by the other electron.

7.5 Excited States of Helium

An excited helium atom can exist in either a singlet state or a triplet state. If one of the electrons is raised to an excited state in such a way that their spins are antiparallel, the excited atom is in a singlet state, and is called **parahelium**. On the other hand, if the spin of the excited electron is parallel to the spin of the ground state electron, the atom is in a triplet state, and is called **orthohelium**. This distinction is made because the observed spectral lines of parahelium and orthohelium are different. The former arise from transitions within the singlet states, and the latter arise from transitions within the triplet states (see Figure 7.7).

The total orbital angular momentum L of the helium atom is the vector sum of the orbital angular momenta of the individual electrons:

$$\mathbf{L} = \mathbf{L}_1 + \mathbf{L}_2 \tag{7.36}$$

with the total orbital angular momentum quantum number L given by

$$L = |l_1 - l_2|, |l_1 - l_2| + 1, \ldots, l_1 + l_2 \tag{7.37}$$

The total orbital angular momentum of helium, or of any atom in general, is designated by capital letters, namely,

$$L = 0, 1, 2, 3, \ldots$$
symbol : $S, P, D, F, \ldots$

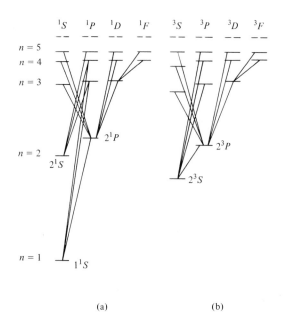

^1S ^1P ^1D ^1F ^3S ^3P ^3D ^3F

$n = 5$
$n = 4$
$n = 3$

2^1P

2^3P

2^3S

$n = 2$
2^1S

2^3S

$n = 1$
1^1S

(a)   (b)

Figure 7.7 (a) Excited states of parahelium. (b) Excited states of orthohelium.

In writing out the electron configuration, the multiplicity of an orbital angular momentum state, which is determined by the total spin quantum number S via $(2S + 1)$, is designated by a left superscript. For example, ^1P and ^3P designate the $L = 1$ singlet and triplet states, respectively.

The energy levels of orthohelium consist of three components because of the spin-orbit coupling, except for the $L = 0$ states. Consequently, transitions between them lead to three component spectral lines. However, transitions to the 1^3S state are strictly forbidden, because orthohelium with spins parallel cannot accommodate two electrons in the $n = 1$ state as a result of the exclusion principle. In parahelium, on the other hand, the total spin is zero. Thus there is no spin-orbit coupling, and the spectral lines have a singlet structure. Transitions to the 1^1S state are, of course, allowed. Transitions between the singlet and the triplet states are forbidden because they would require a change in the total spin. Only those transitions obeying the selection rule, $\Delta S = 0$, are allowed; violations of the selection rule can occur only in the presence of some spin-dependent forces.

The energy of an excited state of helium, for a given principal quantum number n, depends, first of all, on the value of the total angular momentum quantum number L. Subshells of higher values of L have higher energies than the subshells of lower values of L. Second, the energy of an excited state is affected by the average electrostatic interaction between the two electrons. This contributes a positive energy, and thus raises the energy level. Third, the energy of an excited state depends on the alignment of the electron spins. The effect of the spin alignment is rooted in the symmetry character of the spatial wave function. The symmetric character of the spatial wave function of a singlet excited state implies that the electrons tend to be found close together as a result of the exchange effect. This gives rise to a slight increase in the electrostatic repulsion between the electrons, and thus contributes a positive energy. On the other hand, the spatial wave function of a triplet excited state is antisymmetric. The exchange effect then causes the electrons to stay farther apart, and thus gives rise to a slight decrease in the electrostatic repulsion between the electrons. This has the effect of lowering the energy level. Consequently, the energies of the singlet excited states are higher than those of the corresponding triplet excited states. This result is in agreement with observation (Figure 7.8).

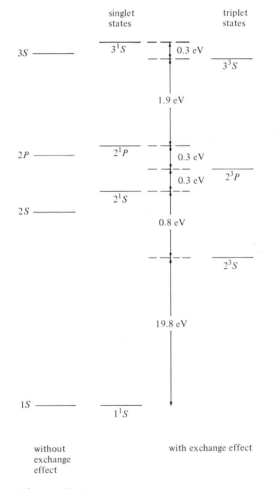

Figure 7.8 The degeneracy between the singlet and the triplet states is removed by the exchange effect.

There is no triplet state corresponding to the singlet ground state. This, of course, is consistent with, and predicted by, the exclusion principle. Historically, the nonexistence of the triplet ground state was first established from observations of the helium spectra, and subsequently led Pauli to formulate the exclusion principle.

7.6   Optical Spectra of Alkali Atoms

Optical spectra are emitted from transitions involving the outermost electrons. The outermost electron of an alkali atom exists outside an inert core. The inert core consists of the nucleus and $(Z - 1)$ electrons, completely filling the subshells. The optical spectra of an alkali atom are thus similar to those of the hydrogen atom.

The nucleus and the filled core of an alkali atom produce a spherically symmetric potential $V(r)$. The potential energy of the outermost electron, however, depends on the angular momentum state of the electron. The lower the value of l, the deeper the electron penetrates the core, thus experiencing a higher effective charge $Z_{eff}(l)$. Therefore, unlike the hydrogen atom, the energy levels of an alkali atom are l-dependent:

$$E_{n,l} = -\frac{Z_{eff}^2(l)13.6\,\text{eV}}{n^2} \qquad (7.38)$$

The total angular momentum of an alkali atom is determined by the outermost electron, that is,

$$\begin{array}{ll} j = l + s & \text{spin-up} \\ j = l - s & \text{spin-down} \end{array} \qquad (7.39)$$

where l is the orbital quantum number of the electron. This is so because the total angular momentum of the filled core is zero. An alkali atom is normally in an S ($l = 0$) state. (It is customary to designate an atomic orbital angular momentum state by a capital letter.)

Sodium has 11 electrons. Its electron configuration is $1s^2 2s^2 2p^6 3s^1$. The ground state of the optically active electron is thus the $3s$ subshell, and has no fine structure. The lowest excited state, namely, the $3P$ state, and all the $l \neq 0$ states, however, display fine structure splittings because of the spin-orbit interaction. The total angular momentum of the $3P$ state, for example, is given by

$$\begin{array}{ll} j = l + s = \frac{3}{2} & \text{spin-up} \\ j = l - s = \frac{1}{2} & \text{spin-down} \end{array} \qquad (7.40)$$

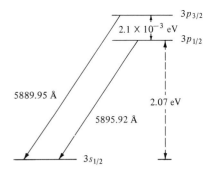

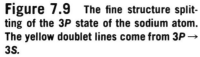

Figure 7.9 The fine structure splitting of the 3*P* state of the sodium atom. The yellow doublet lines come from 3*P* → 3*S*.

The doublet structure of the 3*P* state is designated by $3^2P_{3/2}$ and $3^2P_{1/2}$, where the left superscript indicates the multiplicity of the state and the right subscript indicates the value of *j*. (See Figure 7.9).

The so-called sodium D lines are results of transitions from the doublet 3*P* state to the singlet 3*S* state; their wavelengths differ by $\Delta\lambda = 6.03$ Å. The spectral lines emitted upon transitions from the *P* states to the 3*S* state are referred to as the **principal** series, and are all doublets. Transitions from the *S* states to the 3*P* state give rise to the so-called **sharp** series, which are also doublets. Triplet spectral lines are emitted when electrons make transitions from the *D* (*l* = 2) states to the 3*P* state and from the *F* (*l* = 3) states to the 3*D* state. The former are referred to as the **diffuse** series, and the latter the **fundamental** series. Notice that the initial states leading to the spectral series are designated by the first letters of the series names. All these transitions are governed by the selection rule, $\Delta j = 0, \pm 1$.

The downward transitions of inner electrons lead to x-ray spectral lines. If a 1*s* electron is knocked out, say, by a high-energy electron beam, a vacancy is created in the *K* shell. This vacancy can be filled by an electron in the *L* shell, in which case an x-ray photon, referred to as a K_α line, is emitted. The energy of a K_α line is given by

$$hv_{K_\alpha} = 13.6 \text{ eV } (Z-1)^2 \left(\frac{1}{1^2} - \frac{1}{2^2}\right) \qquad (7.41)$$

The factor $(Z - 1)$ is the effective nuclear charge that the electron making the transition experiences, and thus the nucleus is completely shielded by the unremoved 1*s* electron. Of course, transitions from higher shells also occur, and the emitted x-ray spectral lines constitute the so-called *K* series. Similarly, electrons making transitions to the *L* shell give rise to the *L* series of x-ray spectral lines (Figure 7.10).

Recall now that Moseley was led to the discovery of the atomic number *Z* by a systematic study of the x-ray spectra of the elements. His work established that the periodicity of the elements was a func-

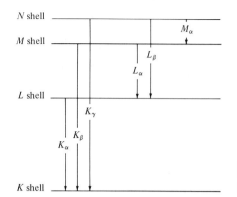

N shell

M shell

M_α

L_β

L_α

L shell

K_γ

K_β

K_α

K shell

Figure 7.10 Characteristic x-ray lines.

tion not of atomic weight, as originally proposed by Mendeleyev, but of atomic number. Now the atomic number determines the number of electrons in a normal atom, and this fact suggested that the physical basis for the periodicity of the elements might be found in the arrangement of the electrons in an atom. Then came the discovery of the exclusion principle by Pauli, in 1925, which, combined with the Schrödinger theory, gave rise to the shell structure of the atom. A rigorous explanation of the periodic table of the elements thus came forth some 50 years after its discovery.

7.7 Summary

An important feature of quantum mechanics is that identical particles are indistinguishable and consequently are described by either a symmetric wave function or an antisymmetric wave function. The wave function describing a system of electrons is antisymmetric, thus yielding the result, referred to as the exclusion principle, that no two electrons can exist in the same quantum state. This result gives rise to the shell structure of the atom and, in turn, to the periodicity of 8 in the main body of the periodic table of the elements. The exclusion principle applies not only to the electrons but to all spin-$\frac{1}{2}$ particles.

The wave function of a system of electrons is the product of a spatial wave function and a spin wave function. The spatial wave function of a singlet state is symmetric; that of a triplet state is antisymmetric. As a result of the symmetry character of the spatial wave function, electrons tend to stay closer together in a singlet state and farther apart in a triplet state. This phenomenon, referred to as the exchange effect, has no classical analog.

The helium atom is a two-electron system. The interaction between the electrons in helium is quite appreciable, and has the effect of partially shielding the nucleus. The normal helium atom can exist only in a singlet state. The excited atom, on the other hand, can exist either in a singlet state or in a triplet state. The spectral lines arising from transitions within the singlet states have a singlet structure, and those from transitions within the triplet states have a triplet structure. Transitions between the singlet and the triplet states are forbidden.

The optical spectra of alkali atoms display a close semblance to hydrogen spectra. However, the energy of the outermost electron of an alkali atom, is, unlike that of hydrogen, nondegenerate with respect to the orbital angular momentum quantum number. This l dependence has its origin in the fact that the potential field of an alkali atom deviates from the strict r^{-1} dependence of the Coulomb potential. As in hydrogen, the energy levels of an alkali atom have a doublet fine structure as a result of the spin-orbit interaction, and are specified by the total angular momentum quantum number. All transitions obeying the selection rule, $\Delta j = 0, \pm 1$, are allowed, and the spectral lines are either doublets or triplets. X-ray spectra are produced by the inner electrons, and their transitions give rise to the characteristic x rays.

Problems

7.1 Verify Eqs. (7.8) and (7.9).

7.2 Argue that a system of particles described by a symmetric wave function does not obey the Pauli exclusion principle.

7.3 Explain why the total angular momentum of He and Ne in the ground state is zero.

7.4 The electrostatic repulsion energy between the electrons in the ground state of helium is given, in a first-order approximation, by

$$\Delta E = \frac{5}{4} Z \left(\frac{\alpha^2 mc^2}{2} \right)$$

where Z is the atomic number of helium, m is the electron mass, and α is the fine structure constant.
   (a) Calculate the ground state energy of the helium atom in this approximation.
   (b) Explain why it takes considerably less energy to ionize a normal helium atom than a singly ionized helium ion.

7.5 Consider a helium atom one of whose electrons is in the $2P$ state and the other in the $3D$ state.
   (a) What are the possible values of its total orbital angular momentum?
   (b) If the total orbital angular momentum vector and the total spin vector of the helium atom are "parallel," give the total angular momentum quantum numbers J for the triplet state of the atom.

7.6 The ionization energies of the $3S$, $3P$, and $3D$ states of the sodium atom ($Z = 11$) are 5.13 eV, 3.0 eV, and 1.5 eV, respectively. Calculate the effective charge "seen" by the electron in each state.

7.7 The observed fine structure splitting of the $3P$ state of sodium is $\Delta E = 2.1 \times 10^{-3}$ eV. By use of the semiclassical expression for the magnetic energy $E_m = -\frac{1}{2}\boldsymbol{\mu} \cdot \mathbf{B}$ (Eq. 5.90), calculate the magnitude of the internal magnetic field of the atom.

7.8 The K_α x-ray spectral line of a certain element has an energy of 6.375 KeV. What is the element?

7.9 (a) Show by referring to problem 7.4 that the energy required to ionize a $1s$ electron of sodium is 1.46 KeV in a first-order approximation.
(b) Calculate the energy of the K_α x-ray line.

7.10 The ionization energy of the least tightly bound electron of the boron atom is 8.3 eV, and the ionization energy of a $1s$ electron is 259.3 eV. If a boron target is bombarded by electrons, calculate the minimum kinetic energy of the electron that will produce a K_α x-ray line.

7.11 Explain why the K_α x-ray spectral line has a doublet structure.

Suggestions for Further Reading

Christy, R. W., and A. Pytte. *The Structure of Matter*. New York: W. A. Benjamin, Inc., 1965. Chapters 22 and 23.

Eisberg, R. M. *Fundamentals of Modern Physics*. New York: John Wiley & Sons, Inc., 1961. Chapters 12, 13, and 14.

Ford, K. W. *Classical and Modern Physics*, vol. 3. Lexington, Mass.: Xerox College Publishing, 1974. Chapter 24.

X rays and crystallography

8.1 Introduction

Ordinary samples of real materials are made up of huge numbers of atoms and molecules. When the sample is in the form of a gas, there are relatively few interactions between the particles. The particles swarm around in random fashion, undergoing chance collisions. The **mean free path** between collisions is large compared to the size of the atom or molecule. Associated with the random motion of each particle is a kinetic energy. The collisions, although elastic, result in interchanges of kinetic energy among the molecules. For a sample in thermal equilibrium, kinetic theory for an ideal gas is able to relate the average molecular kinetic energy to macroscopic quantities,

$$\langle KE \rangle = \tfrac{3}{2} k_0 T \tag{8.1}$$

where $k_0 = 1.381 \times 10^{-16}$ erg-$^\circ K^{-1}$ and T is the absolute temperature.

When two molecules approach each other, as in a collision, forces between them come into play. At moderate distances the force is essentially classical, being that between two dipoles. Ionic molecules have static dipole moments which tend to align within each other's fields. The result of this alignment is an attractive force. Atoms and nonionic molecules do not have a static dipole moment. However, each can be polarized within the field of the other, again resulting in attraction. As the molecules come still closer together, their **electron clouds** begin to overlap. As they do so, the Pauli exclusion principle sets in and the net force becomes one of repulsion. Figure 8.1 is a sketch of the van der Waal's potential, resulting from this interaction. In a gas kinetic energies are sufficiently high that binding cannot take place in the potential well of Figure 8.1. At sufficiently low temperatures, however, the molecules bind together into a solid.

Most solids have crystalline structure. In some cases this structure extends through large enough volumes to make the geometrical shape of the crystal visible to the unaided eye. In other cases the solid may be polycrystalline, that is, made up of many microscopic crystals packed together in somewhat random fashion. There are many techno-

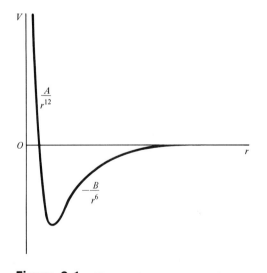

Figure 8.1 The van der Waal's potential.

logical applications that require understanding of crystal structure. Apart from technology, crystals are fascinating on their own account, and the study of them is an important aspect of fundamental physics. Much of our knowledge of crystal structure comes from analyses of x-ray diffraction effects. At the same time the diffraction of x rays provides confirmation of the wave nature of x radiation.

8.2 The Discovery and Production of X Rays

The discovery of x rays is credited to a series of observations by Wilhelm Roentgen in 1895. He found that when a sufficiently high voltage is applied to a well-evacuated discharge tube, a radiation is produced that is capable of **penetrating** paper, wood, and other materials opaque to ordinary light. In his early experiments he was able to detect the radiation by the **fluorescence** it caused in a coating of platinum barium cyanide deposited on a paper screen. After a series of careful observations, he concluded that the radiation originated in parts of the discharge tube that were under bombardment by cathode rays. The nature of the radiation, however, was a mystery, hence the name x rays. Nevertheless, the discovery excited great interest, and within a matter of months x rays were being used as an aid in setting broken bones.

The public interest in x rays centered on their use for "seeing through" opaque materials, but physicists were interested primarily in the nature of the radiation itself. In addition to producing fluorescence in certain materials, the radiation darkened photographic film and quickly discharged electroscopes, presumably by **ionizing** the surrounding air. It was observed that although cathode rays were deflected by the field of a nearby magnet, x rays were not so deflected. On these grounds there were early surmises that the x radiation might be electromagnetic waves of unusually short wavelength. Almost two decades intervened before there was irrefutable experimental evidence in confirmation of these surmises.

Many pioneering x-ray experiments were done with cold cathode gas discharge tubes. In these tubes residual gas is ionized by the application of high voltage. When ions so created strike the **cathode,** secondary electrons are produced which, together with those resulting from the ionization process itself, are accelerated toward the anode. When they strike the **anode,** or even the glass wall of the tube, x rays are produced.

As we shall see in later sections, the penetrating power and other properties of x rays are very dependent upon tube voltage. There is a threshold voltage for the onset of ionization. Below this voltage no ionization occurs. Small increases in voltage above the threshold produce very large increases in ionization current. For this reason satisfactory operation occurs only at voltages somewhat above the threshold. The operating voltage is determined largely by the pressure of the residual gas. Consequently, many cold cathode tubes were operated in conjunction with a vacuum pumping station to control gas pressure. These systems were awkward to operate and not very portable. Other cold cathode tubes were designed to be sealed off at the proper pressure. While portable, these tubes were subject to changes in gas pressure during the life of the tube. Gas ions become embedded in the electrodes and glass envelope. The residual pressure falls and higher and higher voltages are required to maintain the discharge. The radiation becomes **harder,** that is, more penetrating. Finally a point is reached when the voltage requirements exceed the capabilities of the power supply.

The development by W. D. Coolidge in 1913 of the **hot cathode** x-ray tube alleviated most of these operating difficulties. It is the only type in use today. Figure 8.2 is a diagram of such a tube and associated electrical power supplies. During manufacture this tube is highly evacuated and then sealed off. Electrons are provided not by ioniza-

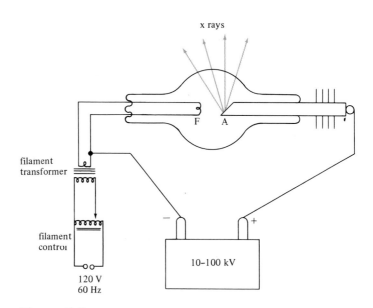

Figure 8.2 The Coolidge hot cathode x-ray tube. Electrons, released from filament F by thermionic emission, are accelerated to anode A by voltage applied to the tube. X rays are produced when electrons strike the anode. A high vacuum is maintained inside the tube. Cooling fins are provided to carry heat away from the anode. In some tubes this function is provided by water or oil cooling.

tion but by emission from a heated filament. The **electron emission**
rate is determined by the temperature of the filament. Since the tube
is at high vacuum, any desired accelerating voltage can be used short
of that which causes direct sparking between electrodes. Individual
control of **electron current** and **accelerating voltage** is a considerable
advantage in many applications.

Electromagnetic radiation is produced when a charged particle
undergoes acceleration. The resulting photon energy depends upon
the magnitude of the acceleration. When electrons strike the target
of an x-ray tube, they undergo many collisions with the atoms there.
These collisions deflect and slow down the electrons. The accompanying changes of velocity constitute accelerations, so radiation is to
be expected. A great many of these collisions are with atomic electrons and are so gentle that the photons produced have too low an
energy to escape from the material of the target. In these instances
the only observable effect is the production of heat within the target.
Some of the collisions are close encounters with atomic nuclei. These
collisions produce much more violent electron accelerations. There is
a continuum in the distribution of accelerations and, consequently, a
continuum of photon energies. The maximum photon energy is produced when the electron is brought to rest in a single collision. Under
these circumstances the entire electron energy is converted to electromagnetic radiation, such that

$$K_e = h\nu_{max} = h\,\frac{c}{\lambda_{min}} \tag{8.2}$$

It follows that

$$\lambda_{min} = \frac{hc}{K_e} \qquad (K_e \text{ in ergs}) \tag{8.3}$$

Upon substituting numerical factors,

$$\lambda_{min} = \frac{(6.626 \times 10^{-27} \text{ erg sec})(2.998 \times 10^{10} \text{ cm/sec})}{(K_e \text{ eV})(1.602 \times 10^{-12} \text{ erg/eV})}$$

$$= \frac{1.2400 \times 10^{-4}}{K_e} \text{ cm}$$

$$= \frac{12.400}{K_e} \text{ Å} \qquad (K_e \text{ in keV}) \tag{8.4}$$

This radiation is called **bremsstrahlung,** a German word which
means **braking radiation.** It results in a continuous spectrum of x rays
whose short wavelength limit is given by Eq. (8.3). The short wavelength limit depends only upon the voltage applied to the x-ray tube

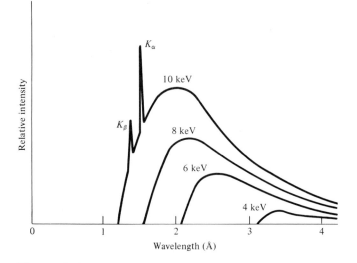

Figure 8.3 X-ray spectrum of copper. Results are shown for four different electron energies.

and not upon the material of the target. The target material does, however, influence the **efficiency** with which the electron energy is converted into radiation. This efficiency increases in an essentially linear fashion with atomic number Z. Figure 8.3 shows relative x-ray intensities that result when a copper target is bombarded with electrons. These are plotted for four different energies, as indicated on the diagram. Following the predictions of Eq. (8.3), the short wavelength cut-off becomes smaller as the bombarding energy becomes higher. The curves for 4 keV, 6 keV, and 8 keV show pure bremsstrahlung spectra.

Superimposed on the curve for 10-keV operation are two sharp spikes labeled K_α and K_β. These discrete **lines** are **characteristic x rays** and formed the basis of Moseley's studies that were treated in Section 5.2.

Characteristic x rays are produced by **transitions** between levels in the target atoms. The process is the same as that which produces emission in the visible spectrum, but in x-ray production much higher energies are involved. The process is quite different from that which produces bremsstrahlen. In order to have a characteristic transition in the x-ray region, one of the inner electrons must first be removed. In an x-ray tube this occurs in some of the collisions between bombarding electrons and target atoms, but only when the incident electron has kinetic energy at least as great as the ionization energy of the pertinent level. The voltage at which the tube operates must be high enough to supply this energy.

It is customary in x-ray studies to use shell numbering and level diagramming systems somewhat different from those used in optical spectroscopy. Electron shells are designated by letters: K for the most tightly bound shell, L for the next, M the next, and so on. These substitute for **principle quantum numbers** as follows:

quantum number, n 1   2   3   4 . . .
x-ray notation K L M N . . .

Subshells are designated by Roman numerals.

The energies of interest involve changes that occur in tightly bound shells. These shells are completely filled when the atom is in its unexcited, ground state. Excitation involves removal of an electron from an inner shell, creating a **vacancy** there. The energy of excitation is greatest when the vacancy is in the K shell, less so when it is in the L shell, and so on. If we direct attention to energy associated with a vacancy, it is sensible to make energy level diagrams showing the K level at the highest position and the others below it. This is the convention adopted for x-ray level diagrams. Under this convention a photon is emitted when a vacancy makes a downward transition from, say, the K shell to the L shell. (Under the convention used for optical spectroscopy we would say an electron has made a transition from the L shell to the K shell, thus filling the vacancy in the K shell and creating one in the L shell. The physical process is exactly the same whichever perspective is adopted.)

Figure 8.4 is drawn on the basis of the x-ray convention. For convenience we show energies on a logarithmic scale, but a linear one could also be used. Optical levels and the ground state are off scale at the bottom. The two K to L transitions that appear in this figure are

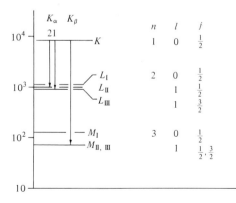

Figure 8.4 X-ray levels for copper. Level designations are shown in both x-ray and atomic notations. Transitions to the 3d and 4s shells are forbidden by the selection rules.

unresolved in Figure 8.3, where together they comprise the K_α line. The K to $M_{II,III}$ transition gives the K_β line. This transition is unresolved in the level diagram as well as in Figure 8.3. Along the right side of Figure 8.4 we show level designations in both x-ray and optical notation. The same selection rules that apply to optical transitions, namely $\Delta l = \pm 1$ and $\Delta j = 0, \pm 1$, also apply to x-ray transitions. We see that K to L_I and K to M_I transitions would require $\Delta l = 0$ and are therefore forbidden. The transition K to $M_{IV,V}$ would require $\Delta l = 2$ and is also forbidden. Transitions between the L level and levels of lower excitation produce L-series x rays. This series is of such low energy (long wavelength) that it does not appear in Figure 8.3.

Table 8.1 shows binding and transition energies for copper x rays. The transition energies are equivalent to differences in binding energies. Although $K_\alpha \sim 8.05$ keV and $K_\beta \sim 8.90$ keV, the bombarding electrons must have more kinetic energy than this to excite the K-series x rays. This is so because in the neutral atom the L and M shells are filled. In order to create a K-shell vacancy, it is necessary to remove the K electron from the atom altogether. Therefore if the bombarding energy were adjusted so λ_{min} were to fall exactly at the K_β wavelength, no characteristic x rays would be observed. Similar information about x-ray wavelengths and energies is available for all elements of the periodic table.

Table 8.1   **X-ray Binding and K-Series Transition Energies for Copper**

Shell	Binding Energy (keV)	K-Series Transition Energies (keV)
K	8.979	
L_I	1.096	
L_{II}	0.951	$K_\beta = 8.905$
L_{III}	0.931	$K_{\alpha_1} = 8.048$
M_I	0.120	$K_{\alpha_2} = 8.028$
$M_{II,III}$	0.074	

8.3 Absorption of X Rays

We have said that x rays penetrate material that is opaque to ordinary light. However, a certain amount of attenuation does occur. This attenuation depends upon the wavelength of the x ray and the material of the absorber. In general, the higher the x-ray energy (shorter wavelength), the more penetrating the radiation. Furthermore, materials of higher atomic number are more effective absorbers than those of low atomic number.

Suppose an x-ray beam of intensity I_0 falls upon an absorber of some particular material. One would expect that as the absorber is made thicker, the intensity of the x rays that penetrate it would be diminished. This is indeed the case, and experiments show that the situation can be described by

$$I = I_0 e^{-\mu x} \tag{8.5}$$

where x is the thickness of the absorber, I_0 the x-ray intensity incident on the absorber, and I that which penetrates through it; μ is called the linear absorption coefficient. Good absorbers are represented by large values of μ. From our remarks in the previous paragraph it is apparent that μ is a function of the atomic number of the absorber and the wavelength, or energy, of the x rays.

A useful modification of Eq. (8.5) is obtained if we recast the exponent.

$$\mu x = \frac{\mu}{\rho} x \tag{8.6}$$

where ρ is the density of the absorber in grams per cubic centimeter. The thickness of the absorber is now expressed in mass per unit area. If we designate it by t, then

$$x\rho = t \text{ g/cm}^2 \tag{8.7}$$

By this rearrangement we introduce a new type of absorption coefficient, μ_m, which is called the **mass absorption coefficient**; it has the units square centimeters per gram.

$$\frac{\mu}{\rho} = \mu_m \text{ cm}^2/\text{g} \tag{8.8}$$

These changes lead to

$$I = I_0 e^{-\mu_m t} \tag{8.9}$$

Since, in general, materials of large atomic number (and consequently large μ) have higher mass-densities than materials of low atomic number, the range in values of μ_m is much narrower than the range in μ. Figure 8.5 shows μ_m as a function of photon energy for absorbers of aluminum and of lead.

Although, in general, absorption coefficients decrease with increasing photon energy, dramatic increases occur at energies corre-

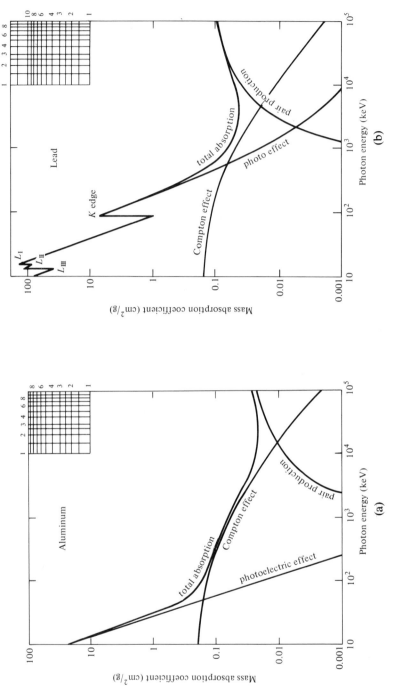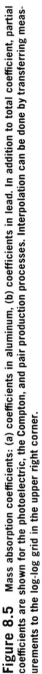

Figure 8.5 Mass absorption coefficients: (a) coefficients in aluminum, (b) coefficients in lead. In addition to total coefficient, partial coefficients are shown for the photoelectric, the Compton, and pair production processes. Interpolation can be done by transferring measurements to the log-log grid in the upper right corner.

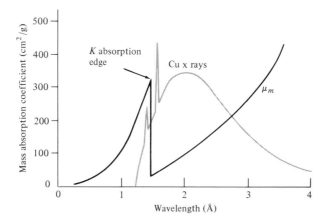

Figure 8.6 Mass absorption coefficient in nickel. The lightly drawn line is the x-ray spectrum of copper superimposed for the purpose of wavelength comparison.

sponding to the binding energies in atomic shells. These increases are spoken of as absorption edges. Figure 8.6 shows the mass absorption coefficient for nickel in the neighborhood of its K absorption edge, which occurs at a wavelength of 1.48 Å. In the regions above and below the absorption edge we see the decrease in absorption coefficient as wavelength becomes shorter. At the absorption edge there is a jump by about a factor of 8 in the coefficient. As we approach the absorption edge from the long wavelength side, photons have sufficient energy to eject L-shell electrons, but not K-shell electrons, from atoms of the absorber. When the energy becomes large enough to eject K-shell electrons, this additional interaction causes a sharp increase in absorption.

The process we have just described is the photoelectric effect as applied to an individual atom. The ionization energy of the pertinent shell is the analog of the work function at the surface of a bulk sample. Two other processes can also take place, the Compton effect and pair production. Both are negligible at the energies we are treating here. In fact, pair production, which is the materialization of a negative and positive electron pair out of the energy of the absorbed photon, is completely impossible for photon energies below 1.02 MeV. The Compton effect, which is a billiard-ball-like scattering of the photon by a free electron, although allowed at all energies, makes only a very small contribution to the absorption with which we are dealing. The relative importance of each process changes with photon energy and atomic number of the absorber. Figure 8.7 shows combinations of energy and atomic number for which each process is dominant. Along boundary lines between regions the absorption coefficients due to the processes on each side are equal.

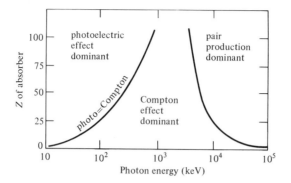

Figure 8.7 Photon absorption processes. Along boundary lines between regions the absorption coefficients due to the processes on each side are equal. [After R. C. Evans, *The Atomic Nucleus* (New York: McGraw-Hill Book Co., 1955). Used by permission of the publisher.]

We have chosen nickel for our illustration because its atomic number is one unit less than that of copper. Consequently its K-shell binding energy is slightly less than the binding energy in copper. It turns out, in fact, that the K absorption edge for nickel lies between the K_α and K_β emission lines of copper. To show this comparison, the copper spectrum at 10-keV bombarding energy is superimposed as the lightly drawn line. By filtering the copper radiation through a nickel foil of suitable thickness, it is possible to nearly eliminate the K_β line while retaining a substantial fraction of the K_α intensity. If the foil is 10^{-3} cm thick, it will pass 65 percent of the K_α radiation, but only 8 percent of the K_β. If it is 0.005 cm thick, about 45 mg/cm², it passes 12 percent of the K_α, but only 3×10^{-4} percent of the K_β. The response to changes in thickness and absorption coefficient is exponential and therefore very rapid. The use of a filter of this sort makes the radiation much closer to monochromatic, an important consideration in many experiments.

8.4 Diffraction

Very soon after the discovery of x rays attempts were made to determine whether the radiation consisted of particles or of waves. As early as 1898 diffraction experiments using narrow slits were attempted. Although there was some evidence of diffraction, the results were marginal and not convincing. To the extent that the results could be accepted, they indicated wavelengths of the order of 1 Å. Although the technology for producing ruled gratings for use in optical spectroscopy was highly developed, it was clearly not feasible to produce gratings with line spacing small enough to use at such short wavelengths.[1] It was not until 1912 that definitive experiments were reported, establishing the wave nature of x radiation. These experiments made use of speculations and surmises regarding the arrangement of atoms in a solid. Since crystals of the same atomic composition exhibit the same geometrical shape, whether the crystal is large or small, and, since crystals cleave along well-defined planes, it was reasonable to presume that the atoms might be arranged in well-ordered layers.

By making some additional simple assumptions, it is possible to calculate the average spacing between atoms in a crystal. To carry out this calculation we need to know the molecular weight and the density of the material. We also need Avogadro's number and a belief that the general shape of the crystal mirrors the arrangement of its atoms. Information of this sort was available at the turn of the century and indicated the spacing between atoms to be a few Ångstroms for most solids.

Example 8.1 Crystals of KCl have a general cubic shape, the molecular weight is 74.55, and the density is 1.984 g/cm³. Calculate the interatomic spacing in KCl.

Solution. Since there are two atoms per molecule, we write

$$2\ \frac{\text{atoms}}{\text{molecule}} \times \frac{1.984\ \text{g/cm}^3}{74.55\ \text{g/mole}} \times 6.022 \times 10^{23}\ \text{molecules/mole} = 3.205 \times 10^{22}\ \text{atoms/cm}^3$$

This leads to 3.120×10^{-23} cm³/atom.

From this we find the interatomic spacing

$$d = (3.120 \times 10^{-23})^{1/3} = 3.148 \times 10^{-8}\ \text{cm} = 3.148\ \text{Å} \quad \bullet$$

Since spacings of the sort found in KCl appeared comparable to the (presumed) wavelengths of x radiation, it seemed possible that a crystal might serve as a three-dimensional grating. This led Max von Laue (1879–1960) to suggest that a beam of x rays, narrowly collimated by pinhole apertures, be passed through a thin crystal. The actual experiment was reported as a collaboration involving W. Friedrich, P. Knipping, and M. Laue. Figure 8.8 shows the arrangement using a photographic film to record x-ray intensities. The results,

[1] There are exceptions to this statement. In 1925 Compton and Doan, and in 1929 Bearden, succeeded in producing x-ray diffraction by reflection at grazing angles from gratings ruled on glass.

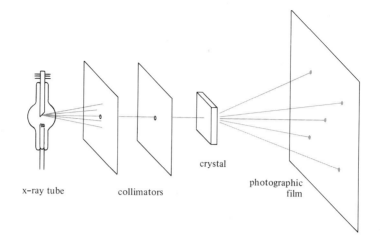

Figure 8.8 Schematic of a Laue diffraction experiment.

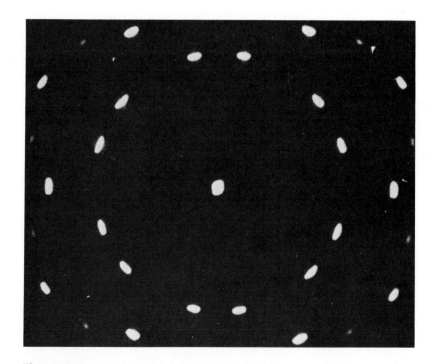

Figure 8.9 Laue diffraction pattern for an Mg crystal. [Reproduced from C. S. Barrett and T. B. Massalski, *Structure of Metals*, 3rd ed. (New York: McGraw-Hill Book Co., 1966). Used by permission of the publisher.]

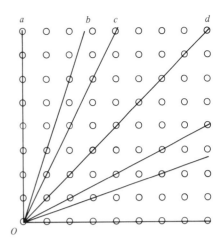

Figure 8.10 A two-dimensional representation of a cubic crystal.

similar to those shown in Figure 8.9, were thoroughly convincing. The symmetrical array of spots could only be interpreted as arising from diffraction effects caused by the three-dimensional array of atoms in the crystal. In this one experiment both the wave nature of x rays and the regularity with which atoms are arranged in a crystal were demonstrated. This turned out to be a truly powerful tool. If one knows interatomic spacings, x-ray wavelengths can be determined from the diffraction patterns. Conversely, known radiation can be used to investigate crystal structure. Since the days of Laue the study of x rays and the study of crystal structure have been intertwined.

Following the announcement of the Laue experiment, W. L. Bragg proposed a mechanism to explain the patterns on the film. Figure 8.10 is a two-dimensional representation of a cubic crystal. It is seen that along certain favored directions there is a high density of atoms. These are analogs of planes with high atomic density in three-dimensional crystals. It was Bragg's suggestion that x rays reflect from these planes, making equal angles of incidence and reflection. This idea gave a generally satisfactory explanation of the Laue patterns. However, the situation is a bit more complicated than simple reflection. Interference effects set limitations upon the angles at which "reflection" can occur. In Figure 8.11 we show a plane wave approaching from the left at angle θ. (Note that in x-ray diffraction work the angle is measured from the surface of the plane rather than from the normal.) The incoming radiation is scattered by the atoms of the crystal. We show ray 1 scattered by atom A and ray 2 scattered by atom B. After scattering, these rays will exhibit constructive interference, provided

$$EB + BF = n\lambda \tag{8.10}$$

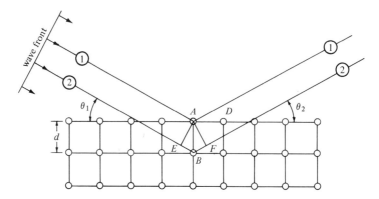

Figure 8.11 Bragg reflection.

with n an integer. Since $EB = BF = d \sin \theta$, we have the Bragg scattering law

$$n\lambda = 2d \sin \theta \tag{8.11}$$

where d is the spacing between the atomic planes.

The careful reader will object that it is not necessary that $EB = BF$ for the condition $EB + BF = n\lambda$ to be met; that is, perhaps it is not necessary that the angle of incidence equal the angle of reflection. However, a third ray scattered from atom D must produce wavelets also in phase with those scattered from atoms A and B. In order to satisfy this requirement simultaneously with that set forth in Eq. (8.10), it is necessary that the angle, after scattering, equal the angle of incidence. It is this equality of angles that leads us to refer to the process as x-ray reflection.

If the radiation is monochromatic, only a great stroke of luck will result in an angle and crystal spacing combination which satisfies Eq. (8.11). It was fortunate that the original Laue experiment was performed with bremsstrahlen, thus providing a wide range of wavelengths. Had monochromatic radiation been used, it is highly probable the diffraction effect would not have been observed. However, out of the continuous x-ray spectrum each set of reflecting planes, with its own angle and interplane spacing, could select the proper wavelength for constructive interference. Since the wavelength selected is not obvious on the film, the spots in a Laue pattern do not give us information about the spacing between atomic planes. The positions of the spots do, however, indicate the angular orientation of these planes.

The Bragg scattering condition, Eq. (8.11) can be used to produce diffraction patterns quite different from Laue patterns. Figure 8.12 is a diagram of a Bragg diffraction apparatus. X rays from a source at the left are collimated into a narrow beam that impinges on the surface of

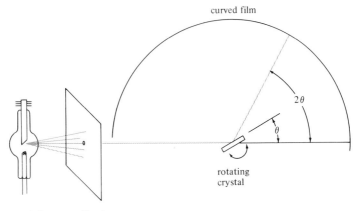

curved film

2θ

θ

rotating
crystal

x–ray tube   collimator

Figure 8.12 A Bragg x-ray spectrometer.

a crystal. A long strip of film is curved into a section of the cylinder whose center is at the crystal. During exposure of the film the crystal is caused to rotate around an axis perpendicular to the plane of the drawing. In this fashion the angle of incidence is continuously varied. Thus, on the film, an angle 2θ measured from the forward direction of the x-ray beam corresponds to the angle θ in Eq. (8.11). Those positions on the film which are strongly exposed correspond to wavelengths of high intensity in the incident radiation. Frequently an ionization chamber, rather than film, is used to detect x rays. The chamber is mounted at the end of an arm that pivots around the rotation axis of the crystal. The arm is moved through an angle 2θ while the crystal turns through an angle θ, thus preserving equality of angles of incidence and reflection. The results of either type of measurement can be presented as an x-ray spectrum similar to that of Figure 8.3.

Example 8.2   A Bragg scattering apparatus uses a KCl crystal. If the radiation comes from a Cu-target x-ray tube operating at 10 keV, calculate the angles 2θ for the short wavelength limit, for the K_β, and for the K_α radiations.

Solution.   We will be using the Bragg scattering condition, Eq. (8.11). From Example 8.1 we have $d = 3.148$ Å. We will need to calculate the wavelengths. By reference to Eq. (8.3) we obtain for the short wavelength limit

$$\lambda_{\min} = \frac{12.40}{10} = 1.240 \text{ Å}$$

The numerical relationship given in Eq. (8.4) is valid for converting photon energy to wavelength, provided we remove the subscript from

λ and express photon energy in keV. Table 8.1 gives the energies of the K_α and K_β transitions. These are the same as the photon energies. We find, therefore,

$$(K_\beta) = \frac{12.40}{8.905} = 1.392 \text{ Å}$$

$$(K_{\alpha_1}) = \frac{12.40}{8.048} = 1.541 \text{ Å}$$

$$(K_{\alpha_2}) = \frac{12.40}{8.028} = 1.545 \text{ Å}$$

Solving Eq. (8.11) for θ, we find

$$\theta = \text{arc sin} \frac{n\lambda}{2d}$$

Assuming $n = 1$, that is, first-order diffraction,

$$2\theta_{\min} = 2 \text{ arc sin} \frac{1.240}{2 \times 3.148} = 22.72°$$

In similar fashion

$2\theta(K_\beta) = 25.55°$
$2\theta(K_{\alpha_1}) = 28.34°$
$2\theta(K_{\alpha_2}) = 28.41°$ ●

Often, for a particular material of interest, it is difficult or impossible to obtain a large single crystal. In that case the technique known as the powder method can be used. A sample is prepared in the form

$2\theta = 180°$ $2\theta = 0°$

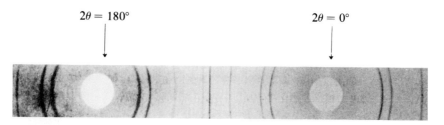

Figure 8.13   X-ray powder refraction pattern for Cu. [Reproduced from B. D. Cullity, *X-ray Diffraction* (Reading, Mass.: Addison-Wesley Publishing Co., 1972.) Used by permission of the publisher.]

of a powder, each grain of which is a tiny crystal of the material to be studied. If need be, the sample may be crushed to produce a powder. The powder is then packed into a thin-walled container of low x-ray interaction, for instance a plastic capillary tube. This is substituted for the crystal in the Bragg apparatus. No rotation of the sample is needed. Since the tiny crystals have random orientation, it is guaranteed that some of them will be found at the proper angle to satisfy the Bragg condition. The resulting radiation leaves the sample in a series of cones. The half angle of each cone corresponds to 2θ of Bragg scattering. Figure 8.13 is a typical x-ray powder refraction pattern.

8.5   Crystal Structure

X-ray diffraction studies have elucidated the crystal structure of a wide variety of materials. The subject is vast and we shall give only an outline of the fundamentals. The internal geometry of a crystal is modeled by a three-dimensional array of **lattice points.** Each lattice point has associated with it an atom or group of atoms. Within the lattice a **unit cell** is designated in such a fashion that the entire lattice can be created by repeated translations of the cell in vector steps which are integral multiples of the **edge vectors** of the cell. Figure 8.14 shows

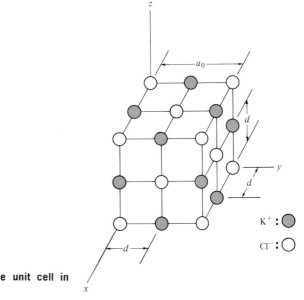

Figure 8.14   The unit cell in sylvite, KCl.

a unit cell for the cubic structure of KCl. The distance d between atom planes is indicated. It is the distance that we calculated in Example 8.1. Because there are two kinds of atoms involved, the edge of the unit cell is $a = 2d$. There is a chlorine atom at each corner of the cube and one in the center of each face. A structure with this geometry is called **face-centered** cubic, or fcc. We could have designated with equal justification a unit cell with potassium atoms at the corners and face-centers of the cube. This would also be an fcc structure. KCl may be said to have two interpenetrating structures, one of chlorine, the other of potassium. However, we shall see shortly another viewpoint that describes KCl in terms of a single fcc lattice.

In the cube as drawn there are 14 chlorine atoms and 13 potassium atoms. Most of these atoms are shared with adjacent unit cells so that the average number per unit cell is much smaller. Of the chlorine atoms each of the six face-centered atoms is shared between two unit cells. Each of the eight chlorine atoms at the corners of the unit cell is shared among eight unit cells. Adding up these fractions, we find a total of four chlorine atoms per unit cell. At the center of each of 12 edges of the cube there is a potassium atom. Each of these 12 is shared among four unit cells and thus altogether three atoms are contributed to each cell. At the body center of the cube is a potassium atom that is not shared with any other cell.

In Figure 8.15 we show the lattice points of the face-centered cubic. Lattice points are geometrical points and do not of themselves

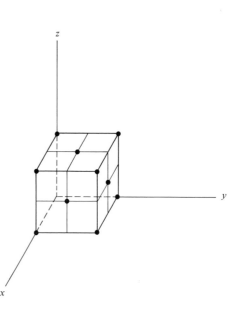

Figure 8.15 Unit cell of the fcc lattice. Lattice points on the three hidden faces are not shown. Since lattice points are shared with adjacent cells, there are, on the average, four lattice points per unit cell.

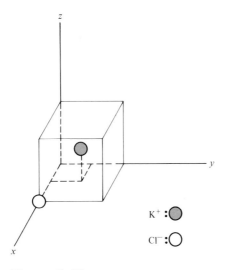

Figure 8.16 A basis in KCl.

represent the locations of atoms. Comparing Figure 8.14 with Figure 8.15, it appears no provision has been made in the latter for the potassium atoms. However, in general each lattice point in a crystal representation may have associated with it more than one atom. This is indeed the case in KCl. Figure 8.16 shows a rectangular coordinate system located so that one of the chlorine atoms has the coordinates $(a_0, 0, 0)$. In units of a_0, the coordinates of the body-centered potassium atom are $(\frac{1}{2}, \frac{1}{2}, \frac{1}{2})$. If, with each lattice point of a single fcc structure, we associate a pair of atoms, one of chlorine and one of potassium, oriented as in Figure 8.16, the location of every atom of the crystal is successfully achieved. When it is necessary to associate a combination of atoms with each lattice point, we call that combination a **basis**. A basis may be quite simple, as in KCl, or quite complex, consisting of hundreds of atoms in the case of a protein crystal.

In addition to **translational periodicity** a lattice will ordinarily exhibit certain **symmetries** with respect to rotation or reflection. The number of lattice shapes that can exhibit these symmetries is limited. We illustrate the point by consideration of the situation in two dimensions. It is possible to construct a lattice with translational periodicity that also exhibits symmetry under rotations of 180°, 120°, 90°, or 60°. Rotation of 72° is not allowed. A regular pentagon exhibits symmetry with respect to rotations of 72°, but it is not possible to construct a lattice with translational periodicity based upon a pentagonal cell. Acceptable two-dimensional lattices are illustrated in Figure 8.17. Lattice (a) is oblique and is the prototype of the other lattices. It exhibits symmetry under **rotation** of 180° but does not exhibit **mirror symmetry**. Lattice (b) is a rectangular lattice that is symmetrical under rotations of 180° and also reflection in the x and y axes. The square

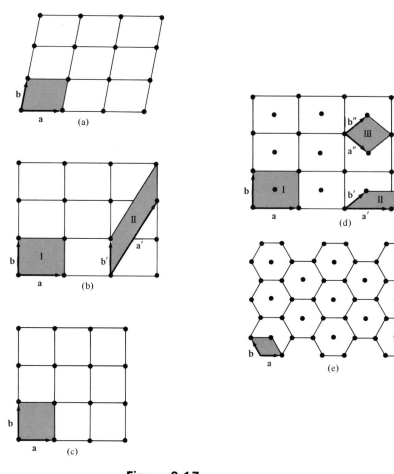

Figure 8.17 The five two-dimensional lattices: (a) oblique, (b) rectangular, (c) square, (d) face-centered rectangular, (e) hexagonal.

lattice (c) is a further specialization that exhibits symmetry under rotation in steps of 90° and also under reflection in the x and y axes. Lattice (d) is face-centered rectangular, and lattice (e) is hexagonal. Lattice (e) exhibits 60° and 120° rotational symmetry.

There is some arbitrariness in the designation of a lattice cell. There can be many different, equally valid shapes. We require only that a cell have such geometry that it is possible to construct the entire lattice by periodic cell translations as described at the beginning of this section. In Figure 8.17(b) cells I and II are equally valid. They both have the same area, $\mathbf{A} = \mathbf{a} \times \mathbf{b} = \mathbf{a}' \times \mathbf{b}'$. However, cell I better displays the symmetry properties of the lattice and would ordinarily be preferred.

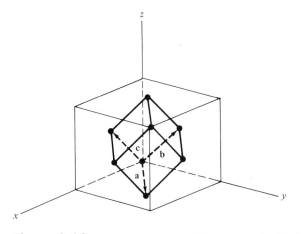

Figure 8.18 Primitive cell of the face-centered cubic lattice.

A cell of minimum area (volume, in the three-dimensional case) is called a **primitive cell.** Both cells in Figure 8.17(b) are primitive. In many cases a unit cell is chosen that is not primitive. In Figure 8.17(d) cell I is the conventional unit cell. It is chosen because it best displays the symmetry properties of the lattice. It is not primitive, however. Cells II and III are primitive cells and have one half the area of cell I. In like manner unit cells for many three-dimensional crystals are not primitive. By way of illustration we return to the face-centered cubic structure. In Figure 8.18 the unit cell of Figure 8.15 is drawn, without showing all its lattice points, as a cube whose edges are parallel to the coordinate axes. The rhombohedron drawn inside the cube is the unit cell. The primitive translation vector **a** is in the xy plane, **b** is in the yz plane, and **c** in the zx plane. Because of sharing with other cells, there is on the average one lattice point per primitive cell. Figure 8.18 shows explicitly only those lattice points associated with the primitive cell. The other lattice points associated with the unit cell may be determined by reference to Figure 8.15. Note that both the primitive cell and the unit cell obey the rules for translational periodicity. The unit cell is usually preferred because it directs attention to the symmetry of the crystal.

In addition to the fcc structure there are two other cubic structures. The **simple cubic,** or sc, has lattice points only at the corners of its cubic cell. In this structure the unit cell is primitive. The other structure is the **body-centered** cubic, or bcc. In addition to lattice points at the cube corners there is one at the center of the body of the cube. This unit cell is primitive. We leave it as an exercise to show that the volume of the primitive cell in Figure 8.18 is one fourth the volume of the unit cell.

The five lattices of Figure 8.17 comprise all possible lattices in

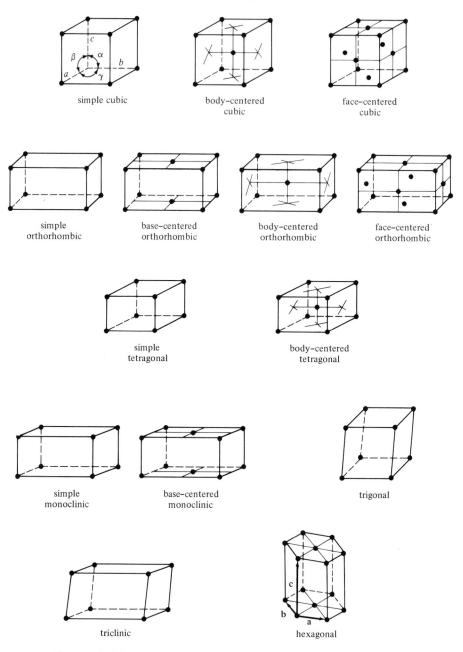

Figure 8.19 The 14 three-dimensional space lattices.

two-dimensional space. Three-dimensional space allows a larger, but still restricted, number of lattice types. There are 14 allowed types. They are shown in Figure 8.19 and some of their features listed in Table 8.2.

Table 8.2   The Fourteen Three-Dimensional Space Lattices

System	Space Lattice	Characteristics of Unit Cell	Rotational Symmetry Axes
Cubic	Simple Body-centered Face-centered	$a = b = c$ $\alpha = \beta = \gamma = 90°$	Four 3-fold (along cube diagonal)
Orthorhombic	Simple Base-centered Body-centered Face-centered	$a \neq b \neq c$ $\alpha = \beta = \gamma = 90°$	Three mutually orthogonal
Tetragonal	Simple Body-centered	$a = b \neq c$ $\alpha = \beta = \gamma = 90°$	One 4-fold
Monoclinic	Simple Base-centered	$a \neq b \neq c$ $\alpha = \beta = 90° \neq \gamma$	One 2-fold
Trigonal	Simple	$a = b = c$ $\alpha = \beta = \gamma \neq 90°$	One 3-fold
Triclinic	Simple	$a \neq b \neq c$ $\alpha \neq \beta \neq \gamma \neq 90°$	None
Hexagonal	Simple	$a = b \neq c$ $\alpha = \beta = 90°$ $\gamma = 120°$	One 3-fold

8.6   Miller Indices

Miller indices are sets of three integers that are used to indicate the orientations of atom planes within a crystal. We shall confine our discussion to an orthorhombic lattice, but the ideas may be carried over to other shapes. The first step in determining the indices of a given plane is to find the intercepts on the three axes of the coordinate system determined by the unit cell. These intercepts are expressed in units of the translation vectors along the axes and are therefore necessarily integers. The second step is to take the reciprocals of these numbers, and the third step is to clear all fractions by multiplying by the lowest common denominator. The resulting set is enclosed in parentheses; the general form is (hkl). Figure 8.20 shows a plane with intercepts at a, $2b$, and c. The associated integers are 1, 2, 1. The reciprocals are 1, $\frac{1}{2}$, 1. Upon clearing the fractions the Miller indices are (212). This is read as two-one-two, and not two hundred twelve. A plane parallel to the yz plane does not intercept the y and z axes. Its indices are (100). Negative Miller indices are allowed and are indicated

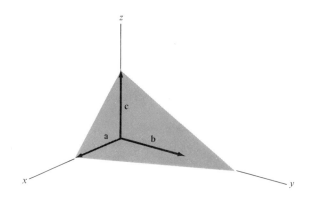

Figure 8.20 A plane with intercepts at *a*, *2b*, *c*. The Miller indices are (212).

by a bar over the top of the index. Thus the plane whose indices are ($\bar{1}$00) has a negative *x* intercept and is parallel to the *yz* plane. In a cubic lattice the planes (100), (010), (001), ($\bar{1}$00), (0$\bar{1}$0), (00$\bar{1}$) are parallel to cube faces and have the same properties. Curly brackets are used to indicate a complete set of planes having the same properties. Thus {100} is equivalent to the previous list of six sets of Miller indices. Figure 8.21 shows three different atom-rich planes in a cubic crystal.

Although the orientation of a coordinate system within a lattice may not be chosen arbitrarily, its location may be so chosen. For this reason Miller indices specify orientation, but not absolute locations of planes. In fact, a given set of Miller indices is usually interpreted as indicating the entire set of planes which have the specified orientation.

Although cubic crystals are very common, this structure does not result in the closest packing of atoms of equal size. This may be seen by considering the stacking of oranges or other spherical objects. The first layer is most closely packed if it is set down in a hexagonal array. (This is the same array that results if pennies are placed on a flat surface in such manner that around each there are six others in contact.) After completing the first layer, the spheres of the second level locate themselves above the curvilinear triangular openings left in the first

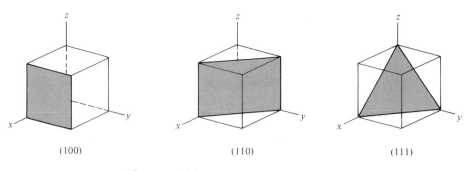

(100)   (110)   (111)

Figure 8.21 Atom-rich planes in a cubic crystal.

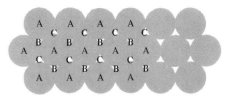

Figure 8.22 Hexagonal packing.

layer. There are two sets of these openings. In Figure 8.22 they are marked with the letters B and C. It is unimportant whether the second layer is placed over B or C. Suppose it is over B. The third layer may center once again over A, or it may center over C. If the sequence continues as ABABAB, the lattice is called hexagon close packed, or hcp. Sequence ABCABCABC produces a different lattice. Surprisingly, this turns out to be identical to our old friend, the face-centered cubic. It is simply the choice of viewpoint that emphasizes the hexagonal or the cubic aspects of the structure. Reference to Figure 8.23 shows the hexagonal symmetry of the lattice points in the (111) plane of an fcc crystal. Seven points and the lattice coordinates of each are shown.

Miller indices are especially useful in analyzing diffraction patterns when the orientation of the crystal cannot be directly inferred from its external geometry. This is very much the case when powder diffraction patterns are produced, but it may also be true for macroscopic crystals. In Figure 8.24 portions are shown of two adjacent planes with Miller indices (hkl). The origin of the coordinate system is located in one of the planes. The length of the normal from the origin to the other plane is equal to the spacing between planes. We see that $OA = a/h$, $OB = b/k$, $OC = c/l$. The direction cosines of the normal are ON/OA, ON/OB, ON/OL. Substituting the interplane spacing d for ON and using Miller indices, the direction cosines become $d/(a/h)$,

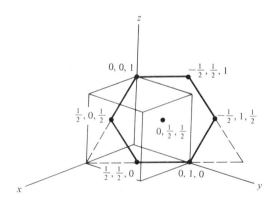

Figure 8.23 Hexagonal symmetry within the (111) plane of a fcc lattice.

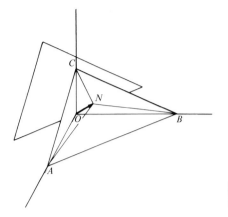

Figure 8.24 Spacing between lattice planes.

$d/(b/k)$, $d/(c/l)$. Since the sum of the squares of the direction cosines of any line equals unity, we have

$$\left(\frac{d}{a/h}\right)^2 + \left(\frac{d}{b/k}\right)^2 + \left(\frac{d}{c/l}\right)^2 = 1 \tag{8.12}$$

from which

$$d = \frac{1}{\sqrt{(h/a)^2 + (k/b)^2 + (l/c)^2}} \tag{8.13}$$

For a cubic crystal $a = b = c$, so

$$d = \frac{a_0}{\sqrt{h^2 + k^2 + l^2}} \tag{8.14}$$

Substitution in Eq. (8.11) gives

$$n\lambda = \frac{2a_0 \sin \theta}{\sqrt{h^2 + k^2 + l^2}} \tag{8.15}$$

High order diffraction ($n > 1$) is much weaker than first order ($n = 1$). It is usually safe to set $n = 1$, and we do so in the following rearrangement.

$$\frac{\sin^2 \theta}{h^2 + k^2 + l^2} = \frac{\lambda^2}{4a_0^2} \tag{8.16}$$

If monochromatic radiation of known wavelength is used, the ratio $\lambda^2/4a_0^2$ is the same for every line in the diffraction pattern. The sum,

$h^2 + k^2 + l^2$, is an integer, since each of the Miller indices is an integer. It is a straightforward task to select, by trial and error, a set of integers, one for each line, such that the ratio $\sin^2 \theta/(h^2 + k^2 + l^2)$ is the same for all. Having done so, we can determine a_0 by use of Eq. (8.15).

8.7   Imperfections

In the preceding sections of this chapter we have idealized crystal structure to a geometrical lattice that has a flawless repetition of the unit cell. Real crystals, however, are subject to a host of imperfections. In this section we shall give brief descriptions of some of the more important ones. **Crystal boundaries** constitute a type of imperfection whose presence is fairly obvious. It must be classified as an imperfection since it represents a failure of the periodic repetition of the unit cell. This failure is not limited to the boundaries of a macroscopic sample. Many materials, although apparently homogeneous, consist of small crystals clustered together with random orientations. These are often spoken of as grains and the surfaces of contact between them as grain boundaries.

Impurity atoms constitute another type of imperfection. When crystals are grown, there are bound to be atoms of other elements in the neighborhood. Some of these atoms incorporate themselves into the crystal. A foreign atom is referred to as a **substitutional** defect if it occupies a site in the crystal normally occupied by an atom of the pure material. If the foreign atom is of about the same size as the atom for which it substitutes, little energy is required to produce this defect. However, appreciable lattice stretching is required to accept significantly oversize atoms. Considerable energy is required to accomplish such stretching, so substitutional defects involving large atoms have low probability.

A **vacancy** is a defect that occurs when an atom is missing from a lattice site. The surrounding atoms tend to move slightly toward the empty spot. This distortion of the lattice is spoken of as **relaxation** and extends outward for several lattice constants around the vacancy. If the vacancy is caused by the removal of an atom to the crystal boundary, the defect is called a **Schottky** defect (Figure 8.25a). If the displaced atom remains in an **interstitial** site within the crystal lattice, the vacancy-interstitial pair is called a **Frenkel** defect (Figure 8.25b). The outward crowding of the atoms that surround the interstitial atom requires much more energy than the creation of a simple vacancy. In copper, for instance, the energy required to form a vacancy is about 1.1 eV, the energy to form an interstitial about 5.6 eV. The fraction of atoms with thermal energy E at Kelvin temperature T, is given by the Boltzmann factor $e^{-E/k_0 T}$. Thus if E_v is the formation energy for vacancies,

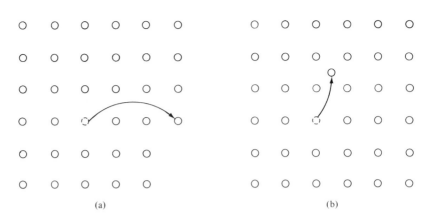

Figure 8.25 Two types of point defects: (a) Schottky defect, (b) Frenkel defect.

$$N_v = Ne^{-E_v/k_0T} \tag{8.17}$$

At room temperature $k_0T \approx 0.025$ eV, so for copper

$$\frac{N_v}{N} \approx e^{-1.1/0.025} = 8 \times 10^{-20}$$

If, in a similar fashion, we calculate the fraction of interstitial formation due to thermal effects, we find

$$\frac{N_I}{N} \approx e^{-5.6/0.025} = 5 \times 10^{-98}$$

This fraction is so small that other mechanisms, such as bombardment with energetic ions, must be resorted to for production of a significant number of interstitial defects.

An **edge dislocation** occurs where a plane of atoms terminates within the crystal. This is illustrated in Figure 8.26, where the plane AB is shown terminating halfway through the lattice. The position of the dislocation is indicated by the symbol $\perp$. Just above the edge B the lattice is under compression. Just below the edge it is under tension. If the crystal is subjected to a shearing stress, for instance by application of a force to the right along the top and a force to the left along the bottom of the crystal, it is relatively easy for an atom at C to shift over underneath the atom at B. This causes the edge dislocation to move to the right. The extra plane now becomes that indicated $A'D$. With continued application of the shearing stress the edge dislocation migrates to the right until it finally reaches the boundary of the crystal. This type of shearing displacement requires much less energy than that needed for an entire plane of atoms in a perfect crystal to slide past the neighboring plane. The sliding of plane EE' past FF', were it to occur, would exemplify the latter type of displacement.

Imperfections profoundly influence the macroscopic mechanical and electrical properties of materials. The electrical resistivity of

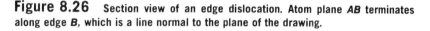

Figure 8.26 Section view of an edge dislocation. Atom plane *AB* terminates along edge *B*, which is a line normal to the plane of the drawing.

metals increases significantly when the fraction of point defects increases. Foreign atoms of specific elements (substitutional defects) may intentionally be introduced into semiconductor materials to increase their electrical conductivities and determine the sign of the charge carriers. Although thermal energy is inadequate to create a significant number of interstitials, bombardment with neutrons, protons, or alpha particles can readily do so. The high flux of particles in a nuclear reactor produces large numbers of interstitials in fuel cladding and other structural parts, thus weakening them and causing serious safety problems. Edge dislocations result in poor elastic properties. However, the open spot beside the dislocation is an opportune site for impurity atoms. Impurity atoms deliberately introduced into the melt from which the material is cast migrate to such sites and increase the strength of the sample by "locking" the dislocation in place. Work hardening of a metal is accomplished by creating a large number of small dislocations with random orientation, thus reducing the possibility of major dislocation migration across the crystal.

8.8   Summary

In an x-ray tube electrons are accelerated through potential differences in the range of 10 keV to 100 keV. Electrons are focused onto a target chosen from a variety of materials. Electron bombardment of a target produces x rays by two fundamentally different processes, namely, bremsstrahlen production and characteristic x-ray production.

A bremsstrahlung spectrum results from the stopping of energetic incident electrons. It exhibits a continuous distribution in wavelength. Its short wavelength (high energy) limit is given by $\lambda_{\min} = 12.400/K_e$ Å.

Characteristic x rays exhibit a line spectrum representing transitions between energy states of target atoms. The wavelength is related to the transition energy by $\lambda = 12.400/h\nu$ Å.

In passing through an absorber, x rays are attenuated exponentially

$I = I_0 e^{-\mu x}$ (linear absorption)
$I = I_0 e^{-\mu_m t}$ (mass absorption)

The absorption coefficients μ and μ_m depend upon the material of the absorber and upon photon energy.

The photoelectric effect, the Compton effect, and pair production are fundamental processes involved in x-ray absorption. The photoelectric effect is dominant for energies of concern to topics of this chapter.

X-ray diffraction is an important tool for the study of crystal structure. The diffraction process is governed by the Bragg law

$n\lambda = 2d \sin \theta$

Laue diffraction, Bragg (rotating crystal) diffraction, and powder diffraction are primary experimental approaches.

A crystal **lattice** is a geometrical model. The lattice is a three-dimensional array of points that is generated in the ideal case by a flawless repetition of a **unit cell**. A unit cell is a small grouping of lattice points which exhibits the symmetry of the structure. A minimum volume cells is said to be **primitive.** Although some conventional unit cells are primitive, many are not. A **basis** is a group of two or more atoms associated with each lattice point. A crystal which can be modeled without a basis has only one atom at each lattice point.

Examples of crystal structures include the simple cubic, the face-centered cubic, the body-centered cubic, and the hexagon close-packed. Miller indices are sets of three integers (hkl) which designate the orientation of a plane of lattice points with respect to conventional axes. In an orthorhombic lattice having a cell with dimensions a,b,c the interplane spacing is

$$d = \frac{1}{\sqrt{(h/a)^2 + (k/b)^2 + (l/c)^2}}$$

In real crystals there are many departures from the ideal geometry of the lattice model. These departures are called defects and play a major role in determining electrical and mechanical properties of the crystal.

In a sample whose Kelvin temperature is T the probability of finding an atom with thermal energy E is given by the Boltzmann factor

$$P(E, T) = e^{-E/k_0 T}$$

Problems

8.1 Calculate the amount of aluminum absorber required to reduce the intensity of 50-keV x rays by a factor of two and by a factor of 1000. Make the same calculations for a lead absorber. Use mass absorption coefficients from Figure 8.5.

8.2 The electron configuration of the copper atom is $1s^2 2s^2 2p^6 3s^2 3p^6 3d^{10} 4s$. Consider the ionization energy of the $4s$ shell negligible with regard to x-ray energies.
   (a) List in x-ray notation the allowed transitions that comprise the L-series x rays of copper.
   (b) What is the highest photon energy emitted in the L series? What is its wavelength?
   (c) What is the lowest photon energy emitted in the L series? What is its wavelength?
   (d) What is the minimum voltage required to excite L-series x-ray emission in copper?

8.3 In the KCl crystal the potassium ion and the chlorine ion have the same number of electrons. They are equivalent in x-ray scattering power which is almost entirely due to electrons. Although the lattice structure of KCl is fcc, it behaves as a simple cubic with regard to x-ray diffraction. Calculate the interplane spacings for the (100), the (210), the (110), and the (111) planes.

8.4 Prove that the volume of the primitive cell in Figure 8.18 is one fourth the volume of the fcc unit cell. *Hint:* Recall that volume $V = \mathbf{a} \times \mathbf{b} \cdot \mathbf{c}$.

8.5 For a pure copper crystal at 1027° C calculate the number of interstitial atoms per cubic centimeter caused by thermal effects. Make the same calculation for the number of vacancies per cubic centimeter at 100° C, at 500° C, and at 1027° C.

8.6 What minimum x-ray tube voltage is required to produce a bremsstrahlung spectrum with a short wavelength limit of 1.0 Å?

8.7 Given the data in the following table, would you choose zirconium or niobium to filter the molybdenum K-series x rays? Give reasons. After filtering, would the radiation be primarily K_α or K_β?

	K-shell Binding Energy (keV)	Photon Energy (keV)	
		K_α	K_β
$_{40}$Zr	17.998	15.7	17.7
$_{41}$Nb	18.986	16.6	18.6
$_{42}$Mo	20.000	17.4	19.6

8.8 From the data tabulated in problem 8.7 determine the L-shell binding energies for $_{40}$Zr, $_{41}$Nb, $_{42}$Mo.

8.9 Explain the discrepancy between the K-shell binding energies and the most energetic K-series x radiations tabulated in problem 8.7.

8.10 From the data tabulated in problem 8.7 determine the wavelengths of the K_α and K_β radiations for $_{40}$Zr, $_{41}$Nb, $_{42}$Mo.

8.11 What is the thickness in centimeters of lead required to reduce the intensity of 1000-keV (1-MeV) x rays by a factor of two? by a factor of 1000? Refer to Figure 8.5 for absorption coefficients.

8.12 In Figure 8.10 order according to atom density the four "planes" which are designated by letters. What is the ratio of the highest density to the lowest density?

8.13 Justify the statement in the paragraph following Eq. (8.11) that in order for wavelets scattered by atom D to be in phase with those scattered by atoms A and B it is necessary that the angle of reflection equal the angle of incidence.

8.14 NaCl has the same structure as KCl. When Mo K_α radiation ($\lambda = 0.713$ Å) is used in a Bragg scattering experiment, x-ray lines are observed at the following angles: 7°16′, 10°20′, 12°40′, 24°50′, and 33°30′. For each angle calculate the interplane spacing and the Miller indices. Determine the distance between the Na and Cl ions and the edge length of the unit cell.

8.15 Comment on the relative merits of photographic versus ionization chamber detection for use in a Bragg rotating crystal spectrometer. What factors would argue in favor of using the first method? What factors would argue in favor of using the second method?

8.16 Suppose the lattice constant for a simple cubic crystal is $a_0 = 3.0$ Å. Determine the number of atoms per Angstrom squared in the (100), the (110), and the (111) planes. (See Figure 8.21)

8.17 For a simple cubic crystal of lattice constant $a_0 = 3.0$ Å determine the interplane spacings for the (010), the (111), the (211), and the (321) planes.

Suggestions for Further Reading

Bragg, Sir L. "X-ray Crystallography," *Scientific American* (July 1968).

Compton, A. H., and S. K. Allison. *X-rays in Theory and Experiment*, 2nd ed. New York: D. Van Nostrand Company, Inc., 1935. Chapter 1. Historical survey of early x-ray experiments.

Kittel, C. *Introduction to Solid State Physics*, 5th ed. New York: John Wiley & Sons, Inc., 1976. Chapters 1 and 2. More mathematically advanced treatment.

Omar, M. A. *Elementary Solid State Physics*. Reading, Mass.: Addison-Wesley Publishing Company, 1975. Chapter 2.

Stern, E. A. "The Analysis of Materials by X-ray Absorption," *Scientific American* (April 1976).

Wert, C. A., and R. M. Thomson. *Physics of Solids*. New York: McGraw-Hill Book Company, 1964. Chapter 2. Good diagrams and photos of crystal models.

Electrical properties of solids

9.1 Introduction

Since the early eighteenth century it has been known that some materials conduct electricity and others do not. Over the years various ideas regarding the nature of electrical conduction have been set forth to explain the difference between conductors and insulators. The band theory of electron states in solids proved to be a key to the explanation.

In this chapter we shall direct our attention to the electronic behavior of solids. We shall discuss two models that establish many analogies between electrons in a metal and molecules in a gas. Apart from these we shall see that, in a solid, electron levels lose their discreteness and become broad bands. The structure of these bands depends upon the crystal structure of the material and upon the electron configuration of its atoms. In addition to conductors and insulators there exists a third class of materials, the semiconductors, now of great technological importance.

9.2 The Classical Electron Gas

From a classical standpoint an **insulator** consists of a material in which the electrons are all tightly bound to their respective atoms. Since the atoms are bound in the crystal lattice, no electric current develops even when an electric field is applied. By way of contrast it is assumed that in a **conductor,** a metal for instance, the outermost electrons of each atom are so loosely bound that they are quite free to jump from atom to atom. In the simplest picture each atom contributes one such **free electron.** These electrons move at random, are scattered by atoms when the approach is sufficiently close, and are reflected at the boundaries of the metal. The picture is that of an ideal gas made up of electrons in thermal equilibrium with the atoms of the conductor.

This model is able to make predictions regarding **specific heats** and **electrical conductivities** of metals. The principle of **equipartition of energy** states that each degree of freedom, that is, mode by which a substance can take up energy, shares equally in the available energy. The energy per degree of freedom is $\frac{1}{2}k_0T$ per molecule, or $\frac{1}{2}RT$ per mole, where $R \approx 2$ cal mole^{-1} °K^{-1}. In the electron gas model of a conductor there should be nine degrees of freedom; three for the potential energy of each atom, three for the kinetic energy of each atom, and three for the kinetic energy of each electron. We would expect the specific heat of a metal to be 9 cal mole^{-1} °K^{-1}. Instead for most metals the specific heat is close to 6 cal mole^{-1} °K^{-1}. This discrepancy is disturbing. However, let us ignore this difficulty for the moment and see what can be done with electrical conductivity.

We shall use λ to designate the **mean free path** and τ the **mean time** between collisions. Since we assume that it is with atoms primarily that the electrons collide, we see that λ is essentially the interatomic distance in the crystal. After each collision the electron changes its direction, and in the absence of an external electric field the average electron velocity, in a vector sense, is 0. As many electrons are going in any one direction as in the opposite direction, and there is no net transport of electric charge (see Figure 9.1). If there is an external electric field, however, produced by the application of a potential

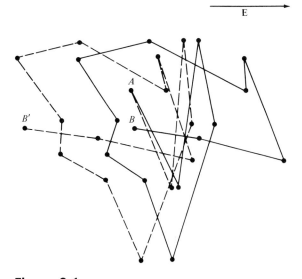

Figure 9.1 Hypothetical motion of an electron through a conductor. The solid line from *A* to *B* represents random motion in the absence of an electric field. The dotted line from *A* to *B'* represents modified motion in the presence of an electrical field *E* to the right. Additional displacement *BB'* is to be associated with drift velocity $\mathbf{v}_d$.

difference V across the ends of a conductor, the situation changes. During the time between collisions the electron is accelerated by the field. This produces a **drift velocity** $\mathbf{v}_d$ superimposed upon the random thermal velocity. Since the electron changes direction violently at each collision, the drift velocity is canceled each time and a new acceleration period must start. The calculation is a straightforward application of Newton's laws. For the acceleration we write

$$\mathbf{a} = \frac{q\,\mathbf{E}}{m} \tag{9.1}$$

where q is the electron charge (negative), $\mathbf{E}$ the electric field, and m the electron mass. It follows that the average drift velocity is

$$\langle \mathbf{v}_d \rangle = \frac{1}{2}\frac{q\mathbf{E}}{m}\tau \tag{9.2}$$

where $\langle\,\rangle$ indicates average value. A common exercise in elementary textbooks on electricity is to show that **electric current** I is given by

$$I = qnA \langle v_d \rangle \tag{9.3}$$

where n is the number of conduction electrons per unit volume and A is cross-section area of the conductor. Rearranging, we have for current density J

$$J = \frac{I}{A}$$

$$= qn \langle v_d \rangle \tag{9.4}$$

$$= qn \frac{1}{2} \frac{qE}{m} \tau$$

Now the mean free path λ can be more readily determined than the mean time τ. The two are related by

$$\tau = \frac{\lambda}{v_{rms}} \tag{9.5}$$

where

$$v_{rms} = \sqrt{\frac{3k_0 T}{m}} \tag{9.6}$$

which is obtained from applying to this model the kinetic theory relation

$$\tfrac{1}{2} m v_{rms}^2 = \tfrac{3}{2} k_0 T \tag{9.7}$$

We then obtain for electrical conductivity

$$\sigma \equiv \frac{J}{E} = \frac{q^2 n \lambda}{2\sqrt{3mk_0 T}} \tag{9.8}$$

We recall that n is the number of conduction electrons per unit volume. Taking this to be the same as the density of atoms and taking λ to be the interatomic distance, we obtain values for conductivity at room temperature that are reasonably close to the experimental values for a number of metals. This seems to be coincidence, however, since Eq. (9.8) also predicts the dependence upon temperature to go as $T^{-1/2}$, whereas experimentally over quite a wide range of temperature the dependence is essentially T^{-1}. We must conclude that the model has overlooked one or more important considerations. We shall see in the next section how the introduction of quantum ideas changes the picture.

9.3   The Fermi Electron Gas

The introduction of quantum ideas changes the response of our model in a number of ways. However, we retain the general picture

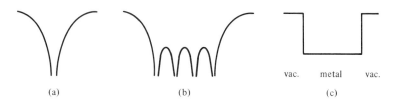

vac. metal vac.

(a)   (b)   (c)

Figure 9.2 Electron potential wells: (a) for a single ion, (b) for four ions, (c) for a very large number of ions, characteristic of a metal crystal.

of a free electron gas confined within the dimensions of a metal sample. Quantum mechanical **confinement** implies the existence of a **potential well.** Since the entire crystal is electrically neutral, there must be as many positive ions as there are free electrons. Figure 9.2 illustrates the progression of well shape from that due to a single ion to that due to the very large number of ions characteristic of a crystal. We know the electrons must obey the Schrödinger equation. In its time independent form this is

$$\frac{-\hbar^2}{2m} \nabla^2 \psi = (E - U)\psi \tag{9.9}$$

Inside the crystal, not too close to the edges, the potential is constant and we may set it equal to zero, giving

$$\frac{-\hbar^2}{2m} \nabla^2 \psi = E\psi \tag{9.10}$$

The solution to this equation, in three dimensions, is of the form

$$\psi(x, y, z) = A e^{i(k_x x + k_y y + k_z z)} \tag{9.11}$$

The k's are subject to the restriction

$$k_x^2 + k_y^2 + k_z^2 = k^2 \tag{9.12}$$

where

$$k^2 = \frac{2mE}{\hbar^2} = \frac{p^2}{\hbar^2} \tag{9.13}$$

and where p is linear momentum. Further restrictions are placed by the boundary conditions. Let us select within the crystal a rectangular parallelepiped of dimensions (x_0, y_0, z_0) where each rectangular dimension subtends sufficient atoms to justify the smoothed-out potential well of Figure 9.2(c). Since the crystal extends beyond the parallelepiped in all three dimensions, we require that ψ at any face be the same as that at the opposite face. Meeting these boundary conditions requires that

$$k_x = \frac{2\pi n_x}{x_0}$$

$$k_y = \frac{2\pi n_y}{y_0} \tag{9.14}$$

$$k_z = \frac{2\pi n_z}{z_0}$$

where

$$n_x = 0, \pm 1, \pm 2, \ldots$$
$$n_y = 0, \pm 1, \pm 2, \ldots \tag{9.15}$$
$$n_z = 0, \pm 1, \pm 2, \ldots$$

Combining Eqs. (9.13) and (9.14) gives the allowed momentum components:

$$p_x = \frac{2\pi \hbar n_x}{x_0} = \frac{h n_x}{x_0}$$

$$p_y = \frac{h n_y}{y_0} \tag{9.16}$$

$$p_z = \frac{h n_z}{z_0}$$

We wish to turn our attention now to the distribution-in-energy of the states represented by the allowed solutions. If, in **momentum space,** we plot each momentum in terms of the components allowed by Eq. (9.16), we obtain a rectangular lattice of points. The first few of these are illustrated in Figure 9.3. The number of states with mo-

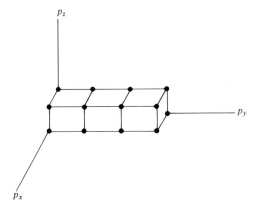

Figure 9.3 Momentum space. Each lattice point represents an allowed combination of three momentum components.

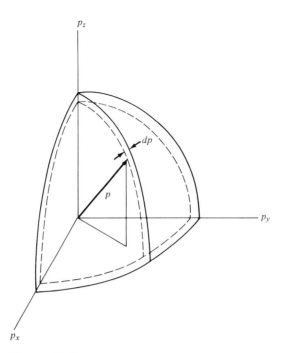

Figure 9.4 Momentum space. One octant of the sphere which encloses all allowed momenta up to magnitude p.

mentum magnitude up to some value p can be determined by counting in momentum space the number of lattice points within a sphere of radius p as shown in Figure 9.4. For macroscopic values of the dimensions (x_0, y_0, z_0) the momentum space volume assigned to each lattice point is extremely small.

$$\text{cell vol} = \frac{h^3}{x_0 y_0 z_0} \tag{9.17}$$

To a very good approximation the number of momentum states is obtained by dividing the volume of the sphere by the volume of the unit cell.

$$N = \frac{4}{3}\pi p^3 \frac{x_0 y_0 z_0}{h^3}$$
$$= \frac{4}{3}\pi p^3 \frac{V}{h^3} \tag{9.18}$$

where V is the physical volume of the parallelepiped in our sample. Since each momentum state can be occupied by two electrons, one with spin up and the other with spin down, the number of electron states becomes

$$N_e = 2\frac{4}{3}\pi p^3 \frac{V}{h^3}$$

$$= \frac{(2mE)^{3/2}V}{3\pi^2\hbar^3}$$
(9.19)

Let us consider a hypothetical way of constructing our electrically neutral crystal. We begin with a crystal consisting of ions from which the free electrons have been removed. We then restore the electrons one at a time. Since electrons are fermions, obeying the Pauli **exclusion principle,** no two electrons can occupy the same state. As the electrons are restored, they fill initially the states of lowest energy and then fill states of successively higher energy. When all electrons have been restored, some state of maximum energy E_f is finally occupied. In a metal there are additional states immediately above E_f allowed but unoccupied. E_f is called the **Fermi energy** and by rearranging Eq. (9.19) and setting $N_e/V = n$ we find

$$E_f = \frac{h^2}{8m}\left(\frac{3n}{\pi}\right)^{2/3}$$
(9.20)

Because n is the number of **occupied** states per unit volume it may also be equated to the number of **free electrons** per unit volume. The assumption that all states below E_f are filled and those above E_f empty is strictly true only at absolute zero. Nevertheless, Eq. (9.20) serves to define E_f at any temperature in a metal. Moreover, as we shall see presently, our assumption is an excellent approximation at ordinary temperatures.

To obtain the distribution of states we return to Figure 9.4 and determine the volume of the spherical shell between p and $p + dp$. Dividing by the volume of the cell in momentum space gives the (differential) number of states.

$$dN = 4\pi p^2 \frac{V}{h^3}\, dp$$
(9.21)

But

$$p^2 = 2mE$$
(9.22)

so

$$p\, dp = m\, dE$$
(9.23)

and

$$p = (2mE)^{1/2}$$
(9.24)

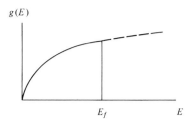

Figure 9.5 Distribution of free electrons in a metal. The cut-off at the Fermi energy is sharp when $T = 0$.

Making these substitutions in Eq. (9.21) and entering the factor of 2 for the spin states, we obtain

$$dN_e = 8\pi(2mE)^{1/2}m\frac{V}{h^3}\,dE$$

$$= V\frac{2^{1/2}m^{3/2}}{\pi^2\hbar^3}E^{1/2}\,dE \tag{9.25}$$

It follows that the distribution function, per unit volume, is

$$g(E) = \frac{2^{1/2}m^{3/2}}{\pi^2\hbar^3}E^{1/2} \tag{9.26}$$

Figure 9.5 shows this distribution rising as $E^{1/2}$ and cut off at E_f.

In describing the filling of the states we assumed all states below E_f to be filled and all those above to be empty. This is valid at $T = 0$. For $T > 0$ some electrons near E_f are promoted into higher energy states, leaving some of the states just below E_f unfilled. The Fermi distribution, which gives the probability that a state at energy E is occupied, is given by

$$f(E,\,T) = \frac{1}{e^{(E - E_f)/k_0T} + 1} \tag{9.27}$$

This is sketched in Figure 9.6. At $T = 0, f(E,\,T) = 1$ if $E < E_f$ and is 0 for $E > E_f$. At higher temperatures the distribution has a toe extending above E_f indicating that, if states exist there, a small fraction of them will be occupied. The same fraction of states below E_f will be unoccupied. Figure 9.7 shows the result of combining $g(E)$ and $f(E,\,T)$ to obtain the complete distribution of electron energies.

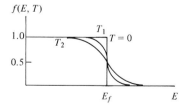

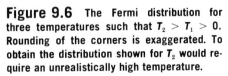

Figure 9.6 The Fermi distribution for three temperatures such that $T_2 > T_1 > 0$. Rounding of the corners is exaggerated. To obtain the distribution shown for T_2 would require an unrealistically high temperature.

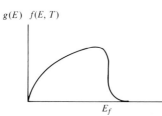

$g(E)$ $f(E, T)$

E_f E

Figure 9.7 Distribution of free electrons in a metal for $T > 0$.

At room temperature the toe of the distribution is undiscernible. A quickly-done calculation shows why this is so. In copper, for instance, Eq. (9.20) leads to $E_f \approx 7$ eV. At room temperature $k_0 T \approx 0.025$ eV. Since $E_f \ll k_0 T$ we see that if E exceeds E_f by only a few percent the exponential term in Eq. (9.27) is very large and $f(E, T) \approx 0$. To make the toe noticeable rather high temperatures are required. For instance, if $E - E_f = 0.7$ eV (0.1 E_f in copper) we must have $T \approx 3700°K$ to make $f(E, T) = 0.1$. Copper melts well below this temperature so such a distribution cannot actually be achieved.

In our discussion we have been speaking of **free electrons**. We recognize, however, that the majority of the electrons are not free to carry electric current. To do so they must be able to move to new sites in the crystal, requiring that they find there states not already occupied. For the vast majority of electrons, those more than a few $k_0 T$ below E_f, all possible states are occupied. Consequently, the most that can happen for such electrons is that a pair of them simultaneously interchange position. No flow of charge results. On the other hand, those electrons near E_f do find unoccupied states available with negligible difference in energy. These we will call **conduction** electrons. In a perfect crystal there would be essentially no resistance to their motion. In a real crystal, however, they are scattered by crystal imperfections, and thereby electrical resistance is produced.

We now see that reinterpreting some of the quantities on the right side of Eq. (9.8) can bring it into reasonable agreement with experimental measurements. λ is no longer the interatomic distance, but is the distance between scattering events. At room temperature most of the scattering is caused by lattice vibrations and these increase with temperature. The decrease in λ is approximately linear. The density of conduction electrons n is very small compared to the density of free electrons. In the denominator the term $k_0 T$ is to be associated with the Fermi energy. This being approximately 7 eV for copper implies a hypothetical temperature of approximately 80,000°K. As a consequence, a change of a few hundred degrees has a negligible effect upon this term. The temperature dependence of conductivity is dominated by the approximately linear decrease of λ with temperature.

With some rather simple assumptions, the Fermi electron-gas model of a metal can show that the contribution of free electrons to its specific heat should be very small. We assume that at a temperature T only those electrons with an energy range $k_0 T$ of the Fermi energy can be thermally excited. The number of such electrons in a metal sample of volume V is

$$N_e' = V g(E_f) k_0 T \tag{9.28}$$

where $g(E)$ is given by Eq. (9.26). If we assume that these electrons behave like an ideal gas with three degrees of freedom, then each electron has the thermal energy $\frac{3}{2}k_0 T$. It follows that the thermal energy associated with the electrons in volume V is

$$Q = \tfrac{3}{2}k_o T V g(E_f) k_0 T \tag{9.29}$$

By resort to Eqs. (9.19), (9.20), and (9.26) and a bit of algebraic maneuvering, it can be shown that

$$V g(E_f) = \frac{3}{2}\frac{N_e}{E_f} \tag{9.30}$$

It follows that

$$Q = \frac{3}{2} k_0 N_e \frac{3}{2} \frac{k_0 T^2}{E_f} \tag{9.31}$$

Since heat capacity is the rate at which thermal energy is absorbed by a sample as its temperature rises, we see that the electron contribution to heat capacity is given by

$$c_e = \frac{dQ}{dT} = \frac{3}{2} k_0 N_e \frac{3 k_0 T}{E_f} \tag{9.32}$$

But molar specific heat is the heat capacity of a 1-mole sample. In a 1-mole sample, N_e is equivalent to Avogadro's number since we assume one electron per atom. Therefore

$$k_0 N_e = R \tag{9.33}$$

The electron contribution to the molar specific heat becomes

$$c_e = \frac{3}{2} R \frac{3 k_0 T}{E_f} \tag{9.34}$$

To estimate the importance of this term, let us apply it to a typical metal. For copper $E_f \approx 7$ eV, and calculations show that at room temperature $c_e \approx 0.03$ cal mole^{-1} °K^{-1}. This being so small, the total molar specific heat is determined almost exclusively by the six degrees of freedom of the atoms themselves. On this basis the molar specific heat is predicted to be ≈ 6 cal mole^{-1} $^\circ$K^{-1}, in reasonable agreement with experimental measurements.

9.4 Energy Bands, Metals, Insulators, and Semiconductors

The energy levels of an **isolated** atom are discrete and well separated. A monatomic gas or vapor is an example of a group of isolated atoms. Most of the time the atoms are so far apart that their electron clouds (wave functions) do not overlap. The optical spectrum of such a sample is a set of discrete lines. On the other hand, when atoms are brought close together their wave functions overlap. The potential in which the electrons move is no longer the same as that of a single atom. As a result the level structure changes. Each level splits into a number of components. If there are two interacting atoms, each level has two components. With three atoms there are three components. In a system of N atoms each level splits into N components. The extent of the splitting depends upon the degree of overlap of the electron clouds, the closer together the atoms the wider the energy region across which the components spread. It is obvious that outer electron shells will overlap more than inner ones for a given distance between atom centers. Therefore higher energy levels exhibit much wider splitting than is found for the lower levels identified with inner electron shells.

Let us apply these ideas to sodium. Sodium is a metal and a good conductor of electricity. Its electron configuration is given by

$$Na = 1s^2 2s^2 2p^6 3s^1 \tag{9.35}$$

The first ten electrons, comprising the $1s$, $2s$, $2p$ states, form closed shells that experience negligible overlap. However, the $3s$ shells overlap to an appreciable extent, causing considerable splitting of that level. Figure 9.8 shows in a schematic way the potential and energy levels for an isolated atom, for two atoms, and for the approach to the solid.

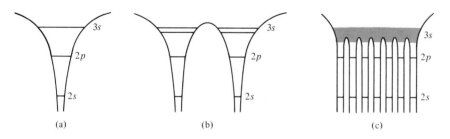

(a)   (b)   (c)

Figure 9.8   Energy levels (schematic) of Na: (a) for an isolated atom, (b) for two atoms, (c) for the "solid."

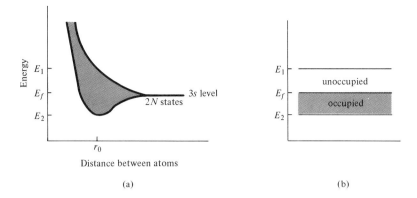

Figure 9.9 The conduction band in metallic sodium.

In the solid the number of atoms is large, $N \sim 10^{23}$. The $3s$ level splits into this very large number of states. The total splitting is but a few electron volts, so states are very close together and form a continuous **band,** which is illustrated in Figure 9.8(c). Figure 9.9(a) illustrates the way in which the band spreads out as the distance between atoms in the solid decreases. It is the behavior of electrons in this band that determines electrical properties. Since each s state can accommodate two electrons, an s band in a solid of N atoms can accommodate $2N$ electrons. In sodium each atom contributes a single $3s$ electron to the band so the band is only **half filled.** Therefore there are unoccupied states immediately adjacent to those which are occupied. We saw in Section 9.3 that this is just the situation required for a material to be a good conductor of electricity. We now see that because it has a single electron in the s shell sodium is a metal. We find that other elements with analogous electron configuration are also metals.

Atoms in a crystal do not exhibit the chaotic motion of atoms in a gas, nor are they free to slide past one another as are atoms in a liquid. Instead they are arranged in a **regular array.** The spacing between atoms is uniform and each atom has its own position in the crystal lattice. This is a position of equilibrium for the atom. If the atom is displaced by a small distance, restoring forces come into play. Each atom site in the crystal is an equilibrium position. The distance between equilibrium positions is known as the **lattice constant** and is indicated by r_0 in Figure 9.9(a). At r_0 the average energy of electrons occupying the $3s$ band is lower than the energy of the atomic $3s$ state. From an energy standpoint this is a more stable arrangement than that of a group of isolated atoms. It is this reduction in energy of the system that is responsible for bonding the atoms together to form the solid.

When the distance between atoms is r_0, the band in Figure 9.9(a) extends from E_1 to E_2. This portion of the band is shown in Figure 9.9(b) and is spread out to suggest that the band extends throughout the crystal. The states in the lower half of the band, shown shaded,

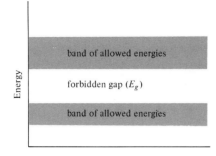

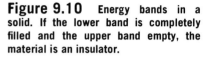

Figure 9.10 Energy bands in a solid. If the lower band is completely filled and the upper band empty, the material is an insulator.

are occupied. Those in the upper half are unoccupied. The occupied states correspond to the shaded band in Figure 9.9(c). The Fermi energy, as defined in Section 9.3, is at the upper edge of the occupied region.

A rudimentary classification of solids divides them into **metallic conductors** and **insulators.** To understand the distinction, further details of band structure must be considered. We have seen the spreading of an occupied atomic energy level into a band, when a large number of atoms are brought close together. The higher, unpopulated atomic levels also spread into bands. Figure 9.10, analogous to Figure 9.9(b), shows two allowed energy bands. It also shows between the bands a **forbidden gap** where no electron states are found. If the lower band is completely filled and the upper band completely empty, the material is an insulator. Electrons in a completely filled band cannot contribute to electric current since there are no empty states into which they can move. Obviously, an empty band can contribute nothing to electric current.

We must supplement these statements by observing that at temperatures other than 0°K there is a possibility of thermal excitation promoting electrons from the lower band across the forbidden gap into the upper band. The probability for this process depends upon the width of the forbidden gap and upon the temperature of the crystal. To a good approximation the probability is given by the Boltzman factor, $e^{-E_g/k_0 T}$. If E_g is more than a few electron volts, the probability is extremely small and the material constitutes an excellent insulator. If E_g is around 1 eV or less, some electrical conductivity results, although it is much less than that of a metal. Such materials are called **semiconductors** and we shall have more to say about them a little bit later.

At first sight it would appear that any material should be nonmetallic under the condition that it be made up of atoms whose outermost electrons occupy closed shells. Contrariwise we might expect materials to be metallic if they do not meet this condition. However, there are numerous examples of materials that contradict these expectations. We must look more carefully at the way in which the band structure is formed.

Consider, for instance, magnesium. It has the electron configuration:

$$Mg = 1s^2 2s^2 2p^6 3s^2 \qquad \textbf{(9.36)}$$

Despite the fact that its outer subshell is filled (two electrons in the s shell), magnesium is a good conductor and is classified as a metal. We must presume that access to the $3p$ band is directly available to electrons from the $3s$ band. To make this possible, the $3p$ band must dip down and overlap the $3s$ band. It turns out that the interatomic spacing in magnesium crystals is correct to accomplish this. In effect, E_g becomes negative and some of the electrons that occupy $3s$ states in isolated magnesium atoms actually occupy $3p$ states in the crystal. This leaves the uppermost states in the $3s$ band unoccupied.

$$\begin{aligned} C &= 1s^2 2s^2 2p^2 \\ Si &= 1s^2 2s^2 2p^6 3s^2 3p^2 \\ Ge &= 1s^2 2s^2 2p^6 3s^2 3p^6 3d^{10} 4s^2 4p^2 \end{aligned} \qquad \textbf{(9.37)}$$

They have in common a pair of s and a pair of p electrons outermost in their atomic configurations. Although the p shells are not filled, these materials are far from being metallic conductors. In its diamond form carbon is an excellent insulator. Silicon and germanium exhibit some conductivity but only enough to be classified as semiconductors. We must examine the band structure for these materials in more detail.

Figure 9.11(a) diagrams the fashion in which the bands develop. At large interatomic distance the levels are those of the atom. Each s

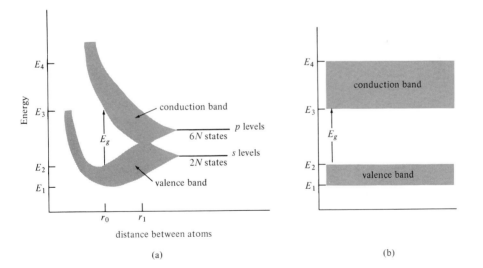

Figure 9.11 Band structure typical of carbon, silicon, and germanium. $E_g \approx 7$ eV for C in the diamond configuration, ≈ 1.2 eV for Si, ≈ 0.7 eV for Ge.

level can accommodate two electrons, and each p level six electrons. As the distance between atoms decreases, the levels spread out into bands, which overlap at the distance of separation r_1. As the separation becomes smaller, the bands cross over and spread apart again. By the time the equilibrium separation r_0 is reached, an energy gap E_g has developed. The lower band is now called the **valence band** and the upper one the **conduction band.** Figure 9.11(b) shows the positions of these bands, extending from E_1 to E_2 and from E_3 to E_4, for the single distance of separation r_0. As the bands cross over, one of the s states remains in the valence band and three of the p states join it. The conduction band is made up of the other s state and the remaining p states. Therefore in a crystal that contains N atoms, there are $4N$ states in each band. But in carbon, silicon, and germanium each atom provides four electrons. As a consequence, the valence band is filled and the conduction band empty.

The electrical properties of a solid depend upon the magnitude of the energy gap. This depends in a rather sensitive way upon the interatomic spacing in the crystal. In the diamond form of carbon $E_g = 7$ eV for the spacing actually found in nature. As we have noted, diamond is an excellent insulator. If the interatomic spacing were greater, it could become a semiconductor or even a conductor. In silicon $E_g = 1.2$ eV and in germanium $E_g = 0.7$ eV. A very small, but electrically significant number of electrons are promoted by thermal excitation across the energy gap, thus making silicon and germanium semiconductors.

Although Figure 9.11 gives a general indication of the way semiconductor properties can arise in certain crystals, it is useful to look a bit further at the electron behavior. For each electron in the conduction band there is an unoccupied state in the valence band. Neighboring electrons are free to move into these unoccupied states, thus leaving states in turn unoccupied at the positions in the crystal from which they came. The region from which an electron is missing has positive charge equal in magnitude to the charge of the electron. Such positive charges can move around in the crystal, in each case in a direction opposite to the motion of the valence electrons. These positively charged regions are called **holes** and behave in the crystal in the same way that a positively charged particle would.

A material with equal numbers of electrons in the conduction band and holes in the valence band is called an **intrinsic** semiconductor. Figure 9.11(a) is drawn for this condition, showing some electrons in the conduction band and an equal number of holes in the valence band. When an electric field is applied, these electrons and holes both contribute to the resulting electric current, the holes moving in the direction of the field and the electrons contrary to it. Figure 9.12(b) is the Fermi distribution of Figure 9.6 rotated 90° to match the convention we are now using of displaying energy in the vertical direction. We have vastly exaggerated the roll-off of the distribution around the Fermi energy in order to make clear the exponential toes that extend

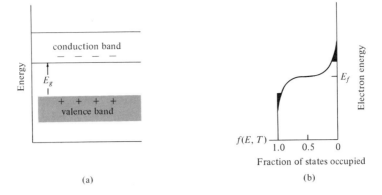

Figure 9.12 Occupation factors for energy bands of an intrinsic semiconductor: (a) band structure, (b) Fermi distribution exaggerated to show the exponential toes.

into the conduction and valence bands. These are shown as the shaded areas in Figure 9.12(b). The shaded region aligned with the valence band represents the electrons **removed** from that band, whereas the shaded area aligned with the conduction band represents the electrons that **appear** there. In an intrinsic semiconductor these areas must be equal. From Eq. (9.27) the fraction of states occupied by electrons at energy E_c in the conduction band is given by

$$f_e(E_c, T) = \frac{1}{e^{(E_c - E_f)/k_0 T} + 1} \tag{9.38}$$

Since a hole represents an electron missing from an otherwise fully occupied band, the fraction of valence states at energy E_v occupied by holes is given by

$$f_h(E_v, T) = 1 - \frac{1}{e^{(E_v - E_f)/k_0 T} + 1}$$

$$= \frac{1}{e^{(E_f - E_v)/k_0 T} + 1} \tag{9.39}$$

Equating f_e and f_h gives

$$E_f - E_v = E_c - E_f \tag{9.40}$$

from which

$$E_f = \frac{E_c - E_v}{2} \tag{9.41}$$

We see that for an intrinsic semiconductor the Fermi energy lies at the **midpoint** of the energy gap between the valence and conduction bands.

When certain impurities are introduced into a semiconductor crystal, its electrical conductivity may be considerably increased. Such a crystal is called an **impurity** semiconductor. Other useful properties are obtained. When such a semiconductor is in electrical contact with other semiconducting material of suitable composition, nonlinear electrical properties result. Such properties form the basis for a wide variety of electronic devices.

In a germanium crystal some of the atoms may be replaced by arsenic atoms. Arsenic has five valence electrons whereas germanium has four. Since the crystal bonding is based on four electrons per atom, the fifth valence electron can be easily removed from the arsenic. In terms of the energy band diagram, the level from which these electrons come is only a few hundredths of an electron volt below the bottom of the conduction band. Thus these electrons have only a small energy gap to cross. At room temperature a substantial fraction of the arsenic atoms lose their electrons to the conduction band. Since they provide electrons to the conduction band, impurities, such as arsenic, drawn from group V of the periodic table are known as **donors.** Figure 9.13 is diagrammatic of this situation. It shows conduction electrons, most of which come from donor levels but a few from the valence band. The 3 to 1 ratio implied in Figure 9.13(a) is not to be taken seriously unless the doping with donor atoms is very light or the temperature of the crystal is very high. For an impurity semiconductor of this type in an electric field most of the current is carried by the electrons in the conduction band. It follows that the **majority carriers** are electrons, the **minority carriers** holes. The positive ions at the donor sites are locked into their crystal positions. They are not free to move and thus do not contribute to the electric current. Since the majority carriers are electrons (negative), crystals doped in this fashion are known as **n-type** semiconductors.

Figure 9.13(b) shows the application of the Fermi distribution to an n-type semiconductor. Since there are many more electrons in the

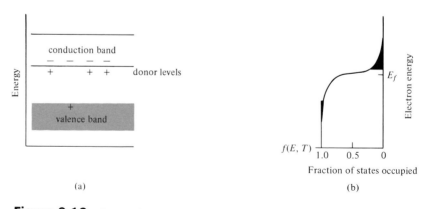

(a)

(b)

Figure 9.13 Occupation factors for energy bands of an *n*-type semiconductor.

conduction band than there are holes in the valence band, the Fermi energy has moved up to the vicinity of the donor levels. Its exact position is dictated by the relative number of donor electrons and electrons from the valence band. Often the number of donor electrons is very large in comparison to the number of electron-hole pairs arising from intrinsic semiconductor properties. Under such circumstances the donor electrons dominate the conduction process. E_f is then found midway between the donor levels and the bottom of the conduction band. However, if the temperature of the crystal is increased, additional intrinsic electron-hole pairs are created. Their contribution may be sufficient to draw the Fermi energy well below the donor levels.

Atoms from group III of the periodic table have three valence electrons. When such atoms are introduced as impurities into a silicon or germanium crystal, they readily accept electrons from nearby atoms in the process of completing the crystal bonding. Such atoms are therefore known as **acceptors.** Acceptor levels are located a few hundredths of an electron volt above the top of the valence band. Since this energy gap is much smaller than the main gap, valence electrons are much more rapidly promoted by thermal excitation to the acceptor levels than to the conduction band. The number of holes in the valence band may be many times greater than the number of electrons in the conduction band. It follows that the majority carriers are positive holes and the crystal is designated a **p-type** semiconductor. The negative ions created at the acceptor sites are fixed in position within the crystal and do not take part in electrical conductivity. Figure 9.14 diagrams the situation for a p-type semiconductor in a fashion analogous to Figure 9.13 for n-type material. Figure 9.14(b) shows the Fermi energy much closer to the valence band than to the conduction band. This must be the case since there are many more valence band holes than conduction band electrons. If the temperature of the crystal is raised, significant numbers of intrinsic electron-hole pairs are created, and the Fermi energy may rise above the acceptor levels, though still below the midpoint of the main energy gap.

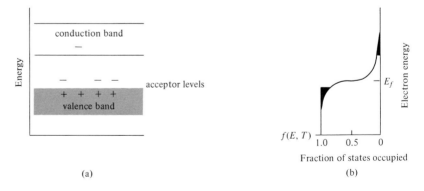

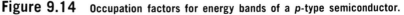

(a)

(b)

Figure 9.14 Occupation factors for energy bands of a *p*-type semiconductor.

9.5 Diodes and Transistors

There are a variety of motivations for the study of semiconductors. One of these is a fundamental interest in the properties and behavior of atoms when bonded together in the solid state. This was the viewpoint of our discussion in the previous section. Another area of interest is the behavior of semiconductors when used as **circuit elements** in electronic devices. It is scarcely possible to overemphasize the important and ubiquitous role these now play in the affairs of mankind. Sophistication of design and construction, of invention and research, the application of microminiaturization and large scale integrated circuitry—all appear to be increasing in exponential fashion. It is impossible to predict the limits to development in this field. Properties of semiconductor diodes and transistors, whether in the form of discrete units or as elements of "solid state chips," are of fundamental importance to modern electronics.

A **diode** is a device through which electric current can readily pass in one direction but not in the other. To understand how semiconductors can accomplish this task, we must trace the interaction of n-type and p-type materials when they are in electrical contact. Figure 9.15 shows the valence and conduction bands and the Fermi energy levels for these materials when they are not in contact. The band structures are those of Figures 9.13(a) and 9.14(a). The donor and acceptor levels themselves are omitted since they do not play direct roles in the conduction of current. For simplicity no minority carriers are shown. It should be noted that the Fermi energy of the n-type semiconductor is higher than that of the p-type. If these materials are brought into electrical contact, a pn **junction** is formed. It is common to use such junctions as diodes.

Actually a pn junction is seldom constructed by placing separate pieces of n-type and p-type materials against one another. There is very little area of contact in such a junction. To produce a junction with intimate contact between the materials, other construction tech-

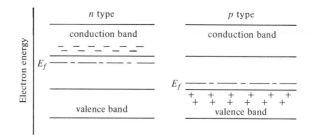

Figure 9.15   Energy bands for electrically unconnected n-type and p-type semiconductors.

niques must be used. One possibility is to start with *p*-type material and diffuse donor atoms into the structure at one face of the crystal. In regions where sufficient donor atoms have been diffused into the structure, the material is converted from *p* type to *n* type. It is at the interface of the two types that the junction itself is located. Another technique involves the control of impurities while a crystal is growing at a liquid-solid interface. A *pn* junction is formed if, during the course of crystal growth, the impurity condition in the liquid is rapidly changed from one dominated by acceptor atoms to one dominated by donor atoms.

At a junction between semiconductor materials, majority carriers from each diffuse into the other. The result is illustrated in Figure 9.16 for a *pn* junction with no external connections. Such a junction is said to be **unbiased.** (An electronic device is said to be **biased** when a steady potential difference is maintained between its terminals. Biasing may be provided by the use of a battery.) The electronic symbol for a *pn* junction is shown below the diagram of the energy bands. In the conduction band there is a much higher electron concentration in the *n*-type than in the *p*-type material; in the valence band the hole concentration is greater in the *p* type than in the *n* type. Because of these concentration gradients, electrons **diffuse** from the *n* type into the *p* type material and holes from the *p* type into the *n* type. The *p*-type material, electrically neutral to begin with, acquires negative charge, whereas the *n* type acquires positive charge. Both effects tend to raise the **electron energy** band of the *p* type above that of the *n* type. An electric field is established that opposes further transfer of charge. The field eventually becomes large enough to stop

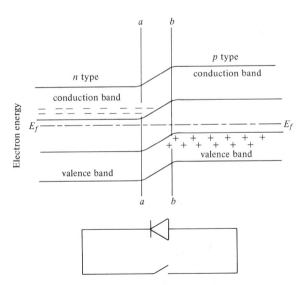

Figure 9.16 An unbiased *pn* junction.

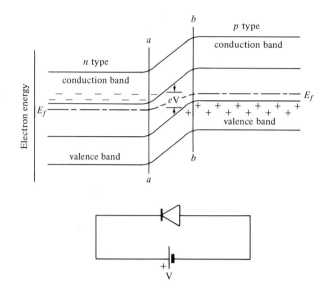

Figure 9.17 A reverse biased *pn* junction.

continued diffusion of carriers across the junction. At this point there is equilibrium between diffusion caused by the concentration gradients and reverse forces due to the electric field. Under these conditions, the Fermi energies of the two materials come to the same level, shown by the horizontal dash-dot line. The junction itself may be thought of as occupying the transition region between the lines *a-a* and *b-b*. Outside the transition region each material is unipotential, with a **contact potential** difference created between them by the differing heights of their electron bands.

Figure 9.17 shows the band structure of a junction biased by a battery. The circuit representation for the junction and the battery is shown below the diagram of the energy bands. When the battery is connected in the direction shown, the junction is said to be **reverse-biased.** With bias voltage *V,* the Fermi energy level of the *p*-type material is raised by an amount *V* above the Fermi level of the *n* type. This makes even higher the potential hill that electrons must surmount to enter the conduction band to the right of the junction. Similarly, holes must surmount a higher potential to enter the valence band to the left of the junction. (In these diagrams increasing electron energy is plotted upward; increasing hole energy is plotted downward. The electrons tend to **sink** to the bottom of their band; holes may be thought of as bubbles that tend to **float** to the top of their band. To keep the electrons from sinking requires that they have extra energy; to keep the holes from floating requires that they also have extra energy.) Because of the higher potentials which they encounter, electrons are prevented from entering the conduction band of the *p*-type materials and holes from entering the valence band of the *n* type. Consequently, neither in the conduction band nor in the valence band does carrier population extend from one face of the diode to the other. For this reason current does not flow through a reverse-biased diode.

If a battery is connected as shown in Figure 9.18, the junction is said to be **forward-biased.** The Fermi level of the *p*-type material is lowered by an amount V below that of the *n* type. The accompanying change in heights of the bands in the two materials allows easy flow of electrons into the conduction band of the *p* type and holes into the valence band of the *n* type. The resulting decrease in electron concentration in the *n*-type conduction band allows additional electrons to enter it from the battery wire at the left contact. The excess electron concentration in the *p* type causes electrons to flow into the battery wire on the right. At the same time that electrons flow to the right, holes flow to the left. The concentration of holes is maintained by their creation at the right contact and neutralization at the left. As long as the forward bias is maintained, there is a sustained current through the device.

The description just given of reverse-biased and forward-biased diodes is somewhat oversimplified. In a real diode there is a small leakage current when the diode is reverse-biased. This current is conducted by minority carriers which are present whether the junction is biased or not. These carriers are created by thermal excitation of intrinsic electron-hole pairs in both types of material. Thus a small number of electrons are to be found in the *p*-type conduction band

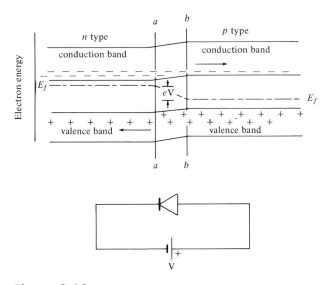

Figure 9.18 A forward biased *pn* junction.

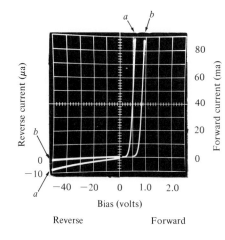

Figure 9.19 Oscilloscope trace of current versus voltage characteristics for semiconductor diodes: (a) germanium diode, (b) silicon diode. Response in the first quadrant is for forward bias; response in the third quadrant is for reverse bias. Note the difference in scale factors for conditions of forward and reverse bias.

and holes in the n-type valence band. They both readily slide down the potential hills that the junction region presents to them. The magnitude of the **reverse current** depends upon the carrier concentration. This concentration depends, through the Boltzman factor, upon the temperature of the diode and the magnitude of the energy gap to be crossed. Since the energy gap in germanium is smaller than that in silicon, we expect the former to exhibit considerably larger reverse current. This is indeed found to be the case. Figure 9.19 shows curves of current versus voltage constructed from measurements made on two typical diodes, one of germanium and the other of silicon. Voltages to the left of and currents below the origin are for the condition of reverse bias. Also, shown to the right of and above the origin is the dependence of current upon voltage in the forward-biased condition. (Note the difference in scale factors for the conditions of forward and reverse bias.) For a given forward current, the germanium diode requires a bias approximately 60% that of the silicon diode. Furthermore, for neither diode is the slope of current versus voltage infinite.

Although the derivation of the current-voltage relationships is beyond the scope of the treatment given here, it can be shown to be

$$I = I_s(e^{\pm |qV|/k_0 T} - 1) \tag{9.42}$$

where q is the electron charge and I_s, called the saturation current, serves as a constant of proportionality. I_s depends upon the physical area of the junction and its semiconductor properties. (The absolute value is used in the exponent to avoid confusion which may arise from the negative charge of the electron. For forward bias the plus sign is to be used in the exponent; for reverse bias the minus sign is to be used.) In a typical junction $I_s \sim 10^{-9}$ amp for Si and $\sim 10^{-6}$ amp for Ge. At room temperature $k_0 T = 0.025$ eV. Therefore the expression for current can be written as

$$I = I\,(e^{40V} - 1) \qquad\qquad (9.43)$$

With forward bias of more than a few tenths of a volt, the exponential term is dominant inside the brackets. To a good approximation

$$I \approx I_s e^{40V} \qquad\qquad (9.44)$$

As one-way valves, diodes have wide application in the **rectification** of ac current to produce dc. They are used to modify pulse shapes, to compare voltage levels, and to perform a variety of other functions in electronic circuits. Some diodes, specially designed to admit energetic particle or photon radiation into the junction region, serve as excellent detectors of such radiation.

A three-layer sandwich of semiconducting material in an *npn* or *pnp* configuration constitutes a transistor. Figure 9.20 shows band diagrams and circuit connections for an *npn* transistor. The transistor is shown unbiased in Figure 9.20(a). The Fermi level is designated by the horizontal dash-dot line. The circuit symbol for an *npn* transistor is shown at the bottom of the figure and labeled with the names of the electrodes. (Reversal of the emitter arrow produces the circuit symbol for a *pnp* transistor.) The *n*-type material on the left is the

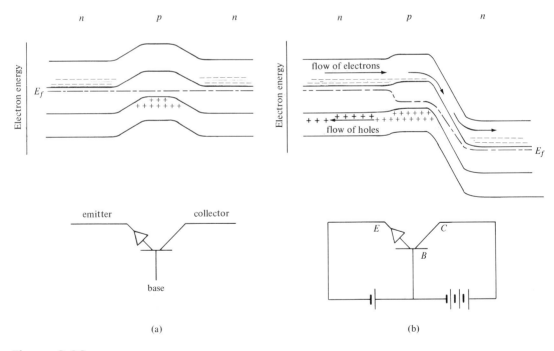

Figure 9.20 Band diagrams and circuit connections for an npn transistor: (a) unbiased, (b) forward biased, normal operation.

emitter, the p type in the center the base, and the n type on the right the collector. Because of contact potentials, the electron energy band in the p-type material is higher than it is in the n types on either side. There are donor electrons in the conduction bands of the **emitter** and **collector** material, and holes in the **base.**

Figure 9.20(b) is for an *npn* transistor with normal biasing. A biasing circuit is shown below the diagram of the energy bands. The collector is made positive with respect to the base by the battery voltage V_{CB}. This reverse-biases the collector-to-base junction located in the region *b-b*. Making the base positive with respect to the emitter by battery voltage V_{BE} forward-biases the base-to-emitter junction. This forward bias permits electrons from the conduction band of the emitter to enter the base region, while holes from the base enter the valence band of the emitter. If the base is physically thin (the distance between *a* and *b*), most of the electrons from the emitter diffuse directly to the collector junction. Very few reach the external base connection. At the collector junction, electrons, which constitute minority carriers within the p-type base, encounter a downward slope of the energy band and readily enter the collector region. Their presence increases the (majority) carrier concentration in the collector material. A slight concentration gradient results, causing electron flow to the right. Electrical contacts with battery wires remove electrons from the conduction band of the collector and provide electrons to the conduction band of the emitter. A sustained current results.

The base-to-emitter bias voltage determines the height of the potential hump that stands between the emitter and the collector. Electrons must cross this hump to carry current between emitter and collector. Small changes in bias voltage between base and emitter produce large changes in current. Furthermore, in a well-designed transistor the external current to the base is very small compared to the current between emitter and collector.

With suitable modifications and giving due consideration to the opposite polarity of carriers, the analysis given here can be applied to *pnp* transistors. It is left as an exercise for the student to design energy diagrams and circuits for unbiased and forward-biased *pnp* transistors.

The power in the collector circuit of a transistor is the product of collector current and voltage. In most transistor applications this product is very large in relationship to the product of base current and voltage. Transistors are effective as **amplifiers** and as **switches.** They serve as elements in electronic logic circuits and in high frequency oscillators. With advanced solid state fabrication techniques they can be made extremely compact; the equivalent of several hundred may be built into a single integrated circuit chip. Although they may be destroyed by electrical insult, they are very rugged mechanically.

9.6 Summary

Energies for isolated atoms are discrete. In a solid the wave functions of the outermost electron shells overlap. This overlap causes the levels to split and spread into bands. The band structure determines whether the material is an electrical conductor, insulator, or semiconductor.

In a metal the uppermost occupied band is partially filled with electrons. There are unoccupied states immediately adjacent to those that are occupied. Electrons have access to unoccupied states at negligible cost in energy. Under these circumstances the material is a good electrical conductor.

In an insulator the uppermost occupied band is completely filled. This is called the valence band. There is a forbidden gap in energy between it and the unoccupied conduction band above. The gap is relatively large, several electron volts, and the number of electrons with sufficient energy to jump the gap is insignificant.

In an intrinsic semiconductor the forbidden gap is relatively narrow, around 1 eV, and enough electrons are able to jump across it to provide measurable conductivity. Pure germanium and pure silicon are examples.

A semiconductor may be doped with foreign atoms to provide donor levels (n-type semiconductor) just below the conduction band, or acceptor levels (p-type semiconductor) just above the valence band. The host material is commonly germanium or silicon. A doped semiconductor is also called an extrinsic or an impurity semiconductor.

The occupation probability for a state of energy E is given by the Fermi distribution

$$f(E, T) = \frac{1}{e^{(E - E_f)/k_0 T} + 1}$$

In a metal the Fermi energy E_f is derived from momentum-space considerations

$$E_f = \frac{h^2}{8m} \left(\frac{3n}{\pi} \right)^{2/3}$$

In an insulator or intrinsic semiconductor the Fermi level is found at the midpoint of the forbidden gap.

In an impurity semiconductor the position of the Fermi level depends upon temperature and impurity concentration.

Problems

9.1 Calculate the electron drift velocity in a copper wire whose diameter is 0.812 mm (No. 12 gauge, typical of residential branch circuit wiring in the USA) carrying a current of 20 amp. The resistance is 33.31 Ω km^{-1} at 20°C. Assume one free electron per atom of copper.

9.2 **(a)** Take the Fermi energy in copper to be 7 eV. Calculate the velocity of electrons with this kinetic energy. (They populate the outer surface of the sphere in momentum space shown in Figure 9.4.)

(b) Assuming the current is carried primarily by these energetic electrons instead of every free electron as assumed in problem 9.1, determine the fraction of electrons which so participate.

9.3 Calculate the Fermi energy for copper, aluminum, and potassium. Assume one free electron for each atom. (The electron interaction with the lattice in some crystals makes it behave as though it has an effective mass different from m_0. In most metals the departure is less than 20%, and in the examples given here the departure is even less.)

9.4 The forbidden gap in silicon is 1.2 eV and in germanium it is 0.7 eV.
(a) Where is the Fermi level for intrinsic silicon and germanium?
(b) Calculate for each of these semiconductors the fraction of states in the conduction band which are occupied when the temperature is 200°K, when it is 300°K, and when it is 500°K.

9.5 A sample of silicon is doped with 10^{16} atoms/cm^3 of As. The donor levels thus created are 0.05 eV below the conduction band.
(a) What fraction of the donor atoms have lost electrons to the conduction band at 100°K? at 300°K? at 500°K?
(b) What fraction of states in the conduction band are occupied by donor electrons at these temperatures?

9.6 For silicon doped as in problem 9.5 what temperature is required so the ratio of donor electrons to intrinsic electrons in the conduction band is 3:1?

9.7 Take the saturation current in a silicon diode to be 10^{-9} amp. Plot a current versus bias voltage curve covering the range from 10 V reverse bias to such forward bias as to produce a current of 0.5 amp.

9.8 Draw energy band and bias circuit diagrams for a *pnp* transistor.

9.9 Complete the algebraic maneuvers needed to arrive at Eq. (9.8).

9.10 For a temperature of 20°C determine the electrical conductances of copper and iron as predicted by the classical electron gas model. What are the ratios of conductances so predicted to those indicated in handbook tabulations?

9.11 Supply the algebraic steps necessary to convert Eq. (9.19) into Eq. (9.20).

9.12 Show that the Fermi function $f(E_f - \Delta E, T) = 1 - f(E_f + \Delta E, T)$.

9.13 Take $E_f = 7.0$ eV in copper. Evaluate the Fermi function at $E = 7.1$ eV, and $T = 300°K$.

9.14 Take $E_f = 7.0$ eV in copper. What temperature is required so that $f(E, T) = 0.01$ for $E = 7.1$ eV?

9.15 Verify Eq. (9.30).

9.16 The Fermi energies in copper and silver are 7.00 eV and 5.48 eV respectively. Determine the electron molar specific heats of these two metals.

9.17 Show that for a semiconductor diode the relationship $V \propto \log I$ applies over a rather wide range of forward currents. (This behavior is useful in analog processing of electric signals. For instance, by adding or subtracting voltages a signal representing the product or quotient of two other signals can be obtained.)

9.18 Verify the final step in Eq. (9.39).

9.19 The work function of tungsten is 4.4 eV and that of platinum is 5.3 eV. Their melting points are 3370°C and 1770°C respectively. According to the Fermi distribution what fraction of the electrons in each of these metals has enough energy for thermionic emission at 1000°C? at 1700°C?

Suggestions for Further Reading

Azaroff, L. V. *Introduction to Solids*. New York: McGraw-Hill Book Company, Inc., 1960. Chapter 12.

Enge, H. A., M. R. Wehr, and J. A. Richards. *Introduction to Atomic Physics*. Reading, Mass.: Addison-Wesley Publishing Company, 1972. Chapter 11.

Thomas, G. A. "An Electron-Hole Liquid," *Scientific American* (June 1976).

Wert, C. A., and R. M. Thomson. *Physics of Solids*. New York: McGraw-Hill Book Company, 1964. Chapters 12 and 13.

Basic properties of the nucleus

10.1 Introduction

In previous chapters we have treated the nucleus of the atom as an extremely small charged particle with a mass very much larger than that of the electrons around it. In fact, in most instances it has been possible to describe atomic structure by treating the ratio of nuclear mass to electron mass to be infinite. Nuclear size and mass, however, are both finite, and nuclei possess a wide range of properties.

In a study of nuclear structure we encounter distances between particles that are much smaller than those between atoms in a solid and characteristic energies that vastly exceed those involved in excitation of the outer electrons of the atom. It is thus desirable to introduce new units of length, energy, and mass, and to review names and symbols for various other quantities. As we shall see presently, the radii of all nuclei are small multiples of 10^{-13} cm. Consequently, this is a convenient unit of length and is called the **fermi.**

1 fermi $= 10^{-13}$ cm

This is five powers of 10 smaller than the Ångstrom (10^{-8} cm), which is appropriate for measuring sizes of atoms. Continuing in this vein we note that nuclear excitation energies are of the order of 10^6 eV, and we introduce the **mega-electron volt,** which is abbreviated MeV, so that

1 MeV $= 10^6$ eV $= 1.60210 \times 10^{-6}$ erg

10.2   Constituents

Through a wide variety of experiments it has been established that nuclei are made up of **protons** and **neutrons.** These are **fermions** with **intrinsic spin** quantum number $s = \frac{1}{2}$. In many respects they appear so similar that the term **nucleon** has been invented to refer to them when we need not distinguish between them. A particular **nuclear species** is made up of nuclei with identical numbers of protons Z and identical numbers of neutrons N. The total number of nucleons in a nucleus is designated by the symbol A. Thus $A = Z + N$ and is called the **mass number** of the nucleus or nuclear species. A nuclear species may be designated by its chemical symbol, thereby implying specification of Z, and a left superscript giving the value of A. To be redundantly explicit, Z may further be specified by a left subscript. The nuclear mass number A is an integer and is to be distinguished from the **exact mass** M of the nucleus or atom. M may be measured in grams or kilograms, but a unit whose magnitude is comparable to the mass of an individual nucleus or atom has wide use. In this system of **unified mass units** the atomic mass of ^{12}C is defined to be exactly 12 u. From this definition

1 u $= 1.66053 \times 10^{-24}$ g

It follows that the energy equivalence of one unified mass unit is

$E = mc^2 = 931.480$ MeV

In an earlier nomenclature A was designated by a right superscript to the chemical symbol. This usage may still be encountered but will not in general create difficulty. A more serious matter and possible source of confusion is the existence of a system of **atomic mass units** in use prior to 1960. This system was based on ^{16}O, setting its atomic mass to be exactly 16 amu. Since the mass of ^{16}O in the system based on ^{12}C is 15.994915 u, it follows that

1 amu = 0.9996822 u

and the energy equivalence of 1 amu is 931.184 MeV. However, **rest energies** and **binding energies,** which will be discussed in the next section, are the same in both systems.

Nuclear species of the same Z but different A are spoken of as **isotopes;** those of the same A but different Z are spoken of as **isobars;** those of the same N are spoken of as **isotones.** Thus $^{12}_{6}C$, $^{13}_{6}C$, and $^{14}_{6}C$ are isotopes of carbon, that is, of the $Z = 6$ system. ^{10}Be, ^{10}B, and ^{10}C are isobars of the $A = 10$ system. ^{14}C, ^{15}N, and ^{16}O are isotones of the $N = 8$ system. The term **nuclide** is used when referring to a **nuclear species.** Thus, ^{14}C is a radioactive nuclide; it is an isotope **of** carbon, but is not properly called "a radioactive isotope" unless the additional phrase, "of carbon," is used.

10.3   Mass Defect and Binding Energies

The measurement of atomic masses has been carried on for many years by both chemical and physical methods. Techniques involving deflection of streams of atomic ions passing through electric and magnetic fields constitute primary measurements of ion masses. Instruments using such techniques are known as mass spectrographs. Results of studies with mass spectrographs, combined with mass-energy changes that occur in nuclear reactions, yield tables of atomic masses for all known species. Uncertainties range from less than a micro mass unit for the best known ones to around a milli mass unit for the poorest known ones.

A given atom, identified in terms of its atomic number Z and its mass number A is understood to consist of Z protons, $(A - Z)$ neutrons, and Z electrons. The difference between the total mass of these components when separate and the mass of the atom is called the **mass defect** ΔM.

$$\Delta M = ZM_p + (A - Z)M_n + ZM_e - M_{Z,A} \qquad (10.1)$$

where

proton mass $M_p = 1.00727661$ u
neutron mass $M_n = 1.0086652$ u
electron mass $M_e = 0.000548593$ u

and $M_{Z,A}$ is the mass of the atom.

The energy equivalence ΔMc^2 of the mass defect is the **binding energy** that would be required to separate the atom into its components. Equation (10.1) may be recast to collect terms in Z.

$$\Delta M = Z(M_p + M_e) + (A - Z)M_n - M_{Z,A} \tag{10.2}$$

but $M_p + M_e$ differs from M_H, the mass of the neutral hydrogen atom, by 13.6 eV, the binding energy of the electron. Since $M_p + M_e = 938.770$ MeV, the 13.6 eV difference can ordinarily be neglected and we can write the mass defect as

$$\Delta M = ZM_H + (A - Z)M_n - M_{Z,A} \tag{10.3}$$

Let $B = \Delta Mc^2$. A quantity of interest is B/A, the binding energy per nucleon. This quantity is plotted in Figure 10.1 as a function of mass number A. A notable feature of this plot is the restricted range of B/A. Except for relatively minor fluctuations in the region of the

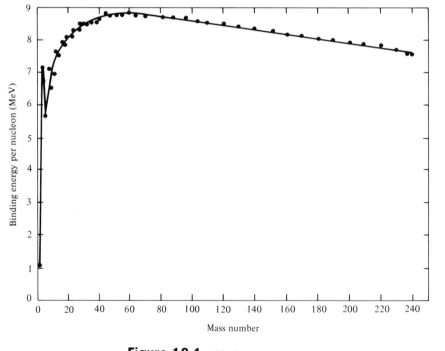

Figure 10.1 Binding energy per nucleon versus the nuclear mass number.

lighter atoms B/A remains between 7 and 9 MeV per nucleon. The nearly constant binding energy per nucleon is evidence that the forces between nucleons exhibit a special behavior called **saturation.**

To illustrate what we mean by saturation, let us contrast the behavior of the electrostatic force, with which we are more familiar. This force links any pair of electric charges without regard to the presence of other charges. That is, the presence or absence of charge q_3 does not change the force between charges q_1 and q_2. If the nuclear force were to behave in this fashion, we should find $A(A-1)/2$ bonding links in a nucleus of mass number A. B/A would then be proportional to $(A-1)$. Instead, after an initial rise in the region of small A, B/A becomes nearly constant. The fact that the binding energy per nucleon does not continue to increase as A becomes large indicates that the nuclear force is effective only between neighboring nucleons. Forces which act only upon near neighbors are said to exhibit saturation.

It is evident from the stability of nuclei that there must be some sort of attracting force between nucleons. Without such a force to bind them together electrostatic repulsion between protons would disrupt all nuclei having $\mathbf{Z} > 1$. Although the gravitational force is a force of attraction, it is much too weak to overcome the electrostatic repulsion. A new force, specifically nuclear in character, is required. It produces a very **strong** attraction between nucleons that are close together, but has quite a **short range.** We also conclude from the gradual fall of B/A, as A increases beyond 60, that in this mass region the additional Coulomb repulsion as we progress to nuclides of higher Z and A is somewhat more effective at reducing the binding energy than the specifically nuclear force is at increasing it.

Example 10.1   Calculate the mass defect and the binding energy per nucleon for $^{34}_{16}\text{S}$.

Solution.   Eq. (9.3) will apply.

$$
\begin{aligned}
ZM_H &= 16 \times 1.007825 = &16.125200 \text{ u} \\
(A-Z)M_n &= 18 \times 1.008665 = &\underline{18.155970 \text{ u}} \\
& &34.281170 \text{ u} \\
-\,(\text{mass of } {}^{34}_{16}\text{S}) &= &\underline{-33.967865 \text{ u}} \\
\Delta M &= &0.313305 \text{ u}
\end{aligned}
$$

$$B = 0.313305 \text{ u} \times 931.480 \text{ MeV/u} = 291.837 \text{ MeV}$$

$$\frac{B}{A} = \frac{291.837}{34} = 8.58 \text{ MeV/nucleon} \quad \bullet$$

Another way of summarizing nucleon relationships as they bear on nuclear stability is shown in Figure 10.2. In this figure each of approximately 280 stable nuclides is represented by a separate loca-

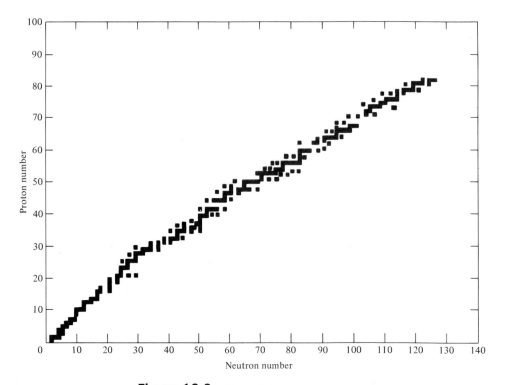

Figure 10.2 Chart of the known stable nuclides.

tion on a chart showing atomic number Z and neutron number N. It is seen that among the lighter nuclides Z tends to be equal to N. This arises from the **charge-independent** character of the nuclear force and the distinguishability of protons and neutrons in the operation of the exclusion principle. For higher Z nuclides the Coulomb repulsion between protons contributes a positive energy proportional to Z^2. It is therefore necessary to dilute the effect with an increasing ratio of neutrons to protons. This is manifested in the slight roll-off of the plotted points at the upper end of the chart.

In addition to the stable nuclides, there are approximately 1120 that are unstable against particle emission. Of these, 46 are in the three naturally occurring radioactive series that extend beyond the band shown in Figure 9.2 up to $Z = 92$, $N = 146$. An additional 17 radioactive nuclides occur naturally and are distributed in the chart from $^{40}_{19}K$ to $^{204}_{82}Pb$. The half-lives of these range from 10^9 yr to 10^{17} yr. There are somewhat over 1050 manmade radioactive nuclides. Some are **transuranic**, with $Z > 92$; others occupy positions off the **band of stability** in Figure 10.2. In general, the farther from the band of stability a nuclide is located, the less stable it is and the shorter its half-life. Also, those below the band are relatively deficient in protons. These nuclides undergo β^- decay, thereby increasing Z. Conversely, those above the band are proton rich and consequently are unstable against

β^+ emission or orbital electron capture. The nature of the various types of radioactive decay will be discussed in a later chapter.

10.4   Rutherford Scattering

In Chapter 5 we briefly described the experiment on the scattering of alpha particles by heavy nuclei conducted by Geiger, Marsden, and Rutherford and the crucial role it played in setting the stage for our present understanding of atomic structure. Historically, scattering experiments of this sort provided the first clues to the sizes of nuclei. **Rutherford scattering** is often referred to as **Coulomb scattering,** because in its pure form it depends on the Coulomb law of force between the charge of a projectile (an alpha particle in the original experiments) and the charge of the target nucleus. Since that force law is well understood, in contrast to the behavior of the specifically nuclear force, the Rutherford scattering law can be calculated with exactness. It thus is a useful standard against which various scattering experiments may be compared. Failure to follow the Rutherford predictions is indication of an interaction other than electrostatic and can be interpreted to give information about the nuclear properties of the target.

Figure 10.3 depicts the path of a projectile of mass m and charge ze approaching a scattering center of charge Ze and mass M. Suppose

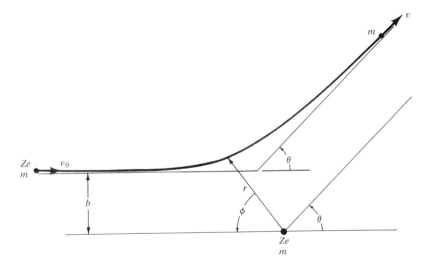

Figure 10.3   Deflection of the path of an alpha particle due to Coulomb interaction with a massive nucleus. The alpha particle, of mass m and charge ze, approaches with an initial velocity v_0 and recedes with a velocity of the same magnitude changed in direction through the angle θ.

the initial velocity of m is v_0 and it is aimed so it would pass the location of M at a distance b if there were no deflection of its path. The distance b is called the **impact parameter.** For simplicity we shall assume that M is so massive that it does not move during the scattering event. This is quite a good approximation for the scattering of alpha particles by medium weight and heavier nuclei. Although we cannot deliberately aim so as to produce a given impact parameter, let us calculate the relationship between b and the scattering angle θ.

By conservation of energy we have

$$\tfrac{1}{2}mv_0^2 = \tfrac{1}{2}mv^2 + \frac{zZe^2}{r} \tag{10.4}$$

where v is the speed of the projectile when it is the distance r from M. Alpha particles from radioactive sources have energies in the neighborhood of 5 MeV. Nonrelativistic calculations will be adequate since v/c for these particles is about 0.05. A useful parameter is D, the distance of closest approach in a head-on collision. In such a collision $b = 0$. As the projectile approaches the heavy nucleus, it slows down, comes to rest at distance D when all its kinetic energy has been converted to potential energy, and then recedes back along the path from which it came. For such scattering $\theta = 180°$.

$$D = \frac{zZe^2}{\tfrac{1}{2}mv_0^2} = \frac{2zZe^2}{mv_0^2} \tag{10.5}$$

In polar coordinates with the origin at M

$$v^2 = \dot{r}^2 + (r\dot{\phi})^2 \tag{10.6}$$

Substitution in Eq. (10.4) and rearrangement yields

$$v_0^2 = \dot{r}^2 + (r\dot{\phi})^2 + \frac{2zZe^2}{mr} \tag{10.7}$$

Since the electrostatic force acts along the line between the charges, there is no torque around the scattering center. As a consequence, angular momentum is conserved

$$mv_0\, b = m(r\dot{\phi})r \tag{10.8}$$

$$\dot{\phi} = \frac{bv_0}{r^2} \tag{10.9}$$

but

$$\dot{r} = \frac{dr}{dt} = \left(\frac{dr}{d\phi}\right)\left(\frac{d\phi}{dt}\right) = \left(\frac{dr}{d\phi}\right)\dot{\phi} = \left(\frac{dr}{d\phi}\right)\left(\frac{bv_0}{r^2}\right) \tag{10.10}$$

Substitution in Eq. (10.7) leads to

$$\frac{1}{b^2} = \frac{1}{r^4}\left(\frac{dr}{d\phi}\right)^2 + \frac{1}{r^2} + \frac{2zZe^2}{mb^2v_0^2r} \tag{10.11}$$

The solution to this differential equation is

$$\frac{1}{r} = \frac{1}{b}\sin\phi + \frac{D}{2b^2}(\cos\phi - 1) \tag{10.12}$$

where D is defined by Eq. (10.5). Equation (10.12) is the polar equation of a **hyperbola** with asymptotes at distance b from an external focus at the origin.

Now, in an actual experiment the scattered particle can only be observed when its distance from the origin is very much greater than the impact parameter. To represent this experimental constraint, we let r approach infinity after the collision. Under these conditions we find

$$\frac{1}{b}\sin\phi = \frac{D}{2b^2}(-\cos\phi + 1) \tag{10.13}$$

In addition $\phi \to \pi - \theta$ so $\cos\phi \to -\cos\theta$ and $\sin\phi \to \sin\theta$.

Making these substitutions leads to

$$\sin\theta = \frac{D}{2b}(1 + \cos\theta) \tag{10.14}$$

Manipulation of trigonometric identities then gives

$$\frac{1 + \cos\theta}{\sin\theta} = \cot\frac{\theta}{2}$$

Thus

$$\frac{2b}{D} = \cot\frac{\theta}{2} \tag{10.15}$$

Equation (10.15) expresses the relationship between the impact parameter and the **scattering angle**. However, since b is not directly observable, we must carry further to obtain an expression that we can compare with experiment.

To guide our approach we remind ourselves that the experiment determines the **fraction** of incident particles that are scattered in a given direction. We are therefore interested in dN_θ, which is the number scattered in directions between θ and $\theta + d\theta$. These particles

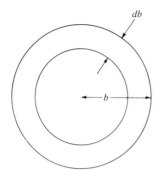

Figure 10.4 The ring of impact parameters between **b** and **b − db**.

are associated with impact parameters between b and $b - db$, as in Figure 10.4. The nuclei that are the scattering centers are contained in a foil of thickness t having n nuclei per unit volume. We assume the foil is thin enough so there is negligible probability that a particle will undergo more than one scattering event in traversing the foil. This is not too severe a condition since the radius of a nucleus is about 10^{-4} times that of an atom. It follows that only about 10^{-12} of the volume of the sample is occupied by nuclei. Most of the space is empty, so the limitation to single scattering is not unreasonable.

The area of the annular ring shown in Figure 10.4 is $2\pi b|db|$. If the frontal area of the foil is A, then there are ntA nuclei and the area occupied by all annular rings of radius b is $dA = ntA\,2\pi b|db|$. The fraction dN_θ/N_0, where N_0 is the number of incident particles, will be equal to the fraction

$$\frac{dA}{A} = \frac{ntA\,2\pi b|db|}{A} = nt\,2\pi b|db|$$

so

$$\frac{dN_\theta}{N_0} = nt\,2\pi b|db| \tag{10.16}$$

But from Eq. (10.15)

$$db = -\frac{D}{4}\frac{1}{\sin^2\left(\frac{\theta}{2}\right)}\,d\theta \tag{10.17}$$

Substitution from Eqs. (10.15) and (10.17) in Eq. (10.16) gives

$$\frac{dN_\theta}{N_0} = nt2\pi\frac{D}{2}\left(\cot\frac{\theta}{2}\right)\frac{D}{4}\frac{1}{\sin^2\left(\frac{\theta}{2}\right)}\,d\theta \tag{10.18}$$

But

$$\cot \frac{\theta}{2} = \frac{\sin \theta}{2} \frac{1}{\sin^2 \left(\dfrac{\theta}{2}\right)} \tag{10.19}$$

Substituting Eqs. (10.5) and (10.19) in Eq. (10.18) leads to

$$\frac{dN_\theta}{N_0} = nt \frac{\pi}{8} \left(\frac{zZe^2}{\frac{1}{2}mv_0^2}\right)^2 \frac{\sin \theta}{\sin^4 \left(\dfrac{\theta}{2}\right)} d\theta \tag{10.20}$$

Equation (10.20) is close to what we want, but there is still a short coming. It shows the fraction of incident particles that scatter into the entire ring of radius a indicated in Figure 10.5, whereas the usual detector counts particles in only a small portion of such a ring. The ring in Figure 10.5 is on the surface of a sphere of radius ρ centered at the scattering site. The ring subtends the differential of solid angle

$$d\Omega = \frac{2\pi a\rho \, d\theta}{\rho^2} = \frac{2\pi \rho \sin \theta \, \rho \, d\theta}{\rho^2} = 2\pi \sin \theta \, d\theta \tag{10.21}$$

Using this in Eq. (10.20) and rearranging leads to

$$\frac{dN_\theta}{N_0} = \frac{nt}{16} \left(\frac{zZe^2}{\frac{1}{2}mv_0^2}\right)^2 \frac{1}{\sin^4 \left(\dfrac{\theta}{2}\right)} d\Omega \tag{10.22}$$

Equation (10.22) is the expression we have been seeking, since it is in a form containing terms which can be tested by experiment. The salient feature is dependence on $\sin^{-4}(\theta/2)$. During 1911 this aspect was verified in the alpha bombardment of silver and gold over the range in angle from 5° to 150°. Data were taken at many angles in this range and showed very satisfactory agreement with the predicted angular dependence. It is interesting to note that this range in angle causes the scattering intensity to change by a factor of 23,000. Agreement was

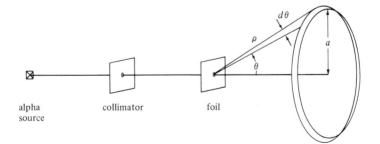

Figure 10.5 The cone of scattering for alpha particles deflected through angles between θ and $\theta + d\theta$.

also noted with the inverse square dependence on kinetic energy and the direct dependence on foil thickness. Since the atomic numbers Z of the heavier atoms were not well known, the Z^2 dependence could not be directly checked. Instead, Eq. (10.22) was used to determine Z. Soon thereafter Moseley's work on x rays gave good measurements of Z. In each instance the Z^2 dependence was confirmed.

Example 10.2 A narrowly collimated beam of 6-MeV alphas bombards a silver foil whose thickness is 300 μg/cm². What fraction of the alphas are deflected through 45° by Rutherford scattering into a detector of 3.0-mm diameter located 10 cm away?

Solution. We will make use of Eq. (10.22). At the outset, however, we note that thickness is given in terms of mass per unit area. In actually measuring a foil, this can be more readily determined than thickness in length units. We see that

$$t = \frac{g/cm^2}{g/cm^3}$$

The density of silver is 10.50 g/cm². Therefore

$$t_{\text{foil}} = \frac{300 \times 10^{-6}}{10.50} = 2.86 \times 10^{-5} \text{ cm}$$

Actually, however, we need not calculate t explicitly because it is the product nt that we need; nt is the number of atoms per unit area, and we proceed as follows.

Since the atomic weight of silver is 107.88,

$$300 \times 10^{-6} \text{ g/cm}^2 = \frac{300 \times 10^{-6}}{107.88} \text{ mole/cm}^2$$

$$nt = \frac{300 \times 10^{-6}}{107.88} \text{ mole/cm}^2 \times 6.02252 \times 10^{23} \text{ atoms/mole}$$

$$= 1.67 \times 10^{18} \text{ atoms/cm}^2$$

We set $z = 2$ (alpha particle); $Z = 47$ (silver); $e = 4.80 \times 10^{-10}$ esu (elementary charge). Then

$$\tfrac{1}{2}mv_0^2 = 6 \text{ MeV} \times 1.60 \times 10^{-6} \text{ erg/MeV} = 9.60 \times 10^{-6} \text{ erg}$$

$$\Delta\Omega = \frac{\text{area}}{\rho^2} = \frac{\pi(0.15)^2}{10^2} = 7.07 \times 10^{-4} \text{ steradian}$$

Combining as indicated in Eq. (10.22),

$$\frac{\Delta N}{N_0} = \frac{1.67 \times 10^{18}}{16} \left(\frac{2 \times 47 \times 4.8^2 \times 10^{-20}}{9.6 \times 10^{-6}}\right)^2 \frac{7.07 \times 10^{-4}}{\sin^4 45°/2}$$

$$= 1.75 \times 10^{-8} \quad \bullet$$

For those target nuclei for which the $\sin^{-4}(\theta/2)$ angular distribution is valid, we can presume that the interaction between the projectile and the target is dominated by electrostatic forces. Conversely, departure from this distribution indicates that some other force is influencing the interaction. In Figure 10.6 we illustrate that both the scattering angle and the distance of closest approach depend on the impact parameter. Equation (10.5) expresses the distance of closest approach for $\theta = 180°$. We leave as an exercise the calculation of closest approach for other angles.

At the time of Rutherford's early experiments alpha particles with energies up to 8.6 MeV were available from ThC' ($^{212}_{84}$Po), and for these D is 26 fermis for gold and 31 fermis for uranium. The predictions of Eq. (10.22) were satisfactory down to these distances of closest approach, but by 1919 anomalies had been noted in scattering from samples of lower Z material such as aluminum, where closer approach distances could be attained. Some of the anomaly could be explained on the basis of the recoil of the target nucleus. In deriving Eq. (10.22) the target was assumed to be sufficiently heavy not to recoil significantly. This assumption is quite good for heavy nuclides, but not so good for light ones. However, even when corrections were made for the recoil effect, anomalies remained at the closer distances of approach. It appeared the alphas were no longer being turned aside by electrostatic forces alone but were reaching the nucleus and interacting with it. It then became a matter of interest to determine the sizes of nuclei.

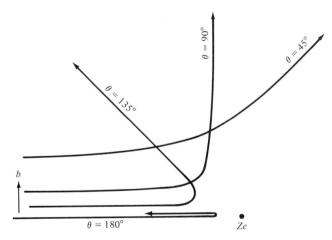

Figure 10.6 The dependence of scattering angle and path shape upon the impact parameter **b**.

10.5 Nuclear Radii

Since the time of the first scattering experiments described in the previous section many measurements of nuclear sizes and shapes have been made. Many different methods have been used.

Details of atomic spectra are influenced by nuclear sizes, since in some cases the electrons penetrate the charge distribution of the nucleus. A portion of the **isotope shift** in the spectra of such atoms can be attributed to this effect, as can also a small but observable change in the **fine structure splitting** in the x-ray spectra of heavy atoms. Similarly, if a muon is captured into a Bohr orbit around a nucleus to form a muonic atom, an x ray is emitted just as would be the case for the capture of an electron into an inner shell. However, the Bohr radius, given by $(n\hbar)^2/Ze^2m_0$, is about 200 times smaller for the muon than the electron. The muon orbit is much more penetrating than the electron orbits and the shifts of **muonic x-ray energies** give good measures of nuclear sizes.

If the nucleus is interpreted as a sphere of charge Ze and radius R, simple electrostatic calculations show it to have a stored **Coulomb energy** $\frac{3}{5}(Ze)^2/R$. When radioactive beta decay occurs, Z changes by one unit, and there is a consequent change in the Coulomb energy. In the case of certain **parent-daughter** combinations the Coulomb energy change can be extracted with good confidence from the energy of the beta decay, thus determining the radius R.

Theoretical calculations of **half-lives** for radioactive alpha decay show very marked dependence on nuclear radius. We have already pointed out that alpha scattering can give a measure of nuclear size. So, too, can the scattering of electrons and neutrons provided projectile energies are sufficiently high so that their de Broglie wavelengths are small compared to the nuclear radius.

Although these various methods of measuring nuclear size differ considerably in theoretical viewpoint as well as experimental technique, they all give a consistent picture. Nuclei are **essentially spherical** with radii proportional to the cube root of the mass number. That is,

$$R = R_0 A^{1/3} \tag{10.23}$$

This equation expresses the characteristic that the density of nuclear matter, stated in nucleons per unit volume, is independent of the number of nucleons that make up the nucleus. The value of R_0 depends somewhat on the method used as a measure of it. It ranges from about 1.1 fermi to about 1.5 fermi. Methods using the electron as a probe lead to the **electromagnetic radius** while those using nucleons give the **nuclear force radius**. In general, the electromagnetic radius seems to be smaller than the nuclear force radius. In the next few paragraphs we shall look in a bit more detail at the experiments involving alpha particles, electrons, and neutrons.

Example 10.3 What is the nuclear radius of ^{56}Fe? of ^{234}Th?

Solution. We adopt $R_0 = 1.4$ fermi; consequently,

$$R_{Fe} = 1.4 \times (56)^{1/3} = 5.4 \text{ fermi}$$
$$R_{Th} = 1.4 \times (234)^{1/3} = 8.6 \text{ fermi}$$

10.6 Alpha Scattering

When an alpha particle approaches a nucleus, the Coulomb repulsion between them can be expressed as a potential energy with a $1/r$ dependence. This r^{-1} dependence will hold until the approach is close enough that the attractive nuclear force becomes effective. The plot of potential energy versus radial separation will then drop abruptly for values of r less than this. This state of affairs is diagrammed in Figure 10.7. The assumption that the Coulomb potential is converted suddenly to a square well potential is obviously unrealistic; it would be

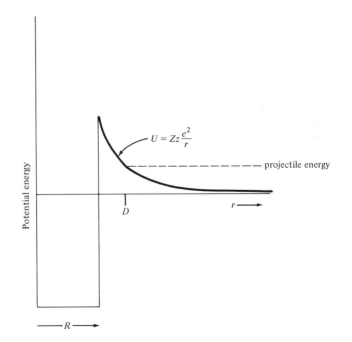

Figure 10.7 The Coulomb potential barrier in alpha particle scattering. *D* is the *collision diameter,* or distance of closest approach in a headon collision.

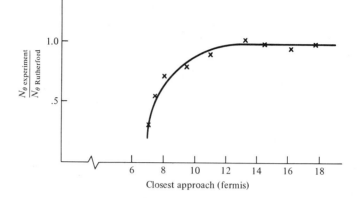

Figure 10.8 Some results of Rutherford's measurement of the scattering of alpha particles by aluminum.

more reasonable if the top of the potential were rounded and the side of the well sloped. There have been many suggestions in the literature for details of this shape, but for our present purposes the shape of Figure 10.7 will suffice.

The simplest interpretation, that Rutherford scattering is valid for all $r > R$, turns out not to be altogether correct. Although classically no alpha particle could reach the interior of the well and interact with the nucleus unless it has sufficient energy to surmount the top of the **potential energy barrier,** the quantum mechanical effect of **barrier penetration** permits a small but nonzero probability of such interaction at lower energies. The probability of barrier penetration depends on both the thickness of the barrier to be penetrated and the height of the barrier above the total energy of the system. Since both the height of the barrier and thickness, $D - R$, depend on R, corrections for barrier penetration can be calculated as a function of R. In Figure 10.8 the results of alpha scattering from aluminum are expressed as a ratio of experimentally observed scattering intensity to that predicted by the Rutherford distribution law, plotted against the distance of closest approach. Definite deviation from Rutherford scattering is observed for $r \gtrsim 10$ fermis. When corrected for barrier penetration and the finite mass of the A1 nucleus, these data are consistent with $R_0 \approx 1.5$ fermi.

Example 10.4 What is the height of the Coulomb barrier for alpha particles incident on $^{234}_{90}$Th?

Solution.

$$U = \frac{zZe^2}{R}$$

From Example 9.3 $R = 8.6 \times 10^{-13}$ cm for ^{234}Th. Therefore

$$U = \frac{2 \times 90 \times (4.8 \times 10^{-10})^2}{8.6 \times 10^{-13}} = 4.8 \times 10^{-5} \text{ erg} = 30 \text{ MeV}$$

10.7 High-Energy Electron Scattering

When fast electrons are scattered by close encounters with nuclei, they experience attractive forces that remain essentially **electrostatic** in nature even when the electron passes through the nucleus itself. The electron "sees" the nucleus as a positive charge distribution, and the potential energy of the system is a function of the distance between the electron and the center of the nucleus. Since the force is one of attraction, there is no potential barrier but only a potential well. In Figure 10.9 the potential function for three simple assumptions regarding distribution of electric charge in the nucleus is shown. For a nucleus with all its charge in a **point** at the center (curve a), $-Ze^2/r$ is valid for indefinitely small values of r. For a nucleus whose charge distribution is in the form of a **spherical shell** of radius R the potential energy is uniform (curve b) inside the shell with the value $-Ze^2/R$. If the charge were distributed with **uniform density** within a sphere of radius R, then the potential energy for $r < R$ becomes

$$U = -\frac{Ze^2}{R} \left(\frac{3}{2} - \frac{1}{2} \frac{r^2}{R^2} \right) \tag{10.24}$$

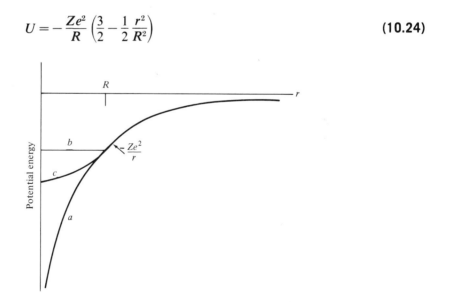

Figure 10.9 Potential function for an electron in the neighborhood of a nucleus. Results are shown for three different assumptions regarding charge distribution in the nucleus. (a) A point charge at the center. (b) Charge distributed on the surface of a sphere of radius R. (c) Charge distributed uniformly throughout the volume of a sphere of radius R.

For all models with spherical symmetry the potential energy is $-Ze^2/r$ when $r > R$.

The discussion in the preceding paragraph treats the electron as though it could be located within a region small compared to the size of the nucleus. This implies a correspondingly small de Broglie wavelength. The total energy and the de Broglie wavelength of the electron are given, respectively, by

$$E^2 = p^2c^2 + (m_0c^2)^2 \tag{10.25}$$

and

$$\lambdabar = \frac{\hbar}{p} \tag{10.26}$$

Anticipating that we shall be dealing with high-energy electrons for which

$$E \gg m_0c^2 \tag{10.27}$$

we find from Eq. (10.25)

$$p \approx \frac{E}{c} \tag{10.28}$$

which then gives

$$\lambdabar \approx \frac{\hbar c}{E} \tag{10.29}$$

Equation (10.27) implies that kinetic energy almost equals the total energy

$$K \approx E \tag{10.30}$$

from which

$$\lambdabar \approx \frac{\hbar c}{K} = \frac{\hbar c}{K}\left(\frac{m_0c^2}{m_0c^2}\right) = \left(\frac{\hbar}{m_0c}\right)\left(\frac{m_0c^2}{K}\right) \tag{10.31}$$

But $\hbar/m_0c$ is the **rationalized Compton wavelength,** and has the numerical value of 386 fermis. For electrons of kinetic energy 150 MeV we find

$$\lambdabar \approx 1.3 \text{ fermi} \tag{10.32}$$

Since the wavelength given in Eq. (10.32) is smaller than nuclear radii, electrons of 150 MeV and higher energy are suitable for probing the nuclear charge distribution.

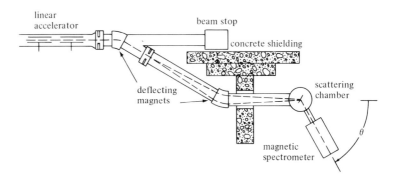

Figure 10.10 Apparatus used by Hofstadter and colleagues at Stanford University for the study of nuclear elastic scattering of high energy electrons.

In an important series of experiments reported during the period 1953 to 1958, R. Hofstadter and his colleagues at Stanford University used an electron linear accelerator to produce controlled beams of electrons of suitable energy. The experimental arrangement is shown in Figure 10.10. The electron beam enters from the left where the last sections of the electron accelerator are shown. The deflecting magnets momentum analyze and focus the beam onto the sample in the form of a thin foil at the center of the scattering chamber. Those electrons that scatter through the angle θ enter the magnetic spectrometer where they are again momentum analyzed to reject electrons scattered inelastically. Elastically scattered electrons are counted with a detector at the exit focus of the spectrometer and a series of measurements taken at various angles.

Figure 10.11 shows angular distributions for electrons of two different incident inergies scattered by gold. The mathematical analysis of the distributions is more difficult than for Rutherford scattering, and we shall not go into the details. It is useful to note, however, that the electron behaves in a fashion similar to light waves impinging on a transparent sphere. The change in momentum of the electron as it enters the region of negative potential energy changes its wavelength in a manner analogous to changes in optical wavelength in a region of varying optical index of refraction. **Diffraction** occurs as well as **refraction** and the shape of the diffraction maxima and minima which are superimposed on the sharply falling curves of Figure 10.11 depend in a sensitive fashion on the charge distribution through which the electrons pass.

Hofstadter et al. found that by using a radial charge density function of the form

$$\rho(r) = \frac{\rho_0}{1 + e^{(r - R) \, (4 \ln 3)/t}} \tag{10.33}$$

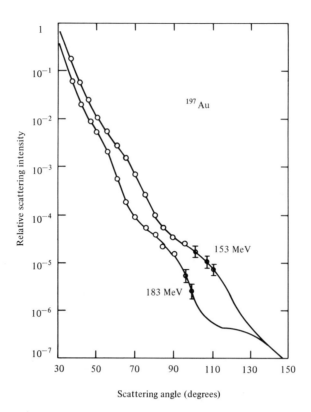

Figure 10.11 Angular distributions in the elastic scattering of electrons by gold.

they could calculate **angular distribution** curves giving very good fit to the experimental data. In Figure 10.11 the plotted points show experimental data and the solid lines the theoretical curves with best fit. In Eq. (10.33) ρ_0, R, and t are treated as adjustable parameters to produce the best fit. They are subject to the requirement that $\int \rho(r)\,dV = Ze$ when taken over the volume of the nucleus. ρ_0 is the charge density at $r = 0$; R is the radius at which charge density is $\frac{1}{2}\rho_0$; t is the surface thickness across which $\rho(r)$ changes from 0.9 to $0.1\rho_0$.

Figure 10.12 shows charge distributions for a number of nuclides from $_6$C to $_{83}$Bi. R and t are marked on the diagram for $_6$C. It is apparent that t is essentially the same for all nuclides and that ρ_0 is nearly so. Since it is the protons that provide the charge on the nucleus, it follows that the number of protons per unit volume is proportional to $\rho(r)$. Assuming that protons and neutrons are uniformly mixed inside the nucleus, the curves of Figure 10.12 are very nearly proportional to the radial distribution of nucleons. We noticed when discussing Figure 10.2 that the ratio N/Z is greater for nuclides of high atomic number than for the low ones. When this change in N/Z is

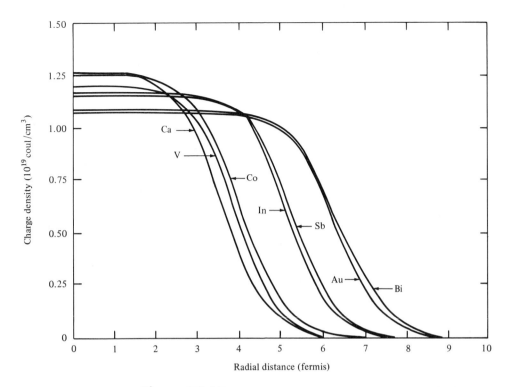

Figure 10.12 Charged density distributions for various nuclides as determined by high energy electron scattering.

taken into account, the nucleon density near the center of the nucleus is even less dependent on Z than is ρ_0. Interpreting R as the nuclear radius, it is found that

$$R = R_0 A^{1/3} \tag{10.34}$$

with

$$R_0 = (1.07 \pm 0.02) \text{ fermi} \tag{10.35}$$

The picture which evolves indicates that nuclei are **mostly spherical,** that the **nucleon density** is very nearly the same in the central region of every nucleus, and that near the surface the nucleon density falls rapidly but not discontinuously to zero. Many nuclides show ellipsoidal distortion, but, except for a very few, the difference between the semimajor and semiminor axes is less than 10% of the average radius that is used in Eqs. (10.23) and (10.34). On the whole, the spherical approximation is quite good.

10.8 Scattering of Fast Neutrons

In addition to alpha particles and electrons, neutrons can be used to investigate nuclear sizes. Being electrically neutral the neutron does not respond to the Coulomb force. We would therefore expect neutron scattering to give more direct information about the **nuclear force radius** than the other two methods. Since the specifically nuclear force, by which the neutron interacts with the nucleus, is of short range (~ 1 fermi), we wish to localize the neutron at least this closely. This implies using neutrons of sufficient energy that $\lambda \leq 1$ fermi. For non-relativistic particles we can write

$$p = mv = \sqrt{2mK} \tag{10.36}$$

and

$$\lambda = \frac{\hbar}{p} = 4.55 K^{-1/2} \tag{10.37}$$

when K is expressed in MeV. From this we see that the neutron energy should be around 20 MeV.

Figure 10.13 is a potential diagram for a neutron-nucleus system. Since the nuclear force is one of attraction, there is no barrier around the potential well. Since the force is of short range, the region outside the well is flat without the long r^{-1} dependence of Figure 10.9. The transition region from zero potential outside the well to negative potential inside the well is short and therefore the well may be approximated by a square well of radius R.

Figure 10.14 is a representation in which two dimensions, x and y, are space dimensions and the third one, U, is potential energy. The well is represented by a cylindrical region dipping below the $U = 0$ surface (i.e., the xy plane). A neutron n is shown approaching the well. If it comes within an interaction distance $(R + r)$ from the center of the well, where r is the neutron radius taken to equal λ, the neutron will "fall into" the well and be **absorbed**. When this happens the neu-

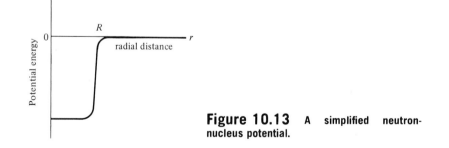

Figure 10.13 A simplified neutron-nucleus potential.

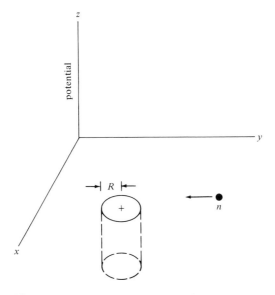

Figure 10.14 Representation of a two-dimensional "square well" for a neutron-nucleus system. The potential is 0 on the xy plane. Inside the nucleus the potential is negative and is represented by the dotted cylinder below the surface. A neutron is shown approaching from the right. It either misses the well altogether or falls into it.

tron is removed from the stream. At passing distances greater than $(R + r)$ it will pass by unaffected.

Of course, Figure 10.14 should really show three space dimensions with still another dimension representing the potential. We shall not attempt that drawing, but instead in Figure 10.15 we show a side view of a grazing encounter of a neutron with a spherical nucleus. We see that the effective area for absorption of the neutron is $\pi(R + \lambda)^2$. In addition to absorption, diffraction of the wave associated with the neutron occurs. This is a scattering that further reduces the number of neutrons that pass straight by. For energies such that $\lambda \ll R$ the result is that the area of interaction for each nucleus becomes $2\pi(R + \lambda)^2$.

Figure 10.15 The interaction radius for a nucleus of radius *R* undergoing bombardment by a neutron of wavelength, and hence effective radius, λ.

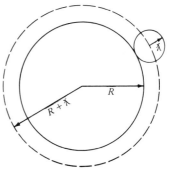

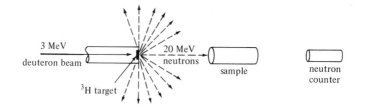

Figure 10.16   Experimental arrangement for a neutron attenuation experiment.

Figure 10.16 is a schematic diagram of an arrangement suitable for measuring the **attenuation** of neutrons passing through a sample. A beam of deuterons of approximately 3 MeV from a Van de Graaff accelerator enters from the left and is focused on a target containing ³H. This bombardment of ³H produces neutrons that leave the target in all directions. The neutrons which travel in the forward direction have kinetic energies of 20 MeV. Attenuation is determined by the ratio of the number of neutrons registered by the counter with the sample in place to that registered with the sample removed.

Consider a portion of the sample of thickness dt along the path of the neutrons, with an area A normal to the beam, and containing n nuclei per unit volume. Let N_0 be the number of neutrons incident on the sample in a given time. This is the same as the number registered by the counter with the sample removed. Let N be the number remaining at an arbitrary depth in the sample. Then

$$-\frac{dN}{N} = \frac{nA\ dt\ 2\pi(R + \lambda)^2}{A} = n2\pi(R + \lambda)^2\ dt \qquad (10.38)$$

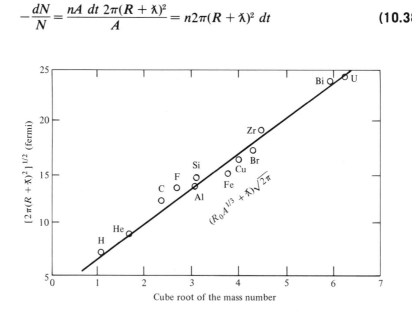

Figure 10.17   Results of a neutron attenuation experiment at 19 MeV.

Integration of Eq. (10.38) and use of the condition $N = N_0$ at $t = 0$ gives

$$N_{\text{thru}} = N_0 e^{-n2\pi(R + \lambda)^2 t} \tag{10.39}$$

The N's are measured by the experiment, n and t are properties of the sample, and λ is known from the kinetic energy of the neutrons. The calculation of R is straightforward.

The results of such an experiment by R. B. Day and R. L. Henkel using 19 MeV neutrons are shown in Figure 10.17. The plotted points are their measurements and the straight line is drawn with $R_0 = 1.4$ fermi and $\lambda = 1.04$ fermi.

10.9   Summary

Atomic nuclei are made up of protons and neutrons bound together by the specifically nuclear force, often called the strong force.

The following nomenclature is used:

Z = number of protons

N = number of neutrons

$A = Z + N$ = mass number

A chemical element X of atomic number Z and mass number A is designated $_Z^A X$.

Some units and quantities useful in nuclear studies are

1 fermi $= 10^{-13}$ cm

1 MeV $= 10^6$ eV $= 1.60210 \times 10^{-6}$ erg

1 u $= 1.66053 \times 10^{-24}$ g $\Rightarrow 931.480$ MeV

The term **nuclide** is used to designate a nuclear species. Other terms are

Isotopes: nuclides of the same Z, different A

Isobars: nuclides of the same A, different Z

Isotones: nuclides of the same N, different Z

The **mass defect** of an atomic species is the difference between the mass of the neutral atom and the sum of the masses of the protons, neutrons, and electrons of which it is made.

$$\Delta M = ZM_H + NM_n - M_{Z,A}$$

The binding energy is ΔMc^2.

Nuclear forces are short range and exhibit saturation.

Radioactive nuclides are unstable against emission of particles, which may be alphas ($^4_2$He nuclei), negatrons (negative electrons), or positrons (positive electrons).

Stable nuclides exist in a **band of stability** extending through the chart of nuclides. In the band of stability $Z \approx N$, but N increases faster than Z in the heavy nuclides.

The Rutherford scattering experiment provided the earliest evidence that atoms have massive nuclei with positive charge. The scattering law is of the form

$$\frac{dN_\theta}{N_0} = \frac{nt}{16}\left(\frac{zZe^2}{\frac{1}{2}mv_0^2}\right)^2 \frac{1}{\sin^4 \theta/2}\, d\Omega$$

To a very good approximation nuclear radii can be expressed as

$$R = R_0 A^{1/3}$$

where

$$\begin{array}{ccc} 1.1 \text{ fermi} & < R_0 < & 1.5 \text{ fermi} \\ \text{(electromagnetic)} & & \text{(nuclear force)} \end{array}$$

Problems

10.1 Rearrange the following list of nuclides into groups of isotopes, isobars, and isotones. Some nuclides may appear in more than one group.

$^{26}_{12}$Mg, $^{19}_{9}$F, $^{13}_{6}$C, $^{60}_{29}$Cu, $^{26}_{13}$Al, $^{9}_{4}$Be, $^{25}_{11}$Na, $^{61}_{28}$Ni

$^{8}_{4}$Be, $^{27}_{13}$Al, $^{26}_{14}$Si, $^{60}_{28}$Ni, $^{10}_{4}$Be, $^{12}_{6}$C, $^{13}_{7}$N, $^{60}_{27}$Co

10.2 Calculate the mass defect and the binding energy per nucleon for the last five nuclides listed in problem 10.1.

10.3 Calculate the nuclear radii for the first five nuclides listed in problem 10.1.

10.4 Calculate the distance of closest approach to $^{197}_{79}$Au for a 5.5-MeV alpha particle which is scattered at 180°. Calculate also the nuclear radius of $^{197}_{79}$Au so the two distances may be compared.

10.5 Derive a general expression for the way the distance of closest approach depends upon the atomic numbers of the projectile and target nucleus, the kinetic energy of the projectile, and the angle of scattering. Evaluate this distance when a 5.5-MeV alpha is scattered by $^{197}_{79}$Au through 60°.

10.6 A gold foil whose thickness is 180 μg/cm^2 is bombarded by a beam of 5.3-MeV alphas collimated to a cross section diameter of 2.0 mm. A detector whose diameter is 5.0 mm is mounted 5.0 cm away from the irradiated spot on the foil. What fraction of the incident alphas undergo scattering through 15°? through 30°? through 60°? through 120°?

10.7 What is the height in MeV of the Coulomb barrier which the first five nuclides listed in problem 10.1 present to alphas? to protons? to neutrons?

10.8 Calculate the wavelength of neutrons whose kinetic energy is 14 MeV, 18 MeV, or 22 MeV.

10.9 A neutron detector which is placed 25 cm from a source of 18-MeV neutrons registers 4205 counts in 100 sec. When an aluminum absorber 1.0 cm thick is placed between the source and the detector, the count drops to 3808 counts in 100 sec. Calculate the radius of the aluminum nucleus from these data and determine the value of R_0.

10.10 The capture of a slow neutron by $^{235}_{92}$U leads to fission into nearly equal fragments. Examine Figure 10.1 to determine the binding energy per nucleon before and after fission. From these data estimate the energy release per fission.

10.11 Calculate the mass defect and the binding energy per nucleon for $^2_1$H, $^4_2$He, and $^3_2$He.

10.12 Calculate the nuclear radius of $^{27}_{13}$Al. What is the height of the Coulomb barrier presented to alpha particles? What is the distance of closest approach if the alpha particle has a kinetic energy of 5.3 MeV? if it has 8.6 MeV?

10.13 An aluminum foil whose thickness is 600 μg/cm^2 is substituted for the gold foil of problem 10.6. If the other parameters remain as described, calculate the fractional scattering at the four stipulated angles.

10.14 Verify by substitution that Eq. (10.12) is a solution to Eq. (10.11).

10.15 Perform the trigonometric manipulations called for immediately following Eq. (10.14) and for verifying Eq. (10.19).

10.16 Determine the electrostatic field inside a sphere of radius R within which there is a uniform charge density ρ. By integration verify Eq. (10.24).

10.17 What is the de Broglie wavelength for an electron of kinetic energy 0.5 MeV? of 2.0 MeV? of 10 MeV? of 100 MeV?

10.18 What kinetic energy protons would be required so their de Broglie wavelength $\lambda = 1.3$ fermi? Discuss whether protons of this energy would be suitable in an experiment comparable to Hofstadter's electron experiment for probing nuclear charge density.

Suggestions for Further Reading

Eisberg, R., and R. Resnick. *Quantum Physics*. New York: John Wiley & Sons, 1974. Sections 15.1 through 15.4.

Evans, R. D. *The Atomic Nucleus.* New York: McGraw-Hill Book Company, Inc., 1955. Chapter 2.

Green, A. E. S. *Nuclear Physics.* New York: McGraw-Hill Book Company, Inc., 1955. Chapter 2.

Marshak, R. E. "The Nuclear Force," *Scientific American* (March 1960).

Nuclear reactions

11.1   Introduction

When a bombarding particle has a sufficiently close encounter with a target nucleus, events other than simple scattering may occur. As we have seen in the discussion of Rutherford scattering, a charged incident particle is turned aside by a Coulomb potential barrier unless the collision diameter given by Eq. (10.5) and the impact parameter are sufficiently small. Thus at low energies scattering predominates, whereas at higher energies specifically **nuclear reactions** can occur. Rutherford scattering conveys information about nuclear charge and size; nuclear reactions convey information about the interior of the nucleus.

In the broadest sense any encounter in which specifically nuclear forces play a role may be called a nuclear reaction. We shall limit our discussion, however, to those situations in which a single projectile interacts with a nucleus, producing an **emitted particle** and a **residual nucleus**. We shall omit discussion of spallation reactions in which the struck nucleus breaks into numerous pieces, as well as three-body breakup. Now, as neutrons and protons may be used as projectiles, so also may heavier nuclei and photons. The use of charged projectiles implies a Coulomb barrier to be surmounted. This barrier becomes higher as projectiles of higher charge are selected. Since the general trend is for $Z \approx \frac{1}{2}A$, heavier nuclei require higher bombarding energies if there is to be appreciable likelihood of producing reactions. Most of our discussion will be centered on reactions using projectiles with $A \leq 4$, but the general principles will apply to those of any mass.

11.2   Typical Reactions

In the typical reaction a **projectile** x interacts with a **target nucleus** X, producing as end products a **residual nucleus** Y and an **outgoing particle** y. With this nomenclature the reaction may be summarized by the equation

$$X + x \rightarrow y + Y \tag{11.1}$$

or by a more compact form which is to be understood as giving the same information:

$$X(x, y)Y \tag{11.2}$$

Usually x and y are nucleons or light nuclei or photons. Figure 11.1 is a schematic representation of the reaction. It is assumed that the target X is stationary before the reaction. In the "before" portion of the figure, x is shown approaching the target. After the reaction, y is shown receding from the site of the reaction at a direction θ relative to the forward direction of x while Y travels in the direction ϕ.

If y is the same as x, the reaction is a **scattering** reaction and it must follow that X and Y are the same. If the sum of the kinetic energies of y and Y is the same as the kinetic energy of x, we have **elastic scattering**. If the total kinetic energy afterward is less than that before, then the reaction is one of **inelastic scattering** and Y is left in an ex-

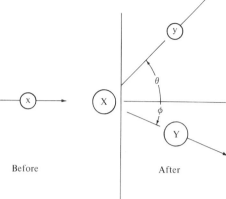

Figure 11.1 Geometrical relationships in a nuclear reaction. A projectile x is shown approaching from the left. It strikes a target nucleus X, particles y and Y leave the scene of the reaction following paths which make angles θ and ϕ respectively with the forward extension of the path of x.

Before

After

cited state. If y is a photon, the reaction is called a **radiative capture** reaction, and if x is a photon, it is called a **nuclear photo** reaction. The residual nuclide may or may not be identical to a nuclear species found in nature. Fundamentally we may say the same thing about the target nuclide, but for practical reasons it must be sufficiently long lived and available in sufficient quantity to permit preparation of an adequate sample.

Table 11.1 shows 24 examples of nuclear reactions expressed in the notation of Eq. (11.2). This is by no means an exhaustive list but is intended only to demonstrate some typical reactions. Some of the

Table 11.1 Representative Nuclear Reactions

Reaction*	Q Value MeV	Reaction*	Q Value MeV
1. $^2_1\text{H}(\gamma, n)^1_1\text{H}$	-2.225	13. $^{10}_5\text{B}(\alpha, p)^{13}_5\text{C}$	4.071
2. $^2_1\text{H}(d, p)^3_1\text{H}$	4.032	14. $^{10}_5\text{B}(\alpha, n)^{13}_7\text{N}$	1.068
3. $^2_1\text{H}(d, n)^3_2\text{He}$	3.268	15. $^{11}_5\text{B}(d, p)^{12}_5\text{B}$	1.136
4. $^3_1\text{H}(d, n)^4_2\text{He}$	17.571	16. $^{11}_5\text{B}(\alpha, n)^{14}_7\text{N}$	0.154
5. $^6_3\text{Li}(n, \alpha)^3_1\text{H}$	4.804	17. $^{12}_6\text{C}(d, p)^{13}_6\text{C}$	2.723
6. $^7_3\text{Li}(p, \alpha)^4_2\text{He}$	17.337	18. $^{12}_6\text{C}(d, n)^{13}_7\text{N}$	-0.280
7. $^7_3\text{Li}(p, n)^7_4\text{Be}$	-1.647	19. $^{13}_6\text{C}(d, p)^{14}_6\text{C}$	5.944
8. $^9_4\text{Be}(\gamma, n)^8_4\text{Be}$	-1.666	20. $^{14}_7\text{N}(n, \gamma)^{15}_7\text{N}$	10.833
9. $^9_4\text{Be}(p, d)^8_4\text{Be}$	0.559	21. $^{14}_7\text{N}(\alpha, p)^{17}_8\text{O}$	-1.198
10. $^9_4\text{Be}(d, t)^8_4\text{Be}$	4.591	22. $^{15}_7\text{N}(d, \alpha)^{13}_6\text{C}$	7.684
11. $^9_4\text{Be}(\alpha, n)^{12}_6\text{C}$	5.708	23. $^{19}_9\text{F}(d, \alpha)^{17}_8\text{O}$	10.042
12. $^{10}_5\text{B}(n, \alpha)^7_3\text{Li}$	2.793	24. $^{27}_{13}\text{Al}(p, n)^{27}_{14}\text{Si}$	-5.593

* In this table γ = gamma-ray photon, n = neutron, p = proton, d = deuteron, t = triton, α = alpha particle.
Q values will be discussed in Section 11.4.

reactions are of historical interest, having been involved in "discovery" experiments. For instance, reaction 21 led to the discovery of artificially induced transmutations by Rutherford in 1919. While studying the scattering of alpha particles by light nuclides he noticed that, with nitrogen as the scattering material, radiation was produced that had a range in air considerably greater than that of the alpha particles themselves. Upon identifying the radiation as protons coming from the region under bombardment and eliminating other possible sources, it was concluded that nitrogen was being transmuted into oxygen according to reaction 21. Some years later an even more penetrating radiation was observed to result from alpha bombardment of beryllium and boron. After considerable effort by groups of experimenters in Germany, France, and England to understand the interaction of this radiation with various absorbing materials, J. Chadwick, in 1932, showed that a consistent and simple interpretation could be based on the assumption that the radiation consisted of uncharged particles whose mass was close to that of the proton. Thus the discovery of the neutron using reactions 11, 14, and 16.

In 1930, Cockcroft and Walton demonstrated that reaction 6 could be made to "go" using as their projectiles protons accelerated by electrical means to energies of a few hundred kiloelectron volts. This was the first transmutation reaction produced by projectiles from an accelerator in the laboratory rather than by those from radioactive materials. Some of the other reactions also are used nowadays as tools for neutron production. Reactions 3, 4, and 7, using accelerator produced protons and deuterons to induce the reactions, are good sources of neutrons. As we shall see presently, a certain amount of control of the neutron energy can be obtained by choice of the reaction and the bombarding conditions. Reaction 11 can be used to produce neutrons without an accelerator. It is the usual source of neutrons in a **neutron howitzer,** the reaction being produced by an intimate mixture of beryllium and an alpha emitting radioactive material encapsulated in a small metal cylinder placed at the center of a barrel of paraffin, polyethylene, or water. A popular source of alpha particles for this purpose is $^{239}_{94}$Pu, with a half-life of 2.4×10^4 yr.

11.3   Conservation Laws

Examination of the reactions displayed in Table 11.1 reveals the operation of two **conservation laws.** Several additional conservation laws also operate in nuclear reactions and knowledge of them is very helpful in understanding various reactions. Let us look first at the ones that are obvious in Table 11.1. As examples we rewrite reactions 6 and 7, using the notation of Eq. (11.1).

$$\ce{^7_3Li} + \ce{^1_1}p \rightarrow \ce{^4_2}\alpha + \ce{^4_2He} \tag{11.3}$$

$$\ce{^7_3Li} + \ce{^1_1}p \rightarrow \ce{^1_0}n + \ce{^7_4Be} \tag{11.4}$$

We see that the **mass number** A is conserved; the total for A on the left side of the equation is the same as the total on the right. This is the same as saying that the total number of nucleons remains unchanged during the course of a reaction. Nucleons may be rearranged into different configurations, but they are neither created nor destroyed. There is an exception. At very much higher energies than we are considering, proton-antiproton production can occur. However, in such reactions the nucleon and antinucleon always appear in pairs and can be regarded as canceling each other in the count of nucleons. Thus we conclude that nucleon number is conserved.

Directing our attention to the subscripts, which designate the **nuclear charge** Z, it is apparent that this total is the same before and after the reaction. The interpretation that the total number of protons remains unchanged in all reactions is too strong, however. Table 11.1 limited itself to reactions produced by bombarding a target nuclide with projectiles of one sort or another. If, however, beta decay is included as a possible reaction, we find that the proton number changes, increasing by one when a negative beta particle is emitted and decreasing by one when the emitted particle is a positive beta. In any of the reactions total charge, due to protons and betas together, is conserved.

These first two conservation laws will specify the complete reaction when any three of the four particles involved are known. Also it is common to omit the left subscript, which designates atomic number, since the chemical symbol gives the same information. Thus if we are told that a reaction under investigation is $\ce{^10B}(\alpha, p)$ we readily conclude that the residual nuclide is $\ce{^13C}$ and that the complete statement of the reaction would be $\ce{^10B}(\alpha, p)\ce{^13C}$. However, the mere fact that a nuclear reaction equation **balances** with regard to nucleon number and charge does not give assurance that the reaction will actually occur. The reaction must be **energetically possible**; that is, it must conserve mass-energy. We also anticipate that **linear momentum** and **angular momentum** will be conserved. We shall use the ideas of **mass-energy** and linear momentum while discussing the energetics of nuclear reactions in the next section.

Both the proton and the neutron have the **intrinsic spin** quantum number $s = \frac{1}{2}$. In addition, their motions in a nucleus around the center of mass contribute **orbital angular momentum** which is quantized with integral quantum number similar to the situation in atomic physics.

Consequently, odd A nuclei have **total angular momenta** with quantum number J half-integer and even A nuclei have J integral. In a nuclear reaction the initial angular momentum arises from the inherent angular momentum of the target nucleus, the projectile and any quantized angular momentum from the collision process itself. The quantum numbers J_X, J_x, and L combine in various orientations to produce integer or half-integer resultants according to the **vector model.** Similarly, the **final angular momentum** of the system is the resultant of the quantized combination of J_Y, J_y, and L_f, where L_f is the **mutual angular momentum** of the products Y and y. The only reactions that can occur are those for which the final angular momentum equals the initial angular momentum.

The basic idea of angular momentum is familiar to us from the study of classical physics and we are not surprised at the requirement that it be conserved. That it should be quantized is also familiar from the study of atomic systems. There are two other quantities, explicitly quantum mechanical in nature, which are also conserved in nuclear reactions. These are the **statistics** and the **parity** of the system. If the wavefunction of a system of two identical particles changes sign upon interchange of the particles, the wavefunction is said to be antisymmetric in the coordinates of the particles. The particles obey **Fermi-Dirac** statistics and are called **fermions.** Nucleons, beta particles, and neutrinos are fermions. If the wavefunction does not change sign, it is symmetric in the coordinates. Particles characterized by such wavefunctions obey **Einstein-Bose** statistics and are called **bosons.** The process of interchanging two nuclei may be viewed as a stepwise process, interchanging pairs of corresponding nucleons until the entire transfer has been accomplished. At each pair interchange the wavefunction will change sign. It is apparent that if an individual nucleus is of even nucleon number, the completed process will restore the wavefunction to its original sign, and contrariwise for odd A nuclei. Thus odd A nuclei obey Fermi-Dirac statistics, whereas even A nuclei obey Einstein-Bose statistics. We have seen that if the number of nucleons is even prior to a reaction, it will be even following the reaction; if it is odd before, it will be odd afterward. It follows, therefore, that the statistics of the system is conserved.

Another operation that can be performed on a quantum mechanical system is to reflect all its particles through the origin of the coordinate system in which it is located. In a rectangular coordinate system this is equivalent to replacing the (x, y, z) coordinates for each particle of the assembly by $(-x, -y, -z)$. If this reflection causes the wavefunction to change sign, the assembly is said to have odd parity; if the sign does not change, the assembly has even parity. For many years it was believed that parity must be strictly conserved in all transformations. In 1956–1957, however, it was suggested, and experimental evidence soon confirmed that parity conservation could be violated in beta decay processes. On the other hand, no evidence has been found for nonconservation of parity in reactions such as those shown in Table 11.1, which involve nucleons only. Such reactions invoke the specifically nuclear force (frequently called the strong force) between nucleons. At this writing it appears that parity conservation is valid in reactions involving strong forces.

There is reason to believe that the specifically nuclear force acts between pairs of nucleons in a fashion independent of the electric charge. That is, the force between two neutrons is the same as the force between a neutron and a proton and in turn the same as that between two protons. If so we can view the neutron and proton as two different states of the nucleon. We should then be able to distinguish these two states by assigning a quantum number that would have one value if the nucleon is in its proton state and a different value if it is in the neutron state. Such a formalism has been developed and the quantum number is called the **isospin.** Other names that have been used for the same quantity are **isotopic spin** and **isobaric spin.** Isospin projection is $t_3 = \frac{1}{2}$ for a proton and $t_3 = -\frac{1}{2}$ for a neutron. No actual spin is involved, but the formalism was developed by people who saw analogies to two possible spin states, $+\frac{1}{2}$ and $-\frac{1}{2}$, of electrons and of nucleons. Much of the formalism for combining the isospins of nucleons to obtain the total isospin of a nucleus is analogous to that for coupling the spins of electrons in a multielectron atom. Since the Coulomb force must be neglected in a pure isospin treatment, we expect this treatment to give its best results in the region of the light nuclides where Coulomb forces play a less important role than they do among the heavy nuclei. Reactions in which the total isospin of the system is the same before and after should be favored. This indeed seems to be the case and in a later section we shall discuss the experimental evidence.

We have discussed eight quantities to be conserved in a nuclear reaction:

1. Nucleon number

2. Total charge

3. Mass-energy

4. Linear momentum

5. Angular momentum

6. Statistics

7. Parity

8. Isospin

In the next section we shall be concerned primarily with conservation of mass-energy and linear momentum while tacitly assuming that the other six conservation requirements are also met.

11.4 Q Values and the Q Equation

In a nuclear reaction produced by bombarding a nucleus X with a projectile x the target nucleus is ordinarily at rest. We express conservation of mass-energy as follows:

$$K_x + M_x c^2 + M_x c^2 = K_y + M_y c^2 + K_Y + M_Y c^2 \qquad (11.5)$$

where the K's are the **kinetic energies,** the M's are the **exact masses** of the particles and the subscripts are defined in accordance with the notation of Eq. (11.1) and Figure 11.1. Rearranging yields

$$(K_y + K_Y) - K_x = (M_x + M_X)c^2 - (M_y + M_Y)c^2 \qquad (11.6)$$

The increase in kinetic energy is defined as the Q of the reaction.

$$Q = (K_y + K_Y) - K_x \qquad (11.7)$$

We see also that Q equals the decrease in the rest energy of the system.

$$Q = (M_x + M_X)c^2 - (M_y + M_Y)c^2 \qquad (11.8)$$

From conservation of linear momentum we can write

$$p_x = p_y \cos \theta + p_Y \cos \phi \qquad (11.9)$$
$$0 = p_y \sin \theta - p_Y \sin \phi \qquad (11.10)$$

Squaring and adding Eqs. (11.9) and (11.10) eliminates the angle ϕ of particle Y.

$$p_x^2 - 2 p_x p_y \cos \theta + p_y^2 = p_Y^2 \qquad (11.11)$$

Although the mass-energy concept is inherently relativistic, we shall limit our discussion to kinetic energies

$$K \ll Mc^2 \qquad (11.12)$$

Under this condition

$$p = (2MK)^{1/2} \qquad (11.13)$$

Making this substitution for each p in Eq. (11.11) and combining with Eq. (11.7) eliminates E_Y. We then obtain

$$Q = K_y \left(1 + \frac{M_y}{M_Y} \right) - K_x \left(1 - \frac{M_x}{M_Y} \right) - \frac{2(M_x M_y K_x K_y)^{1/2}}{M_Y} \cos \theta \qquad (11.14)$$

This expression is known as the Q equation. Through it the energy release in a reaction can be obtained in terms of the directly measurable quantities K_x, K_y, and θ. Although often M_Y may not be known with precision, M_y/M_Y and M_x/M_Y are usually small compared with unity. As a consequence the uncertainty in Q is considerably smaller than the uncertainty in the M's. In fact, it is often possible to get reasonably good values of Q using mass numbers A in place of M.

Example 11.1 When 5-MeV deuterons are used in the $^{14}N(d, p)^{15}N$ reaction, protons emitted at 90° have kinetic energy 12.124 MeV. Calculate the Q of the reaction using exact masses. Compare the result using mass numbers.

Solution. At 90° Eq. (11.14) reduces to

$$Q = K_y\left(1 + \frac{M_y}{M_Y}\right) - K_x\left(1 - \frac{M_x}{M_Y}\right)$$

For the given reaction

$K_y = 12.124$ MeV

$K_x = 5.000$ MeV

$M_x = 2.014102$ u (^2H)

$M_y = 1.007825$ u (^1H)

$M_Y = 15.000108$ u (^{15}N)

$$Q = 12.124\left(1 + \frac{1.007825}{15.000108}\right) - 5.00\left(1 - \frac{2.014102}{15.000108}\right) = 8.610 \text{ MeV}$$

Using mass numbers

$$Q = 12.124\left(1 + \frac{1}{15}\right) - 5.00\left(1 - \frac{2}{15}\right) = 8.599 \text{ MeV}$$

$\Delta Q = 0.011$ MeV

a difference of about 0.1%.   ●

A question arises regarding interpretation of the masses to be used in this treatment. Should the M's be those of **bare nuclei** or those of **neutral atoms**? Equation (11.5) certainly implies that M_x should be the mass of the (usually) bare nucleus or nucleon employed as the projectile. Similarly, y is usually emitted from the reaction in a state bare of electrons. It follows that the M's are those of bare nuclei. However, in the important Eq. (11.8), relating Q to change in mass of the system before and after the reaction, we see that if we add $(Z_x + Z_X)$

electron masses to the reactants before the reaction and $(Z_y + Z_Y)$ electron masses after the reaction, the value of Q will not be changed. The equality of $(Z_x + Z_X)$ and $(Z_y + Z_Y)$ is a consequence of charge conservation. Hence in Eq. (11.8) neutral atomic masses may be used.

It may be objected that the neutral atomic mass differs from the sum of the nuclear mass and that of Z electrons by the binding energy of the electrons in the atom. This is true, of course, but the ratio of electron binding energy to the atomic rest energy is approximately 2×10^{-6} for atoms at the heavy end of the periodic table and ranges down to about 1.5×10^{-8} for hydrogen. Furthermore, it is only **differences** in electron binding energy before and after the reaction that are important and these differences are even smaller. Use of neutral atomic masses is entirely adequate and is the customary approach.

Example 11.2 Calculate the Q value of the $^{16}O(d, p)^{17}O$ reaction on the basis of mass-energy.

Solution. We use Eq. (11.8)

Initial masses	*Final masses*
^{2}H : 2.014102 u	^{1}H : 1.007825 u
^{16}O : 15.994915	^{17}O : 16.999133
18.009017	18.006958

The initial mass is greater than the final mass by 2.059×10^{-3} u. Therefore

$$Q = \Delta M c^2 = 2.059 \times 10^{-3} \text{ u} \times 931.480 \text{ MeV/u} = 1.918 \text{ MeV}$$

The reaction is exoergic.   ●

Although **beta decay** will be discussed at greater length in Chapter 14, it is appropriate to consider at this time the relationships between masses and energy releases. Equation (11.1) becomes

$$_Z^A X \rightarrow {}_{Z-1}^A Y + e^- + \bar{\nu} \tag{11.15}$$

or

$$_Z^A X \rightarrow {}_{Z-1}^A Y + e^+ + \nu \tag{11.16}$$

for negative and positive (negatron and positron) beta decay, respectively. The antineutrino and the neutrino are denoted by $\bar{\nu}$ and ν, respectively. In terms of nuclear masses,

$$_Z M = {}_{Z+1} M + m_0 + \frac{Q}{c^2} \qquad (\beta^- \text{ decay}) \tag{11.17}$$

$$_zM = {}_{z-1}M + m_0 + \frac{Q}{c^2} \qquad (\beta^+ \text{ decay}) \tag{11.18}$$

where m_0 is the electron rest mass.

If Z electrons are added to each side of Eqs. (11.17) and (11.18) and are bound to their respective nuclei, we obtain the atoms and ions of the before and after conditions. The corresponding equations in terms of atomic masses become

$$_zM = {}_{z+1}M^+ + m_0 + \frac{Q}{c^2} \tag{11.19}$$

$$_zM = {}_{z-1}M^- + m_0 + \frac{Q}{c^2} \tag{11.20}$$

where the plus and minus superscripts indicate positive and negative ions. Although the total electron binding energies may be appreciable, their differences between the two sides of the equations are negligible. Now

$$M^+ + m_0 = M + \frac{BE}{c^2} \tag{11.21}$$

and

$$M^- = M + m_0 - \frac{BE}{c^2} \tag{11.22}$$

where BE is the binding energy of the outermost electron. This is of the order of a few electron volts and can ordinarily be neglected. We thus obtain

$$Q = ({}_zM - {}_{z+1}M)c^2 \qquad (\beta^- \text{ decay}) \tag{11.23}$$
$$Q = ({}_zM - {}_{z-1}M)c^2 - 2m_0c^2 \qquad (\beta^+ \text{ decay}) \tag{11.24}$$

Example 11.3 Calculate the transition energy (Q value) in the beta decay of ^{12}B.

Solution. The residual nucleus must be determined. Since Z increases by one unit, the residual nucleus is $^{12}_6$C.

    Initial mass   *Final mass*
^{12}B : 12.014354 u ^{12}C : 12.000000 u
$\Delta M = 0.014354$ u
$Q = \Delta Mc^2 = 14.354 \times 10^{-3}$ u $\times 931.480$ MeV/u $= 13.370$ MeV   ●

In a related process a nucleus may capture an electron from the K or L shell of the atom. In so doing the nuclear charge decreases by one unit and a neutral daughter atom is produced. This atom, however, is in an excited state corresponding to the vacancy in the K or L shell. This energy is released in the form of x rays characteristic of the daughter. The equations become

$$_zX \rightarrow {}_{z-1}Y + \nu + \gamma \tag{11.25}$$

$$_zM = {}_{z-1}M + \frac{BE + Q}{c^2}$$

$$Q = ({}_zM - {}_{z-1}M)c^2 - BE \qquad \text{(electron capture)} \tag{11.26}$$

where γ represents the x-ray photons and BE is the K or L shell binding energy. The nuclear energy release, Q, goes entirely into the energy of the neutrino and the recoil of the residual atom. Since BE may be 100 keV for heavy atoms, it should not be neglected.

Beta decay is energetically allowed whenever Q is positive. For **negatron decay** this occurs when the rest energy of an atom exceeds that of the $(Z + 1)$ neighboring isobar. **Positron decay** involves a parent atom and its $(Z - 1)$ isobar and requires that the rest energy of the parent exceed that of the daughter by $2m_0c^2$, or 1.022 MeV. **Electron capture** (EC) also involves a parent and its $(Z - 1)$ isobar. In this case the difference in mass-energy between parent and daughter need be only 10 to 100 keV. Where positron emission is energetically allowed, electron capture will be also and the two processes may both occur in competition.

We have seen that the energy released in a bombardment reaction can be determined via Eq. (11.4). Equation (11.8) can be rearranged to yield

$$M_Y = M_X + M_x - M_y - \frac{Q}{c^2} \tag{11.27}$$

The masses M_Y of a large number of nuclides have been determined using Eq. (11.27) and carefully measured Q values from Eq. (11.14). This approach is particularly effective when the masses on the right side of Eq. (11.27) are known accurately. For example, reaction 17 in Table 11.1 is $^{12}C(d, p)^{13}C$ and its Q value is 2.723 MeV. Since 1 u = 931.480 MeV the Q value of this reaction is equivalent to (2.723 MeV)/(931.480 MeV u^{-1}) = 0.002923 u. The other masses are

$^{12}C = 12.000000$ u
$^{2}H = 2.014102$ u
$^{1}H = 1.007825$ u

and lead to $^{13}C = 13.003354$ u. It is apparent this technique can be repeated at length relating the masses of many nuclides to ^{12}C and to one another. Masses that have been determined both by Q value measurements and by mass spectroscopy are in good agreement with each other.

Under some circumstances it may be desired to calculate the energy K_y of the emergent particle in terms of the other quantities in Eq. (11.14). This can be done by treating that expression as a quadratic in $K_y^{1/2}$. Its solution leads to

$$K_y^{1/2} = \frac{(M_x M_y K_x)^{1/2} \cos\theta \pm \{M_x M_y K_x \cos^2\theta + (M_y + M_y)[M_Y Q + (M_Y - M_x)K_x]\}^{1/2}}{M_Y + M_y}$$

(11.28)

Physically meaningful solutions require $K_x^{1/2}$ to be positive and real. For exoergic reactions (Q positive) with $M_Y > M_x$ this condition is met for all values of K_x and θ, but only if the plus sign is selected before the second square root term. K_y is single valued.

For endoergic reactions (Q negative) the reaction cannot go at all unless $K_x \geq K_{th}$, the threshold energy which is given by

$$K_{th} = \frac{-Q(M_Y + M_y)}{M_Y + M_y - M_x}$$

(11.29)

K_y is double valued for incident energies K_x such that

$$K_{th} < K_x < K_0$$

(11.30)

where

$$K_0 = \frac{-Q\,M_Y}{M_Y - M_x}$$

(11.31)

Figure 11.2 shows the energetics for the endoergic reaction $^3\mathrm{H}(p, n)^3\mathrm{He}$. For proton energies below K_{th} no neutrons are emitted. For proton energies above K_{th} and below K_0 neutrons are emitted with two energies in a given direction. At threshold neutrons can be emitted only in the forward direction, $\theta = 0°$. As K_p increases, neutrons are emitted in a forward cone whose angle widens with increasing K_p. When K_p reaches K_0, the cone has opened to 90°. With K_p greater than K_0 all angles of emission are possible and the neutron energy becomes single valued.

Example 11.4 The $^3\mathrm{H}(p, n)^3\mathrm{He}$ reaction is a useful source of neutrons. It is endoergic with a Q value of 0.764 MeV. Calculate the threshold energy. Calculate also the neutron energy at 30° when a proton energy of 1.120 MeV is used.

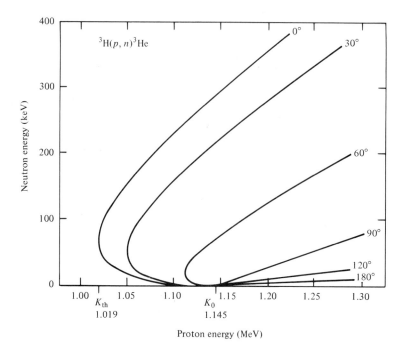

Figure 11.2 Neutron energy versus energy of the bombarding protons in the vicinity of threshold for the ^{3}H(p, n)^{3}He reaction. $Q = -0.764$.

Solution. Equations (11.29) and (11.28) apply. Let us use mass numbers instead of exact masses to simplify the arithmetic. This simplification affects the answer by about 0.1 percent. In Eq. (11.29)

$$K_{th} = -(-0.764)\frac{3 + 1}{3 + 1 - 1} = 1.019 \text{ MeV}$$

In Eq. (11.28)

$$K_n^{1/2} = \frac{(1 \times 1 \times 1.12)^{1/2} \cos 30°}{3 + 1}$$

$$\frac{\pm\{1 \times 1 \times 1.12 \cos^2 30° + (1 + 3)\,[3(-0.764) + (3 - 1)1.12]\}^{1/2}}{3 + 1}$$

$$= \frac{0.9165 \pm 0.7950}{4} = 0.4279 \text{ MeV}^{1/2} \quad \text{or} \quad 0.0304 \text{ MeV}^{1/2}$$

$$K_n = 0.183 \text{ MeV} \quad \text{or} \quad 0.00092 \text{ MeV}$$

The neutron energy is double valued. ●

The energetics of elastic scattering can be represented by Eq. (11.28) upon setting $M_y = M_x$ and $Q = 0$. It follows, of course, that

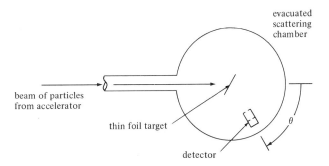

Figure 11.3 Experimental arrangement for a typical Q-value measurement.

$M_Y = M_X$. Inelastic scattering is represented by $M_y = M_x$ and $Q < 0$.

Figure 11.3 is a schematic diagram of the experimental arrangement for a typical Q-value measurement. Energies of bombarding and emitted particles must be measured with precision. The angle θ must be well defined. To accomplish these ends the bombarding particles are energy analyzed by deflection in electric or magnetic fields after acceleration. They are then carefully **collimated** and **focused** onto the target. The target must be very thin so that the bombarding and emitted particles do not lose significant energy by collisions extraneous to the reaction. The detector must be well collimated to define θ and must measure E_y. Although Figure 11.3 suggests a fairly simple detector, and there are semiconductor detectors that serve this purpose very well, the achievement of the highest precision in energy measurement usually requires the use of an electric or magnetic spectrograph at the site of the detector.

11.5 Nuclear Energy Levels

As it turns out, when a target is subjected to a bombardment experiment of the type just described, it is common for many different reactions to occur and great care is needed to sort them out. Not only do reactions occur leading to different end products, as for instance protons, neutrons, and alphas resulting from deuteron bombardment, but in most cases there is a variety of energies observed for the same species of emitted particles. As an example, let us consider the bombardment of a thin carbon target with deuterons of 5.00 MeV energy. We shall confine our attention just to the protons that are emitted at $\theta = 90°$. Table 11.2 shows fifteen different proton energies which appear.

Table 11.2 Proton Energies in (d, p) Reactions on a "Carbon" Target in MeV

9.55	3.58	3.03
6.45	3.41	3.01
5.98	3.28	2.86
5.15	3.12	2.33
3.87	3.08	2.70

Since all these proton energy groups are derived from the (d, p) reaction, we have two choices in an attempt to explain them.

1. The target is contaminated with many nuclear species other than those intended.

2. The (d, p) reaction on a given nuclide can give rise to more than one Q value.

The first explanation is tested by varying the target composition, searching for correlations between changes in yield of the different groups with changes in composition. Changes in chemical composition identify the groups at 5.98, 5.15, 3.08, and 2.33 MeV with oxygen in the target. They arise from the O(d, p) reaction; the remainder are due to carbon. Carbon has two stable isotopes, ^{12}C and ^{13}C. They make up 98.9% and 1.1%, respectively, of natural carbon. By the use of targets enriched in ^{13}C, that isotope is determined to be the source of protons at 9.55, 3.87, 3.41, 3.28, 3.12, 3.01, and 2.70 MeV.

The conclusion reached is that the protons at 6.45, 3.58, 3.03, and 2.86 MeV all arise from the $^{12}C(d, p)$ reaction. Both explanations apply. The Q values that result are displayed in Table 11.3. Since in each instance ^{13}C is formed, we conclude it is formed in a variety of energy states. The reaction with the highest Q value leaves the residual nucleus in its lowest, or most stable, energy state. We call this the ground state. Lower Q values (in the example they become increasingly negative) leave the "missing" energy of the reaction stored as **excitation energy** of the residual nucleus. The degree of excitation of a

Table 11.3 $^{12}C(d, p)^{13}C$ Reaction at 90° and $E_d = 5$ MeV

E_v (MeV)	Q (MeV)	E_{ex} (MeV)
6.45	2.72	0.0
3.58	−0.37	3.09
3.03	−0.96	3.68
2.86	−1.14	3.86

state is simply the difference between the Q value of the reaction leading to that state and the Q value of the ground state reaction. The third column of Table 11.3 shows the ground state and first three excited states of ^{13}C. An asterisk is commonly used to indicate an excited state. Thus $^{12}C(d, p)^{13}C^*$ denotes a reaction leading to an excited state in ^{13}C.

The reaction just discussed was not the "discovery" reaction for observation of excited energy levels in nuclei. That honor belongs to the (α, p) reaction on boron. Studies of alpha-induced reactions by Bothe and Franz from 1928 to 1930 led to the observation of a variety of proton energy groups. In particular, the $^{10}B(\alpha, p)^{13}C$ reaction leads to many of the same states in ^{13}C as does the $^{12}C(d, p)^{13}C$ reaction. It is clear, of course, that finding the same values for energy levels by two independent reactions enhances confidence in the correctness of the assignment of the levels.

With the development of methods for producing excited energy levels in nuclei and for measuring their properties an entire field of nuclear spectroscopy has come into being. Much of the terminology and viewpoint of atomic spectroscopy has been carried over to this new spectroscopy. Discrete, quantized energies exist; transitions from an excited level to a lower one in the nucleus are accompanied by emission of photons, which are called gamma rays; each state has angular momentum that is the resultant of the combined spin and orbital angular momenta of the constituent nucleons; selection rules based on the conservation laws govern the probability of the transitions and reactions. There are important differences, though. The **energy changes** are of the order of MeV's rather than electron volts; the range of **lifetimes** is much larger—from 10^{-15} sec to 10^9 years; levels that are discrete exist at excitation energies high enough to be **unbound** against particle emission; there are **many more mechanisms** for production of excited states; prediction of energies and other properties of nuclear states by quantum mechanics has had some but by no means complete success. More will be said about nuclear spectroscopy in Chapter 12.

Figure 11.4 shows an energy level diagram for ^{13}C. The heights of the levels are given in MeV above the ground state. Along side the ^{13}C levels are shown various reactions that lead to ^{13}C and thereby give information about its excitation and structure. The Q value of the transition to the ground state of ^{13}C is shown for each reaction in MeV. The vertical arrow drawn from the $^{12}C(d, p)$ reaction represents 5 MeV deuteron kinetic energy used in our example of this reaction. This converts to 4.282 MeV kinetic energy of the system, $^{12}C + d$, in **center-of-mass** coordinates.

$$K_{CM} = K_x \frac{M_X}{M_x + M_X} \tag{11.32}$$

K_{CM} is the energy available for excitation; the difference between K_{CM} and K_x is the translational energy of the system as a whole in the

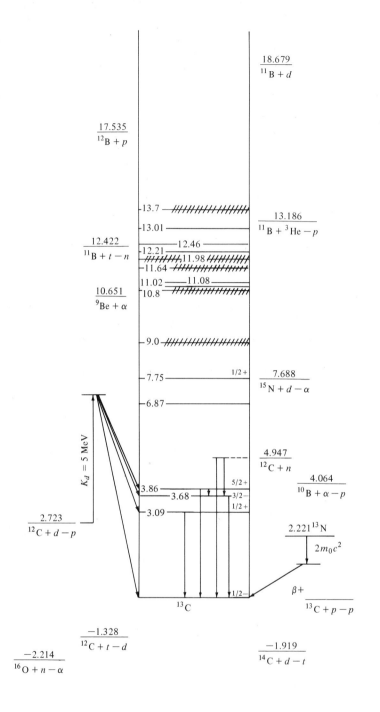

Figure 11.4 Energy levels of ^{13}C. Energies are given in MeV and are measured above the ground state of ^{13}C. Other mass-13 systems are shown beside the ^{13}C level diagram. The differences in mass-energy between each system and that of the ^{13}C ground state are shown in MeV.

laboratory coordinates and is not available for excitation. The diagonal arrows downward represent emission of protons, leaving ^{13}C in the four states listed in Table 11.2. Similar vertical and diagonal arrows for the other reactions involving ^{13}C have been omitted in this figure. Reactions, such as ^{11}B $+ d$, ^{12}C $+ n$, and so on, that do not show an emitted particle are radiative capture reactions.

11.6   The Cross-section Concept

In treating the **yield** or relative probability of a reaction a convention has been adopted that speaks of the **cross section** for the reaction as a measure of this probability. In this view, each nucleus is represented, for the purposes of the reaction, by a disk of area σ normal to the direction of incident particles. If the incident particle strikes the disk, the reaction occurs; if it misses there is no reaction.

Suppose there is a sample that presents a thickness t and area A to a stream of monoenergic incident particles arriving at a rate of N_0 particles per unit time. If there are n nuclei per unit volume, then the fraction of the frontal area shadowed by the hypothetical disks is $nAt\sigma/A$. By the definition of σ, this fraction is the same as the ratio of the number of reactions to the number of incident particles. Let the yield in number of reactions per unit time be N. Then

$$N = N_0 nt\sigma \tag{11.33}$$

The derivation of Eq. (11.33) assumes that the number of incident particles is substantially the same at all depths in the sample. This condition is met when σ is sufficiently small that $N \ll N_0$. Equation (11.33) constitutes the definition of **cross section, σ**.

Frequently the beam of incident particles may produce several types of reactions with a given target nuclide. For instance, with deuterons we may find (d, p), (d, n), (d, α), (d, d), and (d, d') reactions. By measuring the rate of each reaction separately, we define a **partial cross section** for each. The sum of all the partial cross sections equals the **total cross section**. In principle, observing the loss of particles from the incident beam measures directly the total rate of all reactions occurring in the target sample. In practice, this is not practical if the bombarding particles are charged. A target thick enough to produce measurable diminution by nuclear reactions causes overwhelming loss by nonnuclear events, such as ionization. If the incident beam consists of neutrons, however, ionization losses are absent, and total cross sections can be measured in this fashion. In fact, although we did not at that time identify it as such, the experiment on fast neutron scattering discussed in Section 10.8 is a total cross section measurement; the term $2\pi(R + \lambda)^2$ that appears in Eq. (10.39) is the cross section.

A partial cross section is measured by the rate of an individual reaction. This rate can be measured by detecting emitted protons, neutrons, and so on, or by observing the production of the residual nuclide. If the residual nuclide is radioactive, the number of atoms produced can be calculated from the induced radioactivity and is, therefore, usually easy to determine. Cross sections determined in this way are called **activation cross sections** and measure reaction rates without regard to angle between the directions of the incident and emitted particles.

A partial cross section may be a function of the angle of emission of the reaction products. To determine this, we count only that portion $dN(\theta)$ of the particles emitted into the solid angle $d\Omega$ at the angle θ, measured with respect to the direction of the incident particles. We thereby define a **differential cross section** $d\sigma/d\Omega$ by the relation

$$dN(\theta) = N_0 nt \left(\frac{d\sigma}{d\Omega}\right) d\Omega \tag{11.34}$$

Measurements in a scattering chamber, such as that depicted in Figure 11.3, ordinarily determine $d\sigma/d\Omega$. If the total cross section is desired, measurements must be made at enough angles to allow integration over the full solid angle, 4π steradians. Note that in Eq. (10.22) the term

$$\frac{1}{16} \left(\frac{zZe^2}{\frac{1}{2}mv_0^2}\right)^2 \sin^{-4} \frac{\theta}{2}$$

is the differential Rutherford cross section.

For purposes of calculation σ must be expressed in reciprocal units of those used for nt, that is, in square centimeters or square meters. However, to avoid extreme negative powers of ten, a new unit of area, the **barn**, is defined such that

$$1 \text{ barn} = 10^{-24} \text{ cm}^2 \tag{11.35}$$

Total cross section is expressed in **barns**, and differential cross section in **barns per steradian** (Eq. 11.34).

Ranging as they do from microbarns, and less, to over 10^5 barns, cross sections should not be taken as representing the physical sizes of nuclei. The range is much too great. Furthermore, in a given reaction there may be large and rapid changes in cross section as a function of bombarding energy, arguing strongly against interpreting σ as giving the actual size of the target nucleus. Cross section is simply a convenient terminology for indicating reaction probability.

Example 11.5 A layer of NaBr is bombarded with a 2-μamp beam of 10-MeV deuterons. After 10 min of bombardment the activity of ^{24}Na produced by the ^{23}Na(d, p)^{24}Na reaction is found to be 5.08×10^5 disintegrations/sec when summed in a counting geometry of 4π steradians. The NaBr is 3.0 mg/cm^2 thick. What is the cross section for the reaction?

Solution. Equation (11.33) applies. N is the number of ^{24}Na atoms produced that we must find from the radioactivity. To do this we begin by looking up the half-life of ^{24}Na; it is 15 hr $= 5.4 \times 10^4$ sec. From this the decay constant

$$\lambda = 0.693/T_{1/2} = 1.283 \times 10^{-5} \text{ sec}^{-1} \qquad \text{(see Section 13.2)}$$

Therefore

$$N = \frac{\text{activity}}{\lambda} = \frac{5.08 \times 10^5}{1.283 \times 10^{-5}} = 3.96 \times 10^{10} \text{ atoms of } {}^{24}\text{Na}$$

N_0 is the number of deuterons involved. We find this from the beam current and duration of bombardment.

$$N_0 = \frac{2 \times 10^{-6} \text{ amp} \times 600 \text{ sec}}{1.60 \times 10^{-19} \text{ coulomb/deuteron}} = 7.5 \times 10^{15} \text{ deuterons}$$

The molecular weight of NaBr $= 102.9$ g/mole, so

$$nt = \frac{3 \times 10^{-3} \text{ g/cm}^2}{102.9 \text{ g/mole}} \times 6.023 \times 10^{23} \text{ molecules/mole}$$

$$= 1.76 \times 10^{19} \text{ molecules/cm}^2$$

But since Na is monoisotopic, there is one atom of ^{23}Na in each molecule. Rearranging Eq. (11.33), we find

$$\sigma = \frac{N}{N_0 nt} = \frac{3.96 \times 10^{10}}{7.5 \times 10^{15} \times 1.76 \times 10^{19}} = 0.3 \times 10^{-24} \text{ cm}^2$$

$$= 0.3 \text{ barn} \qquad \bullet$$

The reactions ^{23}Na(d, n)^{24}Mg and ^{23}Na(d, α)^{21}Ne occur simultaneously with the (d, p) reaction. Therefore it would be proper to say that the calculation presented in this example applies to the "partial cross section of the ^{23}Na(d, p)^{24}Na reaction."

11.7 Resonances and the Compound Nucleus

Continuing with our example of the $^{12}C + d$ reaction, we show in Figure 11.5 the energy dependence of the cross section for the reaction that emits protons, $^{12}C(d, p)^{13}C$, and for the reaction that emits neutrons, $^{12}C(d, n)^{13}N$. Superimposed on a general upward trend are several sharp peaks. Although the proton and neutron curves do not altogether duplicate each other, it is apparent that many of the peaks in one correspond in energy to those in the other. Response curves of this sort suggest the phenomenon of **resonance,** a particularly strong yield being obtained when the energy provided to a system matches a quantized **characteristic energy** of the system. The existence of such characteristic energies in the $^{12}C + d$ reactions indicates that at least temporarily the target nucleus and the projectile have joined together into a single quantum mechanical system. Inspection of the available nuclear charge and mass identifies this system as $^{14}_{7}N$.

This joining together is spoken of as **compound nucleus** formation. Reactions involving a compound nucleus are viewed as proceeding through a two-stage process. In the first stage the projectile and the target nucleus interact to form the compound nucleus. This process results in the compound nucleus being formed in a highly excited state. The energy which is available for excitation consists of both the center-

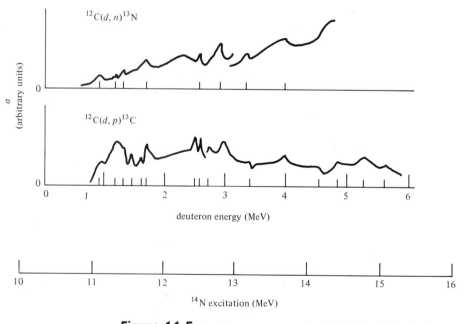

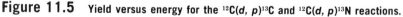

Figure 11.5 Yield versus energy for the $^{12}C(d, p)^{13}C$ and $^{12}C(d, p)^{13}N$ reactions.

of-mass kinetic energy of bombardment, Eq. (11.32), and the mass-energy difference, in the current example, between

$$^{12}_{6}C + {}^{2}_{1}H \qquad \text{and} \qquad {}^{14}_{7}N$$

As the projectile penetrates the target nucleus, its energy is rapidly transferred to and shared among the other nucleons of the new system. This rapid sharing is a consequence of the strong interaction among nucleons and results in the loss of identity of the projectile. There are two energy scales in Figure 11.5. The one at the top shows deuteron kinetic energy in laboratory coordinates; the bottom scale is excitation energy of ^{14}N obtained by adding the center-of-mass kinetic energy to 10.26 MeV, the energy equivalence of

$$^{12}_{6}C + {}^{2}_{1}H - {}^{14}_{7}N$$

Once formed, the compound system usually has several different decay modes, called **exit channels,** open to it by which it can transform to a lower energy state. This transformation is the second stage of the reaction. In the compound nucleus configuration, however, decay does not take place immediately, that is, within the approximately 10^{-21} sec required for a single nucleon transit of the nucleus. Instead, hundreds to thousands of transit times may be required before the excitation energy, shared as it has been throughout the nucleus, sufficiently concentrates in a single nucleon or small grouping of nucleons to permit escape. A compound nucleus may have a lifetime of 10^{-16} to 10^{-19} sec. The hundreds to thousands of nucleon transits before decay result in the system no longer "remembering" how it was formed. Thus **the way in which a particular state in a compound nucleus decays is expected to be independent of the reaction by which that state was formed.**

Figure 11.6 shows an energy level scheme for ^{14}N, indicating numerous reactions which produce these levels. This diagram is entirely comparable to Figure 11.4, which shows levels for ^{13}C. More details about the reactions are shown, however, in Figure 11.6. The curves of Figure 11.5 are repeated in the upper left corner of Figure 11.6, rotated 90° so the energy axis for the curves parallels the energy axis for the main level scheme. The numerical values along the energy axis for the inset are bombarding energies in laboratory coordinates (deuteron kinetic energies in the case of the $^{12}C + d$ reaction) for prominent resonances, but the scale is adjusted to center-of-mass system to align the resonances with corresponding levels in ^{14}N. Also shown are resonance curves for $^{13}C + p$ and $^{10}B + \alpha$, indicating that the ^{14}N compound nucleus can be formed by these reactions as well as the one we have been treating. For each curve the outgoing particle is indicated. With $^{10}B + \alpha$ the curve labeled p_0 is for the reaction leading

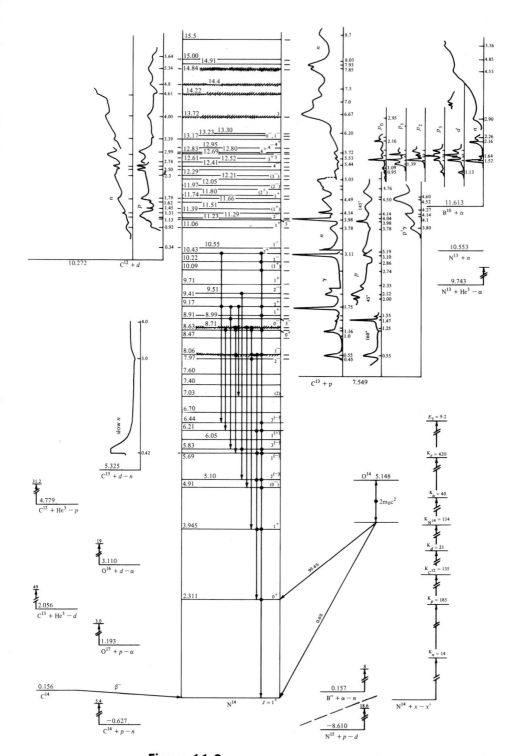

Figure 11.6 Energy levels of ¹⁴N. Energies are given in MeV and are measured above the ground state of ¹³N. Additional mass-14 systems are shown. The differences in mass-energy between each system and that of the ¹⁴N ground state are shown in MeV. [Adapted from T. Lauritson and Fay Ajzenberg-Selove, *Energy levels of light nuclei. Nuclear Data Sheets,* National Academy of Science-National Research Council, Washington, D.C. (1962).]

to ^{13}C in its ground state, p_1 is for that leading to the first excited state in ^{13}C, p_2 to the second excited state, and so on (see also Figure 11.4).

As we have noted, a state in a compound nucleus can decay through several exit channels. Reference to Figure 11.6 shows that in ^{14}N those states whose excitation energy is in excess of 11.61 MeV have sufficient energy to decay by emission of neutrons, protons, deuterons, alphas, and gamma-ray photons.

Entrance Channel	Compound Nucleus	Exit Channel	Reaction	
^{12}C + d $\longrightarrow$	^{14}N	^{13}N + n	^{12}C(d, n)^{13}N	
		^{13}C + p	^{12}C(d, p)^{13}C	
		^{12}C + d	^{12}C(d, d)^{12}C	elastic scattering
		^{12}C* + d'	^{12}C(d, d')^{12}C*	inelastic scattering
		^{10}B + α	^{12}C(d, α)^{10}B	
		^{14}N + γ	^{12}C(d, γ)^{14}N	radiative capture

11.8 Summary

Nuclear reactions are symbolized by the reaction equation

$$X + x \rightarrow y + Y \qquad \text{or} \qquad X(x, y)Y$$

The equation must balance with regard to charge and mass number, but not exact mass.

The difference in the total of the exact (neutral atom) mass before and after the reaction gives the Q of the reaction

$$Q = (M_X + M_x)c^2 - (M_y + M_Y)c^2$$

Also

$$Q = K_Y + K_y - K_x$$

Eliminating K_Y leads to the "Q equation"

$$Q = K_y\left(1 + \frac{M_y}{M_Y}\right) - K_x\left(1 - \frac{M_x}{M_Y}\right) - \frac{2(M_xM_yK_xK_y)^{1/2}}{M_Y}\cos\theta$$

and

$$K_y^{1/2} = \frac{(M_x M_y K_x)^{1/2} \cos\theta \pm \{M_x M_y K_x \cos^2\theta + (M_y + M_Y)[M_Y Q + (M_Y - M_x)K_x]\}^{1/2}}{M_Y + M_y}$$

In endoergic reactions there is a threshold energy

$$K_{th} = \frac{-Q(M_Y + M_y)}{M_Y + M_y - M_x}$$

Also K_y is double valued if

$$K_{th} < K_x < \frac{-QM_Y}{M_Y - M_x}$$

In beta decay

$$Q = ({}_Z^A M - {}_{Z+1}^A M)c^2 \qquad \text{(for } \beta^- \text{ decay)}$$
$$Q = ({}_Z^A M - {}_{Z-1}^A M)c^2 - 2m_0 c^2 \qquad \text{(for } \beta^+ \text{ decay)}$$
$$Q = ({}_Z^A M - {}_{Z-1}^A M)c^2 - BE \qquad \text{(for electron capture)}$$

In a reaction of cross section σ

$$N = N_0 nt\sigma$$

σ is tabulated in barns, where 1 barn $= 10^{-24}$ cm^2

Resonances are indicative of formation of discrete levels in the compound nucleus.

In nuclear reactions the following quantities are conserved:

1. Nucleon number
2. Total charge
3. Mass-energy
4. Linear momentum
5. Angular momentum
6. Statistics
7. Parity
8. Isospin

Problems

11.1 Complete the following reaction equations by filling in the missing symbols where the question marks appear. ${}^7\text{Li}(d, p)$?, ${}^{27}\text{Al}(d, p)$?, ${}^{12}\text{C}(d, ?){}^{10}\text{B}$, ${}^7\text{Li}(p, ?){}^8\text{Be}$, ${}^3\text{H}(p, n)$?, ${}^9\text{Be}(?, n){}^{11}\text{B}$, ${}^{24}\text{Mg}(p, \gamma)$?, $?(d, n){}^{23}\text{Na}$.

11.2 From the exact masses of the nuclides involved, calculate the Q values for the ${}^{10}\text{B}(n, \alpha){}^7\text{Li}$ and ${}^{28}\text{Si}(d, p){}^{29}\text{Si}$ reactions.

11.3 Show that ${}^8\text{Be}$ is unstable against decay into two alpha particles. What is the kinetic energy of each of these alphas?

11.4 What is the energy of the neutrons emitted at 90° in the ^{3}H(d, n)^{4}He reaction when the deuteron energy is 0.150 MeV? What is it at 0°?

11.5 When ^{16}O is bombarded with 5.5-MeV deuterons, several groups of alpha particles are emitted which at 90° have the following energies: 8.217 MeV, 4.288 MeV, 3.017 MeV, 2.267 MeV. What is the residual nucleus? Calculate the four Q values for which there are data. Assuming the most energetic alphas are associated with the ground-state transition, calculate the excitation levels of the residual nucleus associated with the other alpha energies.

11.6 The nuclide ^{64}Cu is radioactive with a half-life of 12.74 hr. Compare its mass with the masses of the neighboring isobars, and thereby predict the possible mode(s) of decay. Calculate the energies of the emitted particles.

11.7 Calculate the threshold energy for the ^{7}Li(p, n)^{7}Be reaction and the energy of the neutrons emitted at threshold.

11.8 When ^{9}Be is bombarded with protons, prominent gamma-ray resonances are observed at proton energies of 0.998 and 1.087 MeV. What is the compound nucleus? What energy levels are excited by these resonances?

11.9 How many atoms of ^{63}Zn are produced by the ^{63}Cu(p, n)^{63}Zn reaction at the end of a 2-min bombardment of a copper foil of natural isotopic composition 0.001-in. thick with a 5-μamp beam of 13-MeV protons? The cross section for the reaction is 0.52 barns.

11.10 $^{239}_{94}$Pu (2.4 × 10^4 yr half-life) emits alpha particles of 5.15 MeV kinetic energy. Calculate the maximum neutron energy produced when these alpha particles bombard $^9_4$Be.

11.11 $^{214}_{82}$Pb (26.8 min half-life) is a source of 3.52 MeV gamma rays. What are the neutron kinetic energies produced when these gamma rays bombard $^2_1$H and $^9_4$Be?

11.12 Perform the algebraic manipulations required to arrive at Eq. (11.14).

11.13 Calculate for the ^{11}B(d, p)^{12}B reaction with 150 keV deuterons the kinetic energy of protons emitted at 45°, at 90°, and at 135°.

11.14 Solve for $K_y^{1/2}$ in Eq. (11.14) to obtain Eq. (11.28).

11.15 Verify that when $Q < 0$ the conditions stated in Eqs. (11.30) and (11.31) are necessary and sufficient for K_y to be double valued.

11.16 Show that in elastic scattering ($Q = 0$) of a projectile by a target nucleus of equal mass (for instance, neutrons scattered by protons) neither particle can be projected into the rear hemisphere (lab coordinates).

11.17 For the ^{7}Li(p, n)^{7}Be reaction determine the minimum proton energy for emission of neutrons at 90°. What is the neutron energy at 90° when the proton kinetic energy is 3.0 MeV?

11.18 In Figure (11.5) there are two energy scales. Make a two-point verification of their comparison by calculating the energy available for excitation of ^{14}N when deuterons of zero kinetic energy (ignore Coulomb barrier difficulties) and of 6.0 MeV kinetic energy are incident upon ^{12}C.

11.19 A nickel absorber 5 cm thick is placed in a beam of 1.0 MeV neutrons. The total cross section is 3.5 barns. What is the ratio of neutron count rate with the absorber in place to that with it removed?

Suggestions for Further Reading

Evans, R. D. *The Atomic Nucleus*. New York: McGraw-Hill Book Company, Inc., 1955. Chapters 4, 13, and 14.

Green, A. E. S. *Nuclear Physics*. New York: McGraw-Hill Book Company, Inc., 1955. Chapter 7.

Eisberg, R., and R. Resnick. *Quantum Physics*. New York: John Wiley & Sons, 1974. Sections 16.7 and 16.8.

Nuclear models

12.1 Introduction

It is natural to ask ourselves what the **internal structure** of the nucleus might be like. Scattering experiments such as those described in Chapter 10 show that the nucleus is not a simple point object but has distributed charge and mass. We envision the nucleons, as carriers of charge and mass, joined together in some sort of dynamic structure. A fundamental approach to the problem of nuclear structure might follow along lines that have successfully unraveled atomic structure. We would thus apply a knowledge of the forces between particles to obtain a **potential energy function** for an assumed **configuration** of particles. We would apply the Schrödinger equation to this function and seek out the **allowed states** and **probability distributions.** We would compare these distributions with our assumed configuration, readjust the latter, and repeat the process until a **self-consistent** set of configurations and distributions were obtained. The resulting quantum mechanical specification of states would constitute our description of the nucleus.

Although a certain degree of success has been achieved along these lines the problem is much more difficult than it is for the atom. There are two reasons for this. The specifically nuclear force itself is less well understood than is the Coulomb force. It is the former that is of primary importance to the binding of the nucleus, the latter to the binding of the atom. Furthermore, in the atom the central-force interaction between electrons and nucleus is dominant; the many-body problem of electron-electron interaction may be treated as a small perturbation. In contrast, the nucleus is inherently a many-body system with no apparent central-force interaction. In order to avoid the intractable many-body problem we invoke a simplifying assumption, namely that each nucleon executes its motion in an average potential field created by all the other nucleons.

In a sense this simplification is an admission of defeat. We forego the complete first-principles quantum mechanical approach and fall back upon descriptive models in our effort to account for observed nuclear properties and behavior. Although very useful, no single model has yet been developed to account for all observations. Some models are more successful in dealing with nuclides from one region of the periodic table than with those from another region; some models are able to deal with certain aspects of nuclear behavior but not with other aspects. We uneasily acknowledge that existing models do not give the full picture of the nucleus, that much work remains to be done. Nevertheless, models do help our understanding and they convey at least partial answers about nuclear structure.

12.2 The Fermi Gas Model

The Fermi gas model is useful for its simplicity and for ideas that can be incorporated into more complex models. A Fermi gas is one made up of **indistinguishable** particles which obey the Pauli **exclusion principle.** Such particles have odd half integer spin and are antisymmetric upon interchange of pairs of particles. Neutrons and protons are such particles and so also are electrons. For the purposes of our model we shall treat the nucleus as an assemblage of two populations of such particles bound within a potential well. We shall, at least temporarily, ignore the Coulomb force between protons and shall assume the *nn, pp,* and *np* forces to be equal. The existence of the well results from the attractive nature of these forces, but we shall not attempt a detailed analysis of the motions of the nucleons. We recognize that discrete energy levels exist within the well but we shall not try to calculate their spacing. Under the assumption of **charge independence** of nuclear force the levels for the proton population will be the same as for the neutron population and a single potential well will serve for both.

Figure 12.1 is a very schematic diagram of such a well. We construct the nucleus by adding one nucleon after another. Each level can hold **four nucleons,** one neutron with spin up and one with spin down, one proton with spin up and one with spin down. As each nucleon is added it is placed in the lowest energy state available to it. Each level is filled before nucleons are added to the next higher one. U_0 is the **depth** of the well; S_n is the **separation energy** required to remove the last nucleon; and E_f is the **Fermi energy,** the energy of the topmost occupied level measured upward from the bottom of the well. In Chapter 9 we treated the conduction electrons in a metal as a Fermi gas confined within a potential well whose spatial extent is determined by the boundaries of the metal. Recalling results developed in that chapter, we write

$$E_f = \frac{\pi^2 \hbar^2}{2 M_n} \left(\frac{3}{\pi} n \right)^{2/3} \tag{12.1}$$

where n is the number of particles per unit volume in the Fermion population and M_n is the mass of each particle. Let us evaluate this

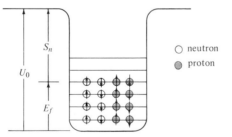

Figure 12.1 Schematic diagram of a square well with independent proton and neutron populations. The operation of the Pauli exclusion principle produces alpha particlelike clustering of four nucleons in each level. Through analogy with a Fermi gas, the uppermost occupied level is called the Fermi level. Its height above the bottom of the well is E_f. S_n is the energy required to remove a nucleon from the shell at the Fermi level.

expression. Since neutrons are distinguishable from protons we treat each population separately. Thus, the neutron density is N/V and the proton density is Z/V, where V is the nuclear volume.

$$V = \tfrac{4}{3}\pi R_0^3 A \qquad (12.2)$$

The assumptions of our simple model dictate that

$$N \approx Z \approx \tfrac{1}{2}A \qquad (12.3)$$

since otherwise nucleons would occupy higher energy states while lower ones remained unfilled. If the excess nucleons are neutrons the system energy can be lowered by converting them to protons via β^- decay and allowing the protons to drop into the unoccupied lower states. If there is a proton excess, then β^+ decay brings the system into a more stable lower energy configuration.

We find then that

$$n = \frac{\tfrac{1}{2}A}{4\pi R_0^3 (A/3)} = \frac{3}{8\pi R_0^3} \qquad (12.4)$$

Setting

$$R_0 = 1.4 \text{ fermi}$$
$$M_n = 1.675 \ 10^{-24} \text{ g}$$
$$\hbar = 1.0545 \ 10^{-27} \text{ erg-sec}$$

leads to

$$E_f \approx 25 \text{ MeV} \qquad (12.5)$$

For other than very light nuclides

$$N \approx 0.6A \qquad (12.6)$$

is a better approximation than the statement in Eq. (12.3). With this we find

$$E_f \approx 28 \text{ MeV} \qquad (12.7)$$

For nuclides in the middle of the periodic table, typically

$$S_n \approx 7 \text{ MeV} \qquad (12.8)$$

and therefore

$$U_0 \approx 35 \text{ MeV} \qquad (12.9)$$

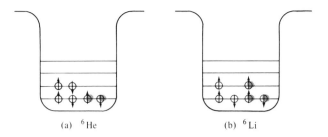

(a) ^{6}He   (b) ^{6}Li

Figure 12.2   Nucleon occupation of levels for $A = 6$ nuclides: (a) ^{6}He and (b) ^{6}Li. ^{6}He is unstable and transforms by beta decay to ^{6}Li. In this transformation one proton in the upper level is converted into a neutron.

In addition to predicting the depth of the potential well for a typical nuclide our model suggests the interesting possibility of nucleons being bound in **alpha particlelike** groups of four. After the filling of a level with four nucleons the next nucleon must go into a higher **level** or **shell,** and as a consequence is less tightly bound. This behavior is indeed characteristic of light nuclides. $^4_2$He is a tightly bound system, but the next nucleon to be added, whether neutron or proton, is so poorly bound that no $A = 5$ nuclide exists in nature. Both $^5_2$He and $^5_3$Li are unstable against nucleon emission even in the ground state. The ^{6}Li$(n, d)^5$He reaction indicates a ground state width in ^{5}He of about 0.8 MeV and the ^{6}Li$(d, t)^5$Li reaction shows the ground state of ^{5}Li to be about 1.5 MeV wide. Through the **uncertainty principle** the lifetimes of these states are found to be about 0.8×10^{-21} sec and 0.4×10^{-21} sec, respectively. These times are shorter than the nucleon

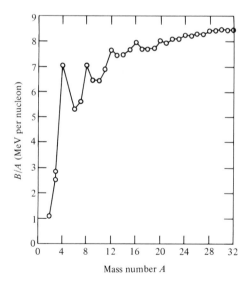

Figure 12.3   Binding energy per nucleon for light stable nuclides. The line connecting plotted points is drawn merely to guide the eye. ^{3}H and ^{8}Be are included, though not strictly stable.

transit time across a nucleus so it is dubious terminology even to speak of these assemblages as nuclides. The addition of a sixth nucleon, as in Figure 12.2, however, enhances binding in the second shell so both of the new nuclides are stable against nucleon emission, although ^{6}He does β^- decay to ^{6}Li.

Further confirmation of the stability of shells of four particles is indicated in the plot of binding energy per nucleon shown in Figure 12.3. This is similar to Figure 10.1 but is confined to the mass region $A \leq 32$ and plotted to a larger scale.

Example 12.1 Calculate the separation energy of the "last" neutron in $^{40}_{20}$Ca and $^{41}_{20}$Ca.

Solution. The separation energy is the difference in rest energy between the original nuclide and the system consisting of a free neutron and the $(A - 1)$ isotope.

Initial mass	*Final mass*
^{40}Ca : 39.962589 u	^{39}Ca : 38.970691 u
	n : 1.008665
	39.979356

$\Delta M = 0.016767$ u $\Rightarrow$ 15.618 MeV

^{41}Ca : 40.962275 u	^{40}Ca : 39.962589 u
	n : 1.008665
	40.971254

$\Delta M = 0.008979$ u $\Rightarrow$ 8.364 MeV   ●

Most nuclides have separation energies of 7–8 MeV. The binding of the last neutron in ^{40}Ca is especially tight. This is one piece of evidence that $N = 20$ is a **magic number,** a property that is discussed in Section 12.4.

12.3   The Liquid Drop Model and Semiempirical Mass Formula

Although there is evidence among lighter nuclides that alpha particlelike clusters of nucleons are especially stable and hence might be of importance in considering nuclear models, it does not appear that this configuration is important to the medium weight and heavier nuclides. Beyond $A = 24$ the curve of binding energy per nucleon no longer shows discernible peaks at multiples of four. We turn to a smoothed-out model that describes general trends of binding energy across the periodic table.

A number of nuclear properties we noted in Chapter 10 are

similar to those found in a **liquid drop:** The mass-density is **independent of size,** the forces binding the constituent particles are **short range** and exhibit **saturation,** and the total binding energy is approximately **proportional to volume.** Using the liquid drop as the basic analogy and adding extra empirical terms, an energy formula is worked out expressing dependence of **binding energy** on nucleon numbers. This approach was first used by C. F. Von Weizsäcker in 1935 and his name is often associated with the formula.

Five main contributions to binding energy are identified. The first and most important term is the **volume energy.** Since the density of nuclear material is constant, volume is proportional to mass number A and we write

$$E_{\text{vol}} = a_v A \tag{12.10}$$

This term assumes that all nucleons are equally surrounded by other nucleons. Because of the saturation of the nuclear force only nucleons that are near neighbors are effective in binding. Nucleons on the surface have fewer near neighbors and consequently the binding energy of the nucleus is reduced by a **surface energy** term proportional to the surface area.

$$E_{\text{surf}} = -a_s A^{2/3} \tag{12.11}$$

The Coulomb force between protons in the nucleus is one of repulsion. The **electrostatic energy** reduces the binding energy. If there are Z protons in a spherical region of radius $R = R_0 A^{1/3}$, a straightforward calculation in electrostatics shows that the stored energy is

$$E_{\text{coul}} = \frac{3}{5} \frac{Z(Z-1)e^2}{R_0 A^{1/3}}$$

The third term in our formula, then, involves **Coulomb energy** and is

$$E_{\text{coul}} = -a_{\text{coul}} \frac{Z(Z-1)}{A^{1/3}} \tag{12.12}$$

We have seen in the preceding section that because of the Pauli exclusion principle nuclides with $Z = N$ are favored. Departure from the condition $Z = A/2$ tends toward lowered stability. The effect is less important for heavier nuclides, going approximately inversely with the mass number. A term expressing this **symmetry effect** becomes the fourth term of the formula.

$$E_{\text{sym}} = -a_{\text{sym}} \frac{(Z - A/2)^2}{A} \tag{12.13}$$

Table 12.1   Abundances of
Natural Nuclides

Z	N	Stable	Long-lived
even	even	155	11
even	odd	53	3
odd	even	50	3
odd	odd	4	5

In studying nuclear abundances as found in nature it becomes apparent that nuclides with Z and N both even are considerably favored and those with Z and N both odd are disfavored. Table 12.1 shows this distribution. We conclude that the binding energy in even-even nuclides is increased over that in even-odd or odd-even nuclides and that for odd-odd it is reduced. A fifth term to express this **pairing energy** is introduced. Empirically a satisfactory form for this term is

$$E_{\text{pair}} = a_{\text{pair}} A^{-1/2} \tag{12.14}$$

where a_{pair} is positive for **even-even** nuclides, zero for **even-odd** and **odd-even** nuclides, negative for **odd-odd** nuclides.

Combining Eqs. (12.10) through (12.14) the expression for binding energy becomes

$$B = a_v A - a_s A^{2/3} - a_{\text{coul}} \frac{Z(Z-1)}{A^{1/3}} - a_{\text{sym}} \frac{(Z - A/2)^2}{A} + a_{\text{pair}} A^{-1/2}$$
$$\tag{12.15}$$

Now binding energy is just the same as mass defect ΔM, which was given in Eq. (10.3). The only distinction is an emphasis on energy units for the one and mass units for the other. If the a's are given in mass units it follows that nuclear masses can be expressed by the following formula:

$$M_{Z,A} = Z M_{\text{H}} + (A - Z) M_n - a_v A + a_s A^{2/3} + a_{\text{coul}} Z(Z-1) A^{-1/3}$$
$$+ a_{\text{sym}} \frac{(Z - A/2)^2}{A} - a_{\text{pair}} A^{-1/2} \tag{12.16}$$

We have now constructed the **semiempirical mass formula.** It is empirical in the sense that the constants, the a's, are evaluated by comparison with measured masses rather than by a first-principles theoretical approach. On the other hand, ideas about a particular nuclear model guide the construction with respect to dependence on atomic number Z and mass number A.

Application of Eq. (12.16) to five nuclides of known mass suffices to evaluate the constants. Repeating with several different sets of

Table 12.2 Representative Values of Parameters in the Semiempirical Mass Formula

	Millimass Units	MeV
M_H	1007.825	938.769
M_n	1008.665	939.551
a_v	16.91	15.75
a_s	19.11	17.80
a_{coul}	0.763	0.711
a_{sym}	101.75	94.778
a_{pair}	±12. *	±11.2 *

* See note accompanying Eq. (12.14) regarding sign of a_{pair}.

nuclides and averaging gives a set of constants that produce an excellent fit to the general trend of masses over a wide range of the periodic table. Different investigators arrive at somewhat different sets of values, depending upon the reference nuclides chosen, but in general the masses fit to within a few percent. Table 12.2 shows a representative set of parameters.

The formula does not work very well for light nuclides. This is not surprising since it is based on analogy with a liquid drop. A nucleus with only a few nucleons cannot be closely approximated by a liquid drop which has a very large number of particles (atoms or molecules) within its volume. As an example, however, let us apply Eq. (12.15) to calculate the binding energy of $^{59}_{27}$Co. This is an odd-even nucleus, so $a_{pair} = 0$. Substituting the other parameters in Eq. (12.15) gives

$$\Delta M = 16.91 \times 59 - 19.11 \times 59^{2/3} - 0.763 \times \frac{27 \times 26}{59^{1/3}} - 101.75$$

$$\times \frac{\left(27 - \frac{59}{2}\right)^2}{59} = 559.704 \text{ mu} = 521.353 \text{ MeV} \qquad \textbf{(12.17)}$$

The exact measured mass of ^{59}Co is 58.933190 u. Using this value, along with the values of M_H and M_n from Table 12.2, we can obtain from Eq. (10.3)

$$\Delta M = 555.365 \text{ mu} = 517.311 \text{ MeV} \qquad \textbf{(12.18)}$$

The calculated value from Eq. (12.17) agrees with the measured value from Eq. (12.18) within 0.8 percent. We note also that $B/A = 8.77$ MeV per nucleon, consistent with trends alluded to in Chapter 10.

The example we have just computed augers well for the usefulness of the semiempirical mass formula. If we can calculate exact nuclear masses by formula we can predict the Q values of nuclear reactions along the lines discussed in Section 11.3 and, in particular, indicated in Eq. (11.5). Unfortunately we must not get our hopes that high. Even an error of less than 1%, as in this example, means an error of several MeV since we are dealing with hundreds of MeV total mass-energy. When we take sums and differences the errors compound and are much larger than those ordinarily tolerated in experimental determinations of Q values.

All is not lost, however. The formula does predict **general trends** reasonably well. If our demands are not too stringent, if we only require a picture of overall characteristics and not detailed ones, the liquid drop model is satisfactory. Furthermore, as we shall see in the next section, comparison of measured binding energies and beta-decay energies against the predicted general trends show regularities which can guide our thinking about more sophisticated models.

12.4   The Nuclear Shell Model

As early as the first part of the 1930s attempts were made to formulate descriptions of nuclear structure along lines of **shell models** that had been so successful in describing the structure of atoms. Following the discovery of the neutron in 1932 by Chadwick it was recognized that there are two populations of nucleons and that the exclusion principle would apply to these populations separately. Thus periodic effects arising from closing of shells should be sought in terms of the count of protons or of neutrons in a nucleus, that is, Z or N, rather than in terms of the total nucleon count A. Assuming that quantization proceeds in a fashion analogous to that for electronic structure in atoms it is expected that there will be subshells characterized by orbital angular momentum quantum number l and that each subshell will hold $(2l + 1)$ protons or neutrons, ignoring spin, or $2(2l + 1)$ when spin is considered. The energies of the subshells and their order of filling will depend on the particular potential in which it is assumed the nucleons move.

Despite some initial success with such calculations the model fell into disrepute from the middle 1930s to the late 1940s. A shell model requires us to think in terms of particles moving in **defined orbits.** This implies that even for "inner" nucleons the mean free path in the nucleus is large compared to the nuclear radius. Since the force between nucleons is very large and nucleons are very tightly packed in a nucleus it seems contradictory to ask for a long free path. It seems that collisions must surely take place with great frequency and destroy the orbital character of the motion. Under this view the liquid drop model is much more promising than the shell model.

However, as the body of empirical date grew larger **discontinuities** in a variety of nuclear parameters such as **binding energies, quadrupole moments, beta decay energies,** and **neutron separation energies** became apparent for nuclides with Z or N equal 2, 8, 20, 28, 50, and 82, and for $N = 126$. Discontinuities in these parameters could not be reconciled with models then in vogue. It was all very puzzling and at the same time important and exciting. The numbers came to be called **magic numbers.** The terminology was somewhat tongue-in-cheek, but it has continued and is now common usage.

Although the reasons for abandoning previous efforts toward a shell model were persuasive, the new evidence was compelling. There could be little doubt that shell closures occur at the magic numbers. A mechanism must be found to permit **independent nucleon orbits** without contradicting the evidence for strong interactions between nucleons when a projectile collides with a nucleus. This is done by adopting the following picture of the nucleus. It still consists of a swarm of nucleons close together. Each nucleon moves in an orbit within a potential established by the $A - 1$ other nucleons. All lower quantum states are filled. Thus, except for the highest energy, least tightly bound, last nucleons to be added in the assembly of the nucleus, there is "no place for a nucleon to go" following a collision or other interaction within the nucleus. Since there is no place to go, such interactions are **forbidden** and do not occur. Because of the Pauli **exclusion principle** acting in this fashion the vast majority of nucleons orbit undisturbed. Projectiles from outside the nucleus, however, are not members of bound states. They bring in sufficient energy to raise nucleons in target nuclei to unoccupied states in the compound nucleus or to eject them altogether. Strong interactions of this kind are not forbidden. Finally, it remains to be seen whether suitable nucleon interactions and potentials can be found to operate within the nucleus such that calculations based on them agree with experimental facts.

The atomic case is dominated by the nuclear Coulomb potential in which the orbital electrons move. Except for inner electrons of small angular momentum, which may penetrate the nucleus during a portion of their paths, this potential goes as r^{-1}. The electrons also exert Coulomb forces upon one another, but these forces are small compared to the dominant central force and produce only small perturbations of the main central potential. For the nucleus the situation is quite different. Inside the nucleus the potential will be quite flat, as a function of r, since we predicate that nucleons have low probability of interaction when safely inside. On the other hand, the sides of the potential well must be quite steep in the neighborhood of the nuclear radius, R. We also expect the potential to be spherically symmetric.

A shape that meets the preceding criteria in a reasonable fashion is the **square well** potential:

$$W(r) = \begin{cases} -U_0 = \text{constant} & \text{for } r < R \\ 0 & \text{for } R < r \end{cases} \qquad \textbf{(12.19)}$$

To be sure, we do not expect this potential to be completely realistic. The corners should be rounded, there probably should be some departure from infinite slope at the walls, and for protons a Coulomb portion rising above $U = 0$ just outside the nuclear radius. Nevertheless, as we saw in Eqs. (12.8) and (12.9), U_0, the depth of the well, is relatively great and our results should not be badly distorted by ignoring these fine points.

In discussing the treatment we shall use a variety of quantum numbers, which we now review.

n: The **principal quantum number.** It can take on only positive integer numbers greater than zero. $n = 1, 2, 3, 4, \ldots,$. In a Coulomb potential as encountered in the description of atomic structure n is a good quantum number and is the primary determinant of the energy of a level. In a square well potential n is not a good quantum number and we shall give more emphasis to the radial quantum number, ρ.

l: The **orbital angular momentum quantum number.** This can take on positive integer numbers, and zero, from zero to $(n - 1)$.
$l = 0, 1, 2, \ldots, (n-1)$. Just as in atomic spectroscopy, the letters $s, p, d, f, g, h, i,$ and so on, stand for l-values of 0, 1, 2, 3, 4, 5, 6, and so on. See Table 5.4.

ρ: The **radial quantum number.** This is related to n and l according to $n = \rho + l$. ρ can take only positive integer values greater than zero.

s: The **spin quantum number.** For fermions such as protons, neutrons, and electrons $s = \frac{1}{2}$.

j: The **total angular momentum quantum number.** For a single fermion it can take on either of two values: $j = (l + s) = (l + \frac{1}{2})$ or $j = (l - s) = (l - \frac{1}{2})$. Except that if $l = 0$, then $j = s = \frac{1}{2}$.

m_l: The **magnetic orbital quantum number.** It is the quantized component of l in a particular direction, usually determined by the direction of a magnetic field. It takes on $(2l + 1)$ integer values between l and $-l$.
$m_l = l, (l - 1), \ldots, 1, 0, -1, \ldots, -(l - 1), -l$.

m_s: The **magnetic spin quantum number.** It is the component of s in a particular direction and can take on two values: $m_s = \frac{1}{2}$ or $m_s = -\frac{1}{2}$.

m_j: The **magnetic total angular momentum quantum number.** It is the quantized component of j in a particular direction and takes on $(2j + 1)$ half-integer values between j and $-j$.
$$m_j = j, (j-1), \ldots, \tfrac{1}{2}, -\tfrac{1}{2}, \ldots, -(j-1), -j.$$

In a **spherically symmetric** potential the wave function can be separated into the product of a radial wave function and an angular wave function.

$$\psi(r, \theta, \phi) = R(r)\, Y_{l,m_l}(\theta, \phi) \tag{12.20}$$

R is given by the solution to the radial wave equation

$$\frac{1}{r^2}\frac{d}{dr}\left(r^2 \frac{dR}{dr}\right) + \frac{2m}{\hbar^2}\,[E - U(r)]\,R = l(l+1)\frac{R}{r^2} \tag{12.21}$$

where E is the total energy of the system. In the formal mathematical treatment the l and m_l are introduced as **separation constants** to accomplish the separation suggested in Eq. (12.20). To obtain well **behaved solutions** for R and Y the constants must have the characteristics ascribed to them in the last paragraph and E must be one of a particular set of discrete values, the **eigenvalues** of the wave equation.

All that we have just said applies to any spherically symmetric potential. It applies, for instance, to the Coulomb potential

$$U(r) = -\frac{Ze^2}{r^2}$$

of a one-electron hydrogenlike atom. In this system the states are **degenerate** with respect to m_l and l. That is, the E's depend on the principal quantum number n only and not on m_l and l. Things are quite different, however, with a square well potential or its cousins with rounded corners, finite slopes, and so on. The degeneracy with respect to l is lifted; the energies do depend on l. Furthermore, the dependence on n is not monotonic as in the case of the hydrogenlike atom. In Table 12.3 we show the ordering of states with their spectroscopic notations for the square well potential. Energy is increasing—that is, tending to less tight binding—to the right.

Table 12.3 **Ordering and Spectroscopic Notation for Energy Levels of a Finite Square Well**

1s	1p	1d	2s	1f	2p	1g	2d	3s	1h	2f	3p	1i
										in terms of ρ		
1s	2p	3d	2s	4f	3p	5g	4d	3s	6h	5f	4p	7i
										in terms of n		

For comparison the notation is presented both in terms of the radial quantum number ρ and the principal quantum number n. It is seen that the ρ value gives the order in which a particular l value appears. Thus the third column is the first appearance of a d term and the fourth column the second appearance of an s term. Although degeneracy in l is removed, that in m_l is not. Each of the levels has a $(2l+1)$-fold degeneracy associated with l. Folding in the two orientations of m_s gives $2(2l+1)$ states with the same energy. The next step is to look for noticeable gaps between energy levels, interpret them as occuring just after shell closures, and compare the number of states below the gaps against the known magic numbers.

The left side of Figure 12.4 shows the results of following the procedure just outlined. In addition to the spectroscopic notation for each level we show in parentheses the degeneracy of the level and within brackets the totality of states below promising gaps. Each state is assigned a particular set of quantum numbers, (ρ, l, m_l, m_s). According to the Pauli exclusion principle no two neutrons, nor any two protons, can exist in the same state with all quantum numbers the same. Because of the degeneracy in m_l and m_s, only ρ and l are indicated. Although the first three shell closures reproduce the magic numbers 2, 8, 20, and thereby give us considerable encouragement, there is no indication of closures at the higher numbers 28, 50, 82, 126. Clearly, something is wrong. If the basic idea of the shell model is to be saved, some adjustments must be made to split the levels in a somewhat different fashion. Tinkering with the shape of the potential, changing the rounding of the corners, the slope of the sides, and soon, makes minor changes in the positions of the levels but does not change the splitting and does not improve the match with the higher magic numbers.

In arriving at the left side of Figure 12.4 it was assumed that **coupling** between the spin of the nucleon and its orbital motion could be ignored. In atoms this coupling arises from interaction of **magnetic dipole moments** associated with the electron spin and orbital motion. Nuclear magnetic moments are smaller than atomic moments, whereas the interaction energies sought are much larger. For a long time it was believed that **spin-orbit coupling** could not introduce significant energy terms. However, in 1949, Maria Goeppert Mayer in the United States and Haxel, Jensen, and Suess, in Germany, published independently and essentially simultaneously a model using strong spin-orbit coupling. Not only was the coupling much stronger than had been inferred by analogy with atomic fine structure splitting but the sign was reversed. That is, levels with $j = l + \frac{1}{2}$ are lower (that is, more tightly bound) than levels with $j = l - \frac{1}{2}$. This model leads to the system of levels shown on the right side of Figure 12.4. By proper adjustment of the strength of coupling it predicts all of the magic numbers and a wide range of other observations.

At the time of its introduction the notion of significant spin-orbit

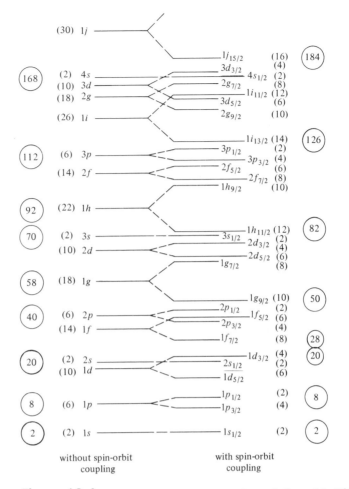

Figure 12.4 Level systems of the nuclear shell model. Without spin-orbit coupling the model fails to predict shell closures at magic numbers above 20; with strong spin-orbit coupling excellent agreement is achieved between shell closures and magic numbers. The degeneracy of the levels is given by numbers in parentheses. The number of nucleons required for each completed shell is shown by numbers in circles.

coupling was largely an ad hoc device to produce level splitting in the hope of accounting for the magic numbers. Within a few years, however, double scattering experiments established that, at proper resonance energies and angles, protons and neutrons can be **polarized** by scattering from helium. Polarization, in this situation, means a preferred, rather than random, **alignment of spin directions** in a beam of particles. To produce polarization by scattering it is necessary that there be an interaction between the intrinsic spin of the projectile and the angular momentum of the projectile-target-nucleus system as the projectile goes by. The sign of the interaction turned out to agree with that needed for the shell model. The magnitude also was about right. Strong spin-orbit coupling is now well established.

We have spoken of the orbital angular momentum quantum number l, the spin angular momentum quantum number s, and the total angular momentum quantum number j. The angular momenta associated with these quantum numbers are the vectors $\mathbf{l}$, $\mathbf{s}$, and $\mathbf{j}$, where

$$|\mathbf{l}| = \hbar \sqrt{l(l+1)}$$
$$|\mathbf{s}| = \hbar \sqrt{s(s+1)} \tag{12.22}$$
$$|\mathbf{j}| = \hbar \sqrt{j(j+1)}$$

If there is strong spin-orbit coupling $\mathbf{j}$ is given by the vector sum

$$\mathbf{j} = \mathbf{l} + \mathbf{s} \tag{12.23}$$

$\mathbf{l}$ and $\mathbf{s}$ precess around their resultant $\mathbf{j}$, as in Figure 12.5. $\mathbf{j}$ becomes a constant of motion. j is a good quantum number. If there is a special orientation direction, for instance, established by a magnetic field in the z direction, $\mathbf{j}$ precesses about it, subject to the quantization conditions on m_j.

The interaction energy that shifts the levels is proportional to $\mathbf{s} \cdot \mathbf{l}$. By the law of cosines in the triangle of the vector sum, Eq. (12.23)

$$2\mathbf{s} \cdot \mathbf{l} = \mathbf{j} \cdot \mathbf{j} - \mathbf{l} \cdot \mathbf{l} - \mathbf{s} \cdot \mathbf{s} = \hbar^2 [j(j+1) - l(l+1) - \tfrac{3}{4}] \tag{12.24}$$

Upon substitution of $j = l + \tfrac{1}{2}$ and $j = l - \tfrac{1}{2}$ in the right side of Eq. (12.24) two values for the energy shift are obtained. Their difference is the energy splitting of the level and is proportional to $(2l + 1)$. Although the constant of proportionality is empirically adjusted, the progression of splitting with increasing l is not. The sublevels are further apart when l is large than when it is small. For $l = 0$, j is single valued and there is no splitting. In spectroscopic notation the j value is given as a right subscript following the l designation. Thus $1f_{7/2}$ is the $j = \tfrac{7}{2}$ member of the first $l = 3$ level. Each sublevel has a $(2j + 1)$-fold degeneracy. With spin-orbit coupling the four quantum numbers that describe a state become: ρ, l, j, m_j

In a system with more than one particle, for instance, a multi-electron atom or a nucleus with many nucleons, strong spin-orbit

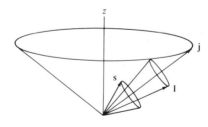

Figure 12.5 Total angular momentum is the resultant of vector coupling of spin angular momentum and orbital angular momentum.

coupling preserves the identity of **j** for each particle. Each **l** and **s** continues to precess around its own **j**. The **j**'s couple together in vector fashion to produce a resultant, designated by **J**. This type of coupling is designated *jj* coupling, and the model introduced by Mayer and Jensen is often referred to as the *jj* coupling model. On the other hand, if spin-orbit coupling is weak the **l**'s couple to form a resultant **L** and the intrinsic spins to form a resultant **S**. Thus scheme is designated *LS* coupling. For given ρ, l the total number of states is the same in either coupling system. This is borne out by comparison of the two columns of Figure 12.4. The energies of the levels, however, are profoundly influenced by the type of coupling. It is the splitting of the $1f$ level in the *jj* coupling model that achieves shell closure at neutron (or proton) number 28. The splitting of the $1g$, $1h$, and $1i$ levels produces shells to match the other magic numbers.

Figure 12.4, as drawn, is extended upward to show an expected shell closure at $N = 184$. No nuclide has yet been made and detected with that many neutrons, but there is speculation that such a nuclide could be quasistable if it were also proton magic. However, the level spacing shown in Figure 12.4 is not completely valid as applied to protons when Z is large. The large Coulomb repulsion makes for lowered stability and raising of the levels. The effect is larger for states of low l than for those of high l since on the average the former penetrate deeper into the Coulomb potential. As a result the $2f$ and $3p$ proton levels are raised so much that the $2f_{7/2}$ level is close to the $1i_{13/2}$ level. This produces a shell at $Z = 114$. Further adjustments based on liquid drop and collective model views predict that $Z = 110$ would optimize stability. It is thought that nuclides located at an **island of stability** in the neighborhood of $Z = 110$, $N = 184$ might have lifetimes as long as 10^8 years.

The shell model has been very successful in dealing with **ground state energies.** In its simplest form it is an **extreme single particle** shell model and in this form it has had good success in predicting **ground state angular momenta.** The single particle model makes the following assumptions:

1. If any shell is occupied by an even number of neutrons (or protons) the individual j's will pair with opposing angular momenta so the total angular momentum sums to zero within the shell. It follows that in even-even nuclides the total nuclear angular momentum $J = 0$.

2. In a shell with an odd number of nucleons, it is the odd nucleon that carries the angular momentum. The other (even number of) nucleons pair off as before. For even-odd and odd-even nuclides the angular momentum will be the half-integer value of the odd nucleon. For odd-odd nuclides the total angular momentum is the vector sum of that for the proton shell and that for the neutron shell.

$$\mathbf{J} = \mathbf{J}_p + \mathbf{J}_n = \mathbf{j}_p + \mathbf{j}_n$$

The allowed values of the quantum number J are integers within the limits

$$|j_p - j_n| \leq J \leq j_p + j_n$$

3. ·A closed shell is spherically symmetric. Its angular momentum is zero since all closed shells have even nucleon number. In other configurations departure from symmetry is due to nucleons added outside a closed shell or missing from (therefore nucleon holes within) such shell.

Let us apply these rules to a few examples and see how we fare, using the level scheme of Figure 12.4:

^6_3Li is an odd-odd nuclide with $Z = 3$, $N = 3$. The level scheme calls for $j_p = \frac{3}{2}$ and $j_n = \frac{3}{2}$. According to rule 2 J should fall between zero and 3. Its measured value is $J = 1$.

$^{14}_7\text{N}$ is also odd-odd with $Z = 7$, $N = 7$. $j_p = \frac{1}{2}$ and $j_n = \frac{1}{2}$. The measured value of J is 1.

$^{17}_8\text{O}$ is even-odd with $Z = 8$, $N = 9$. It has one neutron outside a core of neutron and proton shells, both closed at the $1p_{1/2}$ shell. The closed shells make no contribution to J. The odd neutron should be in the $1d_{5/2}$ shell. Since $j_n = \frac{5}{2}$ we expect $J = \frac{5}{2}$. Its measured value is $J = \frac{5}{2}$.

$^{27}_{13}\text{Al}$ is odd-even with $Z = 13$, $N = 14$. According to rule 1, $j_n = 0$. The odd proton is in the $1d_{5/2}$ shell. The measured $J = \frac{5}{2}$ again agrees with the predicted value.

Comparisons of this sort show that shell model predictions are borne out for a great majority of the nuclides for which ground state J values have been measured. This is a more sensitive test of the validity of the shell model than the matching of magic numbers. Despite these successes, however, the shell model does have its limitations. Its predictions of energy levels of excited states often contradict experimental observations. Furthermore, in some instances ground state J's do not agree with the model. Nevertheless, in general it is an impressive and powerful model.

12.5   The Collective Model

The **collective model** combines features of both the shell model and the liquid drop model. Like the shell model, the collective model views the nucleus as consisting of a core made up of nucleons in closed shells around which one or more nucleons move in well defined orbits. In the extreme single-particle shell model, the core is presumed to be spherically symmetric and quite rigid. The potential in which the outer nucleons move is, as a consequence, spherically symmetric. In the collective model, however, the core is more like a liquid drop. It is deformable and responds to forces exerted on it by the nucleons orbiting around it. The result is a **tidal wave** bulge of the core beneath the nucleon orbit. In these circumstances the core and the potential attendant upon it depart from spherical symmetry. In both models the radius of the outer orbit is taken to be essentially that of the nucleus. In other words, the orbiting nucleon just grazes the surface of the core. The size of the orbit relative to the core is quite different from that of a planetary model.

Distortion of the core makes it possible to match quantitatively a number of nuclear parameters for which the shell model predictions are badly in error. One of these is the **magnetic dipole moment** of the nucleus. Another is the **electric quadrupole moment.** Both quantities interact with atomic electrons, resulting in small energy shifts that manifest themselves in an effect known to atomic spectroscopy as **hyperfine structure.**

The magnetic dipole moment of a nucleus arises from combining the intrinsic dipole moments of its constituent nucleons with the moment due to charge circulation within the nucleus. (The magnetic dipole moment of the proton is in the same direction as its intrinsic spin; that of the neutron is opposite to its spin.) In the shell model the magnetic moment of an odd A nucleus should be due exclusively to the outermost unpaired nucleon. We remember that the nucleons in the core pair with opposing angular momenta. Their magnetic moments therefore cancel. As a consequence, the total magnetic moment of an odd A nucleus should arise from the moments associated with the orbital motion and spin of the odd nucleon. As we have noted, the predictions based on this model do not agree with experimental measurements. For instance, for odd Z nuclides, having $j = l + \frac{1}{2}$, the observed magnetic moment is smaller than the shell model prediction, whereas for odd N nuclides, also having $j = l + \frac{1}{2}$, it is greater (less negative).

The collective model, however, presents us with an additional parameter, the tidal wave in the core. There is, therefore, a prospect for modification of the preceding predictions. We recognize that there is additional circulation of charge because protons as well as neutrons are involved in the tidal wave following beneath the orbiting nucleon. Furthermore, because the tidal wave is caused by the specifically nuclear force it will be present whether the odd nucleon is a proton or a neutron. If the odd nucleon is a proton the magnetic moment that results from combining its moment with the effects of the tidal wave is less positive than that predicted by the shell model. If the odd nucleon is a neutron the result is more positive. These modifications are in the correct direction in each instance and can be adjusted in magnitude to fit magnetic moments determined by experiment.

We now turn our attention to the static electric quadrupole moment. This moment arises from a charge distribution that is non-spherically symmetric. In electrostatics the quadrupole moment is defined by the expression

$$Q = \int \sigma(x, y, z)[3z^2 - (x^2 + y^2 + z^2)] \, d\tau \qquad (12.25)$$

where σ is the volumetric charge density and $d\tau$ is the element of volume. In this configuration there is symmetry about the z axis. In similar fashion the nuclear quadrupole moment is taken to be

$$Q = \int \sigma[3z^2 - (x^2 + y^2 + z^2)] \, d\tau \qquad (12.26)$$

where σ is the time averaged density of protons per unit volume and the integration is to be carried out over the volume of the nucleus. In this usage Q has the units of area and is commonly quoted in barns. The symmetry axis is taken along the direction about which the total angular momentum $\mathbf{J}$ precesses. It is usually assumed that σ is uniform. Therefore Q is positive if $\langle z^2 \rangle > \langle x^2 \rangle = \langle y^2 \rangle$ and negative if $\langle z^2 \rangle < \langle x^2 \rangle = \langle y^2 \rangle$, where $\langle \ \rangle$ denotes average value. The first case represents a **prolate** (cigar shaped) ellipsoid of revolution, the second an **oblate** (saucer shaped) ellipsoid. If the nucleus is spherically symmetric it follows that $\langle z^2 \rangle = \langle x^2 \rangle = \langle y^2 \rangle$ and $Q = 0$.

Measured values of Q are positive for odd A nuclides just below magic numbers and negative for those just above. For nuclides with Z odd and N even the shell model is able to provide a qualitative explanation. A nucleus with one proton more than a magic number is modeled as a closed shell core with a **proton cloud,** representing the odd proton orbit, around its equator. The core is spherically symmetric and makes no contribution to Q. The proton cloud makes the overall charge distribution oblate and thus Q is negative. One proton less than a magic number may be thought of as a **proton hole** in an otherwise closed shell. The hole orbits at the surface of the core, producing a deficiency of positive charge at the equator. The result is a **prolate** charge distribution and a positive quadrupole moment. For a number of nuclides, however, the magnitudes calculated for Q with this model are smaller than measured ones by factors of as much as ten. Furthermore, the model is incapable of explaining the fact that nuclides with Z even and N odd behave about the same as those with odd Z and even N.

In the collective model the charge asymmetry, the source of the quadrupole moment, is attributed primarily to distortion of the core itself. To a first approximation the tidal wave bulge, or depression for a nucleon hole, is independent of whether the odd nucleon is a proton or neutron, thus explaining the lack of separation in Figure 12.6 between odd-Z and odd-N nuclides. Also, larger charge asymmetry,

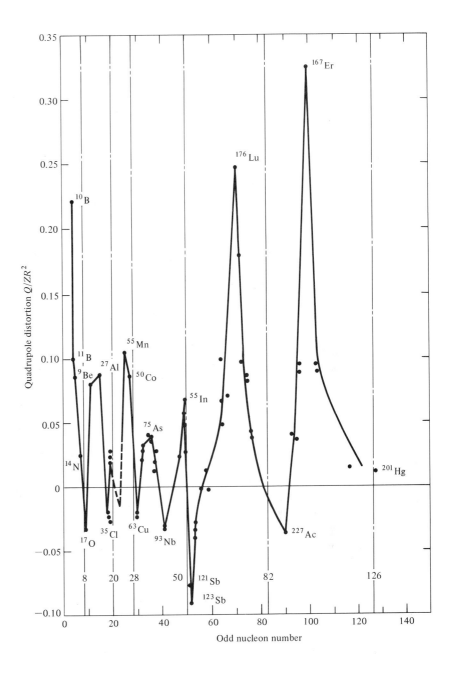

Figure 12.6 Quadrupole distortion of odd-*A* nuclides. The abscissa is the proton number for odd-*Z* nuclides. It is the neutron number for odd-*N* nuclides. There is no apparent separation between the two categories. *Q* is positive just below magic numbers indicated by dash-dot lines, and negative just above.

and with it larger quadrupole moment, can be produced by core distortion than by single particle orbits, thus improving the fit to measured values.

12.6   The Optical Model

The shell model and the collective model are designed to treat bound states of nuclei. These models hypothesize a central potential in which the nucleons move. In a **scattering** or other **bombardment experiment** we may think of the encounter between the projectile and the target nucleus as that of an unbound nucleon incident upon the potential well that represents the nucleus. The simplest situation is that of neutron bombardment since no Coulomb barrier is involved. We expect the potential well to be pretty much the same as that predicated for the previous two models. If we represent the incoming particle as a **plane wave,** we find that its **wavelength changes** upon entering the potential well of the nucleus. At each encounter with the potential boundary there is **partial reflection and transmission.** If the approach is not head on, **refraction** occurs at the two surfaces. **Diffraction** and **interference** effects govern the amplitude of the wave representing the outgoing particle, with the result that the scattering cross section $d\sigma/d\Omega$ may show maxima and minima when plotted against the scattering angle. This description of a nuclear encounter is very similar to one that might be given of a plane light wave impinging upon an optically transparent sphere. As a consequence, the name **optical model** has been adopted.

Figure 12.7 is a wave-mechanical representation of the encounter

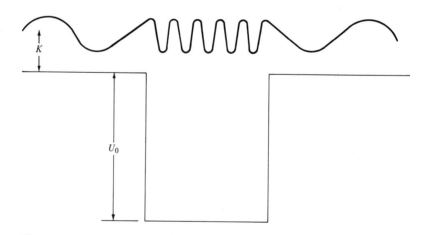

Figure 12.7   The wavelength of an unbound neutron is shorter when it is in the region of a nuclear potential well than it is outside.

of an unbound neutron of kinetic energy K with a potential well of depth U_0. It shows the outgoing particle as having the same kinetic energy as the ingoing particle. This representation, which is compatible with the discussion in the preceding paragraph, limits the treatment to elastic scattering. This is a severe limitation since there are many situations in which the incident nucleon interacts strongly with the target nucleus. For instance, the interaction may be **simple capture** followed by gamma emission, it may be **compound nucleus formation** followed by emission of a different particle, or it may be a **direct reaction** in which all of the energy of the incident particle is delivered to one of the nucleons in the nucleus, thereby ejecting it directly. In each instance the incident particle is lost to the outside world. Consequently, the model needs some modification to introduce **absorption.**

A way to introduce absorption by mathematical maneuvers is to substitute a complex potential $(U_0 + iU)$ for U_0 in Figure 12.7. If there is positive kinetic energy K outside the well, then inside the well the wave equation becomes

$$-\frac{\hbar^2}{2m}\frac{d^2\psi}{dx^2} = (K + U_0 + iU)\psi \tag{12.27}$$

The solution is of the form

$$\psi = e^{ik_1 x}e^{-k_2 x} \tag{12.28}$$

If $U \ll (K + U_0)$, then

$$k_1 = \frac{\sqrt{2m(K + U_0)}}{\hbar} \tag{12.29}$$

and

$$k_2 = \frac{Uk_1}{2(K + U_0)} \tag{12.30}$$

Equation (12.28) is that of a wave traveling in the positive x-direction while decaying exponentially. $|\psi|^2$ gives the probability of finding the incident neutron. The mean free path inside the nucleus is

$$\frac{1}{2k_2} = \frac{K + U_0}{Uk_1} \tag{12.31}$$

A nucleus with $A = 125$ has a radius $R \approx 7.25$ fermis [Eq. (10.23)]. Typical values for the complex potential are $U_0 = 42$ MeV and $U = 0.05W_0 = 2.1$ MeV. Substitution in the preceding equations shows that a neutron with $K = 14$ MeV has a 30% chance of surviving a single passage across a nuclear diameter, once it is inside the nucleus.

Obviously the actual nuclear potential is three dimensional, rather than one dimensional as in our example, and no doubt should have rounded corners as we have suggested before. When both scattering (reflections and refractions at the surface) and absorption (nuclear reactions) are considered, the model makes quite good predictions of total cross sections, except where individual resonances matching sharp levels in the compound nucleus are involved. This model has been applied extensively to neutron total cross-section studies. It predicts broad resonances as a function of neutron kinetic energy and nuclear mass number, which are features seen in the experimental data. The initial work with the model was done by Feshbach, Porter, and Weisskopf in 1954, and it is from their work that the values for R_0, U_0, and U are taken in our example. Other authors have tried various potential shapes, depths, and ratios of imaginary to real parts. Spin-orbit effects, guided by the shell model, have been included, as well as Coulomb barriers for charged particles such as protons. Remarkably good fits, including diffraction effects seen in angular distributions as well as the slower dependence on energy and nuclear mass, have been obtained. Because the model deals with both refraction and absorption it is often referred to as the **cloudy crystal ball.**

12.7 Summary

There are a variety of nuclear models. Some are invoked to describe certain aspects of nuclear behavior, other models to describe other aspects. Although no single model is successful in dealing with all aspects, the use of models helps our understanding and conveys at least partial answers about nuclear structure.

The Fermi gas model predicts extra stability for nuclides made up of alpha particle clusters. Experiments confirm this prediction among light nuclides, but not among medium weight and heavy ones. The Fermi energy is given by

$$E_\mathrm{F} = \frac{\pi^2 \hbar^2}{2M_n} \left(\frac{3}{\pi} n \right)^{2/3}$$

Adding to this the separation energy of the last nucleon gives an estimate of the depth of the nuclear potential well.

The semiempirical mass formula is developed from the liquid drop model.

$$M_{Z,A} = ZM_\mathrm{H} + (A - Z)M_n - a_v A + a_s A^{2/3} + a_\mathrm{coul} Z(Z - 1)A^{-1/3}$$
$$+ a_\mathrm{sym} \frac{(Z - A/2)^2}{A} - a_{pair} A^{-1/2}$$

The single particle shell model with strong spin-orbit coupling is highly successful in predicting the magic numbers 2, 8, 20, 28, 50, 82, and 126, which are experimentally indicated by the existence of especially stable nuclides. It makes good predictions also of ground state energies and angular momenta.

Spectroscopic notation (l, s, j, m_l, m_s, m_j) is borrowed from atomic spectroscopy. For a square well potential (nuclear model), however, the radial quantum number ρ is applicable rather than the principle quantum number n, which is more suitable for potentials with a r^{-1} shape (atomic model). Thus, for instance, $1f_{7/2}$ designates the $j = \frac{7}{2}$ member of the first $l = 3$ level.

Neutron shells and proton shells are treated independently. In even-N and even-Z shells $\mathbf{J}(= \Sigma \mathbf{j}) = 0$. $J = 0$ for the ground state of all even-even nuclides. J is half-integer for even-odd and odd-even nuclides. For odd-odd nuclides $|j_p - j_n| \leq J \leq j_p + j_n$.

The shell model is unsuccessful at predicting energies of excited states, magnetic dipole moments, and electric quadrupole moments.

The collective model combines features of the liquid drop and shell models. A closed shell core behaves like a liquid drop. Nucleons outside closed shells bulge the core by producing tidal waves. With the core thus distorted, the fit to experimentally determined magnetic dipole and electric quadrupole moments is improved.

The optical model describes a scattering reaction in terms of refraction and diffraction of a plane wave representation of the bombarding particle. A complex nuclear potential is introduced to account for absorption, that is, events other than elastic scattering. Since refraction, diffraction, and absorption take place, this is sometimes called the cloudy crystal ball model.

Problems

12.1 Apply the viewpoint of the Fermi gas model to determine E_f for the neutron population and for the proton population of ^{109}Ag. Calculate the separation energies of the last neutron and the last proton in ^{109}Ag and add to E_f to get the well depths.

12.2 By means of the semiempirical mass formula calculate the masses of ^{40}A, ^{133}Cs, and ^{197}Au.

12.3 Select five nuclides with $A > 20$. Use their exact masses in a set of five equations to determine your own set of a's for Eq. (12.16). Avoid nuclides with magic neutron or proton number. Test your a's by using them to predict the masses of several other nuclides.

12.4 On the basis of the single particle shell model, make predictions regarding the angular momenta of the ground state configurations of ^{28}Al, ^{37}Cl, ^{60}Co. In some cases you may have to be satisfied with a range of possible values.

12.5 Verify the conclusions about the probability of interaction of a 14-MeV neutron stated immediately after Eq. (12.31).

12.6 On the basis of the optical model, what is the probability of interaction of a 20-MeV neutron in a head-on collision with ^{109}Ag? Use U_0 from problem 12.1 and let $U = 0.05\ U_0$.

12.7 The nuclides ^{10}Be, ^{10}B, ^{10}C, are neighboring isobars. Which has the highest Coulomb energy; which the lowest? By how much do they differ according to Eq. (12.12)? Which of the three nuclides do you expect to be most stable according to this calculation?

12.8 Other terms in addition to Coulomb energy may be of importance to the triad of nuclides in problem 12.7. Identify and evaluate these terms for all three nuclides. Which member of the triad do you now expect to be most stable? Compare with tabulated nuclide masses.

12.9 Calculate the last proton separation energy for the proton-magic nuclides $^{16}_{8}$O, $^{40}_{20}$Ca, $^{58}_{28}$Ni, $^{120}_{50}$Sn, $^{208}_{82}$Pb, and for their neighboring isotones.

12.10 What are the ground state angular momenta predicted by the shell model for ^{29}Si, ^{43}Ca, ^{60}Ni?

12.11 Consider a hypothetical nucleus modeled as a right circular cylinder whose length is 20 fermis and diameter 15 fermis. Assume it to contain 3.5×10^{37} protons/cm^3. Calculate its quadrupole moment.

12.12 Calculate the de Broglie wavelength of a free neutron of 20-MeV kinetic energy. What is its wavelength upon entering a region of nuclear potential energy $U_0 = -50$ MeV? What (optical analog) index of refraction for the nucleus is implied by these wavelengths?

Suggestions for Further Reading

Cranberg, L. "Fast Neutron Spectroscopy," *Scientific American* (March 1964).

Eisberg, R., and R. Resnick. *Quantum Physics*. New York: John Wiley & Sons, 1974. Sections 15.5 through 15.11.

Enge, H. A. *Introduction to Nuclear Physics*. Reading, Mass.: Addison-Wesley Publishing Co., Inc., 1966. Sections 4.6 and 13.10; Chapter 6.

Evans, R. D. *The Atomic Nucleus*. New York: McGraw-Hill Book Company, Inc., 1955. Chapter 11.

Green, A. E. S. *Nuclear Physics*. New York: McGraw-Hill Book Company, Inc., 1955. Sections 9.1 through 9.8; 10.10 through 10.11.

Peierls, R. E. "The Atomic Nucleus," *Scientific American* (January 1959).

Radioactivity and alpha decay

13.1   Introduction

Pioneering studies with radioactive materials late in the nineteenth and early twentieth centuries indicated that penetrating radiations are emitted from these substances. Three different types of radiation were identified. Suppose a small piece of active material is placed at the bottom of a narrow hole drilled in a block of lead, as in Figure 13.1. The radiation that escapes is thereby **collimated** and allowed to strike

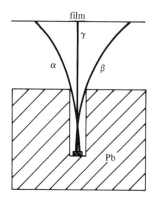

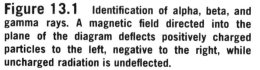

Figure 13.1 Identification of alpha, beta, and gamma rays. A magnetic field directed into the plane of the diagram deflects positively charged particles to the left, negative to the right, while uncharged radiation is undeflected.

a photographic film placed a short distance above the block. The entire assembly is placed in a vacuum chamber to avoid any absorption of radiation by intervening air. Upon application of a magnetic field normal to the plane of the diagram and between the block and film, three different spots on the film are blackened, one from radiation deflected to the left, one to the right, and one undeflected. If the sense of the field is inward to the diagram, the radiation deflected to the left must be composed of positively charged particles and that to the right negative ones. The undeflected radiation must either be uncharged or of such high energy that its deflection cannot be noticed. The name **alpha** radiation was given to the positive component, **beta** radiation to the negative, and **gamma** radiation to the neutral (as it was subsequently identified) radiation. The alpha particles prove to be no different from nuclei of $_2^4$He; beta particles are swift electrons; gamma rays are electromagnetic radiation, that is, photons. Alphas are easily stopped in matter, a sheet of paper being sufficient. Betas are more penetrating and may require several millimeters of aluminum, for instance, to stop them. Gamma radiation is by far the most penetrating and may require many centimeters of lead for adequate attenuation.

If all nuclides found in nature were stable the history of atomic and nuclear structure studies would have been vastly different. In fact, the clues presented by the observation of **natural radioactivity** were so important that it is difficult to surmise how, in their absence, these fields would have developed. The existence of alpha and beta decay enticed early workers into long series of investigations in efforts to understand the source of the radiation. The Rutherford scattering experiment was one such investigation and opened the door to our present understanding of the atom. That there might be structure to the nucleus was suggested by the emission of alpha particles. The first nuclear reactions and the first induced radioactivity were produced by bombardment with alpha particles from radioactive sources. The nature of the radiations and the time dependence of the activities were salient features to be studied. Today we recognize there is no essential difference between natural radioactivity and "artificial" radioactivity induced by manmade experiments.

13.2 The Exponential Decay Law

Chemical treatment of radioactive minerals established the fact that some elements are far more intensely radioactive than others. It also became apparent that for many the activity diminishes with time after chemical separation. For others, any such diminution of activity is too slight to be measured. For a given element the activity is completely unaffected by chemical treatment,[1] by the application of heat, pressure, electric fields, or other gross physical assault. The simplest cases follow an **exponential decay law.**

$$A(t) = A_0 e^{-\lambda t} \tag{13.1}$$

where $A(t)$ is the activity at time t, A_0 is the activity at the start of the time measurement, and λ is the **decay constant.** This is the fundamental decay law. Activities that deviate from it are caused by the presence of more than one radioactive nuclide. Each nuclide individually obeys Eq. (13.1), but with separate decay constants the activities may combine to give an overall activity that does not fall off as a simple exponential. Some examples of such combinations will be discussed later.

Only very simple assumptions are needed for a derivation of Eq. (13.1).

1. Radioactive decay results from the **emission** of an alpha particle, a beta particle, or a gamma ray by a parent nucleus. In the case of artificial radioactivity, which in its broadest sense may be considered to include nuclear reactions, additional radiations may be involved, including positron (β^+) emission, neutrons, protons, and various nucleon clusters.

2. The **probability** of a decay event occurring is directly proportional to the number of parent nuclei in the sample. Corollary to this statement is the implication that the probability for decay of any one nucleus is exactly the same as that for any other of the same species; the time elapsed since formation of the nucleus does not influence its decay probability.

[1] The rates of decay of certain radionuclides that decay predominantly by orbital electron capture (Section 14.3) or by internal conversion (Section 14.11) are affected by electron density in the neighborhood of the nucleus. This density may be influenced somewhat by the molecular structure in which the radioatom is situated. Only a few radionuclides exhibit this effect and in general the change in activity is quite small.

3. The quantity called **activity** is just the number of disintegrations per unit time.

$$A = -\frac{dN}{dt} \tag{13.2}$$

where N is the number of nuclei of the parent species present in the sample. The minus sign is due to the fact that $(dN)/(dt)$ is itself negative since the number of parent nuclei decreases.

From assumptions 2 and 3 it follows that

$$\frac{dN}{dt} = -\lambda N(t) \tag{13.3}$$

where λ is simply a proportionality constant introduced to implement assumption 2. That is,

$$A = \lambda N \tag{13.4}$$

As we shall see shortly, this use of λ is not inconsistent with its use in Eq. (13.1).

Rearranging and integrating Eq. (13.3) leads to

$$\ln N = -\lambda t + C \tag{13.5}$$

where C is a constant of integration. Further rearrangement leads to

$$N = e^C e^{-\lambda t} \tag{13.6}$$

Setting $N = N_0$ at $t = 0$ determines $e^C = N_0$. Finally, then,

$$N = N_0 e^{-\lambda t} \tag{13.7}$$

We see that N, the **population number**, falls off exponentially. Differentiating gives

$$\frac{dN}{dt} = -\lambda N_0 e^{-\lambda t} \tag{13.8}$$

Identifying λN_0 with the initial activity A_0 and making use of Eq. (13.2) leads us back to Eq. (13.1).

Using Eq. (13.7) the **mean-life** τ is defined as the time interval for which $N/N_0 = e^{-1}$. That is $\tau = \lambda^{-1}$. In a similar fashion the **half-life** $T_{1/2}$ is defined as the time interval for which $N/N_0 = 0.5$. It follows that

$$T_{1/2} = \frac{0.693}{\lambda} = 0.693\tau \tag{13.9}$$

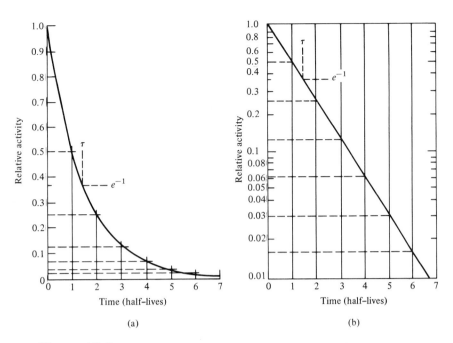

Figure 13.2 Exponential decay of a radioactive sample: (a) a linear plot, (b) a semilog plot. In both plots the mean life τ is indicated. It is the time during which the activity falls to e^{-1} (36.8 percent) of its original value.

During each half-life interval the population, and hence the activity, of a sample falls to 50% of the value it had at the beginning of the interval. Although the mean-life is the more appealing of the two lifetimes from a mathematical stand point, the half-life is easier to use for quick calculations and is the one usually found in tabulations of nuclear properties. Figure 13.2 shows two ways of diagramming the exponential decay. Figure 13.2(a) is a **linear** plot with the ordinate proportional to the activity. Figure 13.2(b) is a **semilog** plot with ordinate height proportional to the logarithm of the activity, but the numerical values indicated are the actual relative activities. For generality the abscissa in each case is calibrated in units of half-life. Dashed lines indicate activities of 50%, 25%, 12.5%, and so on. Also indicated is the activity at $A = e^{-1}A_0$ and the mean-life $\tau = 1.44T_{1/2}$. The decay curves in Figure 13.2(a) are theoretical exponential decays. Curves fitted to experimental data face the problem of dealing with plotted points that spread above and below the true decay curve. It is apparent that the semilog plot with its predicted straight line behavior makes the fitting much easier. The half-life can be read directly from either type of plot.

Determination of half-lives by direct plotting as in Figure 13.2 is limited to nuclides with suitable rates of decay. If the half-life is too long the activity will be essentially constant during any tolerable period of observation. At the other extreme are half-lives so short that the activity dies below detectable levels before adequate measurements can be made. With fast electronic circuits lifetimes of the order of 10^{-8} sec can be measured directly. Faster and faster circuits may be expected to lower the time limits. Indirect means can be used for very short lives. For instance, if the radioactive nuclei can be formed into a beam of swift particles the **distance** they travel before decay can be measured and the time inferred. In other circumstances, upon measuring the **energy spread** of the radiation the uncertainty principle may be invoked in the form

$$\Delta E \, \Delta t \approx \hbar \tag{13.10}$$

to give an estimate of the lifetime of the parent state.

For long-lived nuclides Eq. (13.4) is used directly. For example, consider a 1-g sample of ^{238}U. By use of Avagadro's number we see that the number N of atoms in the sample is 2.53×10^{21}. If the absolute activity is 1.24×10^4 disintegrations/sec, then

$$\lambda = \frac{A}{N} = 4.90 \times 10^{-18} \text{ sec}^{-1} \tag{13.11}$$

$$T_{1/2} = \frac{0.693}{\lambda} = 0.141 \times 10^{18} \text{ sec} = 4.48 \times 10^9 \text{ yr} \tag{13.12}$$

Obviously, even during an observation period of 50 yr, in an attempt to apply the approach of Figure 13.2, there would be no measurable decline in activity.

Although correct in principle this example ignores difficulties associated with measuring the absolute activity. In the first place the alpha particles, which form the main radiation from ^{238}U, have **very short range** in solid material. Unless the ^{238}U is in the form of an extremely thin film or foil most of the alphas will be stopped inside the sample and will not be detected. A milligram, or less, would be a more realistic amount of uranium. Even so, alphas that are aimed nearly parallel to the foil surface may lose too much energy in the foil to be detectable outside. Also, the **overall efficiency** of the detector must be well known. Both the **solid angle** for collection of the radiation and the **intrinsic** efficiency of the detection process itself are important. With care, however, the difficulties can be overcome and good results obtained.

13.3 Series Disintegration

Series disintegration occurs if the **daughter** product of the radioactive decay of a **parent** nuclide is itself radioactive. The next daughter (granddaughter?) may in turn be radioactive, and so on. In fact, most naturally occurring radionuclides are members of radioactive chains. The uranium series, for instance, runs to 19 members. The differential equations, extensions of Eq. (13.3), which govern the transformations in long chains were treated by Bateman in 1910 and are often identified with his name. To illustrate the procedure let us analyze the situation for a parent and first daughter. Let N_1 be the number of parent atoms present at any time and N_2 be that of the daughter. Let us assume there has been chemical separation so we start with a sample that has only parent atoms. Then

$$N_1 = N_{1,0} e^{-\lambda_1 t} \tag{13.13}$$

$$\frac{dN_1}{dt} = -\lambda_1 N_{1,0} e^{-\lambda_1 t} \tag{13.14}$$

But every decay of a parent atom creates a daughter atom. These atoms in turn decay with their own disintegration constant. The net rate of change in the number of daughter atoms is

$$\frac{dN_2}{dt} = \lambda_1 N_{1,0} e^{-\lambda_1 t} - \lambda_2 N_2 \tag{13.15}$$

The solution to this differential equation is

$$N_2 = \frac{\lambda_1}{\lambda_2 - \lambda_1} N_{1,0} (e^{-\lambda_1 t} - e^{-\lambda_2 t}) \tag{13.16a}$$

and

$$A_2 = \lambda_2 N_2 = \frac{\lambda_2}{\lambda_2 - \lambda_1} \lambda_1 N_{1,0} (e^{-\lambda_1 t} - e^{-\lambda_2 t}) \tag{13.16b}$$

In general, N_2 starting from zero initially goes through a period of growth, reaches a maximum, and then falls off when the rate at which it is replenished by decays of the parent is no longer sufficient to maintain the daughter. If $N_{2,0} \neq 0$, then an extra term $N_{2,0} e^{-\lambda_2 t}$ is to be added to the right side of Eq. (13.16a). However, in most cases of interest $N_{2,0} = 0$ and we shall confine further discussion to that situation.

There are several cases that merit special attention. If $(T_{1/2})_1 \ll (T_{1/2})_2$, $(\lambda_1 \gg \lambda_2)$ and if $(T_{1/2})_1$ is also very short compared to the time of observation, then substantially all of the parent is promptly converted into daughter atoms and the subsequent behavior of the daughter is given by

$$N_2 = N_{1,0} e^{-\lambda_2 t} \tag{13.17}$$

A condition known as **transient equilibrium** can be attained if $(T_{1/2})_1 > (T_{1/2})_2$ and if $(T_{1/2})_1$ is comparable to the time of observation.

Since $\lambda_1 < \lambda_2$, the exponential term in λ_2, which appears in Eqs. (13.16a,b), becomes negligible after a sufficiently long time, that is, for $t \gg (T_{1/2})_2$. The expression for N_2 approaches

$$N_2 = \frac{\lambda_1}{\lambda_2 - \lambda_1} N_{1,0}\, e^{-\lambda_1 t} \tag{13.18a}$$

and

$$A_2 = \left(\frac{\lambda_2}{\lambda_2 - \lambda_1}\right) \lambda_1 N_{1,0}\, e^{-\lambda_1 t} = \frac{\lambda_2}{\lambda_2 - \lambda_1} A_1 \tag{13.18b}$$

When these conditions hold, transient equilibrium has been reached and the daughter activity falls at a rate determined by the half-life of the parent. The activity of the daughter is larger than that of the parent in the ratio $\lambda_2/(\lambda_2 - \lambda_1)$. Prior to equilibrium, Eqs. (13.16a,b) express the behavior of the daughter. Manipulating this expression it can be shown that the **maximum activity** in the daughter occurs at a time

$$t_{m2} = \frac{1}{\lambda_2 - \lambda_1} \ln \frac{\lambda_2}{\lambda_1} = \frac{\tau_1}{k-1} \ln k = \tau_2 \frac{k}{k-1} \ln k \tag{13.19}$$

where

$$k = \frac{\lambda_2}{\lambda_1} = \frac{(T_{1/2})_1}{(T_{1/2})_2}$$

At this same time the activities of the parent and daughter are equal. The **total activity** of the sample is the sum of two activities.

$$A = A_1 + A_2 \tag{13.20}$$

For a time it actually increases above the initial activity of the pure parent. It reaches a maximum at a time somewhat earlier than does the activity of the daughter. The maximum combined activity occurs at the time

$$t_m = \frac{1}{\lambda_2 - \lambda_1} \ln \frac{\lambda_2^2}{2\lambda_1\lambda_2 - \lambda_1^2} = \frac{\tau_1}{k-1} \ln \frac{k^2}{2k-1}$$
$$= \frac{k}{k-1} \tau_2 \ln \frac{k^2}{2k-1} \tag{13.21}$$

As the activity of the daughter approaches that expressed in Eq. (13.18) the total activity approaches

$$A = A_{1,0}\, (e^{-\lambda_1 t}) \frac{2\lambda_2 - \lambda_1}{\lambda_2 - \lambda_1} \tag{13.22}$$

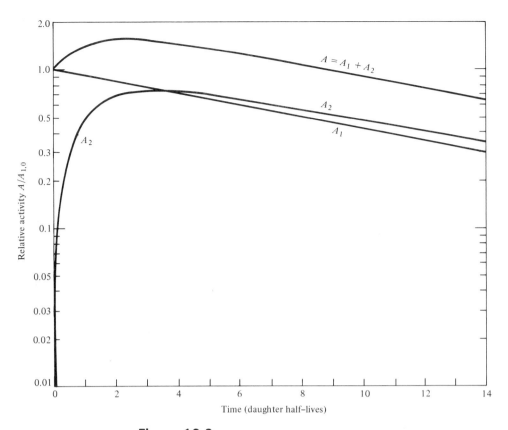

Figure 13.3 Transient equilibrium. A hypothetical series decay in which the half-life of the parent is eight times that of the daughter. The abscissa is labeled in units of the half-life of the daughter. A_1 is the activity of the parent; A_2 is the activity of the daughter; A is the total activity of the sample.

Figure 13.3 shows curves pertaining to transient equilibrium for a hypothetical parent-daughter pair for which $k = 8$.

If the half-life of the parent is very long in respect to the period of observation, there is no appreciable decline in its activity and the activity of the daughter approaches a constant value such that

$$\lambda_2 N_2 = \lambda_1 N_{1,0} \tag{13.23}$$

The daughter disintegrates at the same rate as it is produced. The total amount of the daughter remains constant even though its own half-life may be relatively short. We speak of this condition as **secular equilibrium**. The very long half-life of the parent implies that $\lambda_1 \sim 0$, so $e^{-\lambda_1 t} \sim 1$. Making this substitution in Eq. (13.16), and recognizing that $\lambda_1 \ll \lambda_2$ in the denominator, we find for the daughter

$$N_2 = \frac{\lambda_1}{\lambda_2} N_{1,0}(1 - e^{-\lambda_2 t}) \tag{13.24a}$$

$$A_2 = \lambda_2 N_2 = \lambda_1 N_{1,0}(1 - e^{-\lambda_2 t})$$ (13.24b)

Starting with a sample containing only the long lived parent, nuclide 1, the activity of the daughter grows and asymptoticly approaches that of the parent. The approach is governed by the half-life of the daughter. After five half-lives, for instance, the daughter has achieved about 97% of the activity it can ever achieve.

Example 13.1 Choosing suitable time intervals, calculate the activities of parent and radioactive daughter for each of the following decay sequences. The initial activity of the parent in each case is 10^4 disintegrations/sec. The symbol above the arrow designates the type of decay. The number below the arrow is the half-life. Comment on the nature of each sequence.

(a) $^{30}S \xrightarrow[\text{1.4 sec}]{\beta^+} {}^{30}P \xrightarrow[\text{2.5 min}]{\beta^+} {}^{30}Si$ (stable)

(b) $^{114}Pd \xrightarrow[\text{2.4 min}]{\beta^-} {}^{114}Ag \xrightarrow[\text{4.5 sec}]{\beta^-} {}^{114}Cd$ (stable)

(c) $^{122}Xe \xrightarrow[\text{20 hr}]{\beta^+} {}^{122}I \xrightarrow[\text{3.5 min}]{\beta^+} {}^{122}Te$ (stable)

Solution. The validity of the approximate expressions, Eq. (13.18b) and (13.24b), is restricted to certain time spans. For the purposes of this example we prefer to avoid such restrictions. We shall use Eq. (13.1) for the parent and Eq. (13.16b) for the daughter.

(a)

Time (sec)	Activity ^{30}S	^{30}P
0	10,000	0
1.4	5,000	46
5.0	842	84
15	6	88
30	0	82
90	0	62
180	0	41
330	0	20

$\lambda_1 = 0.495 \text{ sec}^{-1}$

$\lambda_2 = 0.462 \times 10^{-2} \text{ sec}^{-1}$

$\dfrac{\lambda_2}{\lambda_2 - \lambda_1} A_{1,0} = -94.21 \text{ sec}^{-1}$

At time $t = 30$ sec the activity of the parent is negligible. Thereafter the daughter, of much lower activity, decays with its own half-life of 150 sec.

(b)

Time (sec)	Activity	
	^{114}Pd	^{114}Ag
0	10,000	0
2.25	9,892	2911
4.5	9,786	4940
10	9,530	7624
30	8,656	8833
80	6,806	7025
144	5,000	5161
288	2,500	2581

$\lambda_1 = 0.481 \times 10^{-2} \text{ sec}^{-1}$
$\lambda_2 = 0.154 \text{ sec}^{-1}$

$\dfrac{\lambda_2}{\lambda_2 - \lambda_1} A_{1,0} = 10,322 \text{ sec}^{-1}$

Transient equilibrium. After ~ 80 sec the two activities decay at a common rate, determined by the half-life of the parent.

(c)

Time (min)	Activity	
	^{122}Xe	^{122}I
0	10,000	0
3.5	9,980	4994
7.0	9,960	7481
10.5	9,940	8715
14.0	9,919	9322
17.5	9,899	9615

$\lambda_1 = 9.625 \times 10^{-6} \text{ sec}^{-1}$
$\lambda_2 = 0.33 \times 10^{-2} \text{ sec}^{-1}$

$\dfrac{\lambda_2}{\lambda_2 - \lambda_1} A_{1,0} = 10,029 \text{ sec}^{-1}$

Secular equilibrium. The activity of the daughter is (nearly) approaching $A_{1,0}$ in asymptotic fashion.

13.4   Secular Equilibrium—Some Examples

The constant rate of production, which in the case of the radioactive parent is denoted by $\lambda_1 N_{1,0}$, can be provided by "artificial" means such as nuclear reaction in an accelerator or in a reactor. The production rate depends on the cross section for the reaction, the number of target atoms exposed to bombardment, and the number of incident particles per unit time. Letting R stand for the rate of production, the growth of the "daughter" is given by

$$N_2 = \frac{R}{\lambda_2} (1 - e^{-\lambda_2 t}) \tag{13.25}$$

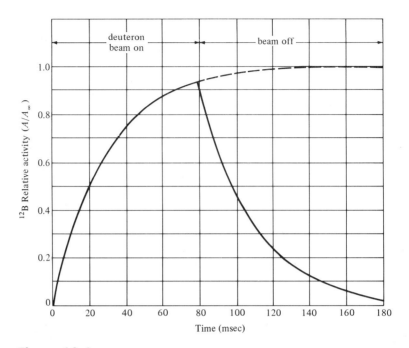

Figure 13.4 Secular equilibrium. ^{12}B is produced by deuteron bombardment of a target containing ^{11}B. In this illustration bombardment proceeds for four half-lives, followed by a decay period of five half-lives. The dashed line indicates the way the ^{12}B activity would grow if bombardment continued uninterrupted.

After a long time (many half-lives) $N_2 \approx R/\lambda_2$. When the target is removed or the accelerator turned off the nuclide created in the reaction decays according to its own half-life.

Figure 13.4 depicts the **growth** and **decay** of ^{12}B produced in an accelerator by the ^{11}B$(d, p)^{12}$B reaction. ^{12}B has a 0.020 sec half-life, decaying to ^{12}C by β emission. Because of its short half-life the supply of ^{12}B must be frequently replenished during studies of its beta decay or measurements of its decay rate. This is done by pulsing the accelerator beam, turning it on for several half-lives to build up the quantity of ^{12}B and then turning it off. During the beam-off periods the rate and nature of decay of the sample can be studied free of the copious prompt radiations produced while the target is undergoing bombardment. During each succeeding half-life of bombardment the net increase in ^{12}B activity is 50% of the difference between the activity at the beginning of the period and the activity which could be achieved by bombardment for infinite time. Because long bombardment times produce diminishing returns, it is usually felt that three or four half-lives provide a practical compromise.

Those radionuclides found in nature with half-lives short compared to the age of the elements exist by virtue of being in secular equilibrium with very long-lived parents. For instance, $^{234}_{90}$Th decays by beta emission and has a half-life $T_{1/2} = 24.1$ days. It is clear the amount now remaining would be indistinguishable from zero if there were no production process. However, ^{234}Th is the daughter, via alpha decay, of $^{238}_{92}$U, which has a half-life $T_{1/2} = 4.5 \times 10^9$ years. The daughter of ^{234}Th is $^{234}_{91}$Pa, which in turn is beta active with a half-life $T_{1/2} = 1.18$ minute. In fact, these three nuclides are at the beginning of the 19-member uranium series that ends with stable $^{206}_{82}$Pb. The properties of this series are listed in Table 13.1 and those of other series in Tables 13.2 through 13.4. If a mineral containing ^{238}U, ^{235}U, or ^{232}Th, each of which is at the head of one of the natural radioactive series, has for a long time (i.e., long compared to the half-life of the second-longest-lived member) been undisturbed by any process that would separate out members of the series, then the entire series with the ex-

Table 13.1 The Uranium Series (4n + 2)

Historical Name	Nuclide	Type of Disinte-gration	Half-life	Disinte-gration Constant (sec^{-1})	Particle Energy (MeV)
Uranium I (UI)	$^{238}_{92}$U	α	4.50×10^9 y	4.91×10^{-18}	4.20
Uranium X$_1$ (UX$_1$)	$^{234}_{90}$Th	β	24.10 d	3.33×10^{-7}	0.193
Uranium X$_2$ (UX$_2$)	$^{234}_{91}$Pa	β	1.1 m	9.87×10^{-3}	2.29
Uranium Z (UZ)	$^{234}_{91}$Pa	β	6.75 h	2.85×10^{-5}	0.49
Uranium II (UII)	$^{234}_{92}$U	α	2.44×10^5 y	9.00×10^{-14}	4.773
Ionium (Io)	$^{230}_{90}$Th	α	7.7×10^4 y	2.85×10^{-13}	4.684
Radium (Ra)	$^{226}_{88}$Ra	α	1600 y	1.37×10^{-11}	4.784
Ra Emanation (Rn)	$^{222}_{86}$Rn	α	3.824 d	2.10×10^{-6}	5.490
Radium A (RaA)	$^{218}_{84}$Po	α,β	3.05 m	3.79×10^{-3}	α:6.003 β:0.33
Radium B (RaB)	$^{214}_{82}$Pb	β	26.8 m	4.31×10^{-4}	1.04
Astatine-218 (^{218}At)	$^{218}_{85}$At	α	~2 s	0.4	6.695
Radium C (RaC)	$^{214}_{83}$Bi	α,β	19.8 m	5.83×10^{-4}	α:5.616 β:3.28
Radium C' (RaC')	$^{214}_{84}$Po	α	1.64×10^{-4} s	4.23×10^3	7.687
Radium C'' (RaC'')	$^{210}_{81}$Tl	β	1.30 m	8.89×10^{-4}	1.9
Radium D (RaD)	$^{210}_{82}$Pb	β	22.3 y	1.13×10^{-9}	0.017
Radium E (RaE)	$^{210}_{83}$Bi	β	5.01 d	1.60×10^{-6}	1.161
Radium F (RaF)	$^{210}_{84}$Po	α	138.3 d	5.80×10^{-8}	5.305
Radium E'' (RaE'')	$^{206}_{81}$Tl	β	4.20 m	2.75×10^{-3}	1.534
Radium G (RaG)	$^{206}_{82}$Pb	stable			

ception of the final stable nuclide will be found to be in equilibrium. Equation (13.23) can be extended in a run-on fashion.

$$\lambda_1 N_1 = \lambda_2 N_2 = \lambda_3 N_3 = \lambda_4 N_4 = \cdots \qquad (13.26)$$

The neptunium series, Table 13.4, is not found in nature. Its longest-lived member, $^{237}_{93}$Np, is alpha active and decays with a half-life of 2.2×10^6 yr. This is short compared to the age of the earth, which is estimated at about 6×10^9 yr. Any neptunium originally present has long since decayed away. In contrast, the long-lived heads of the chains for the other series have half-lives of 4.5×10^9 yr, 7.1×10^8 yr, and 1.4×10^{10} yr, respectively. These are sufficiently long that measurable quantities can still be found. The scarcity of ^{235}U may be attributed to its half-life being the shortest of the three just listed.

Table 13.2   The Actinium Series (4n + 3)

Historical Name	Nuclide	Type of Disinte-gration	Half-life	Disinte-gration Constant (sec^{-1})	Particle Energy (MeV)
Actinouranium (AcU)	$^{235}_{92}$U	α	7.04×10^8 y	3.12×10^{-17}	4.398
Uranium Y (UY)	$^{231}_{90}$Th	β	25.52 h	7.54×10^{-6}	0.303
Protoactinium (Pa)	$^{231}_{91}$Pa	α	3.25×10^4 y	6.76×10^{-13}	5.013
Actinium (Ac)	$^{227}_{89}$Ac	α,β	21.77 y	1.01×10^{-9}	α:4.95 β:0.44
Radioactinium (RdAc)	$^{227}_{90}$Th	α	18.72 d	4.28×10^{-7}	6.038
Actinium K (AcK)	$^{223}_{87}$Fr	α,β	22 m	5.25×10^{-4}	β:1.15 α:5.34
Actinium X (AcX)	$^{223}_{88}$Ra	α	11.43 d	7.02×10^{-7}	5.870
Astatine-219	$^{219}_{85}$At	α,β	0.9 m	1.28×10^{-2}	α:6.27
Ac Emanation (An)	$^{219}_{86}$Rn	α	3.96 s	0.175	6.819
Bismuth-215	$^{215}_{83}$Bi	α,β	~7.4 m	1.56×10^{-3}	2.2
Actinium A (AcA)	$^{215}_{84}$Po	α,β	1.78×10^{-3} s	3.89×10^2	α:7.386
Actinium B (AcB)	$^{211}_{82}$Pb	β	36.1 m	3.20×10^{-4}	1.38
Astatine-215 (^{215}At)	$^{215}_{85}$At	α	1.0×10^{-4} s	6.93×10^3	8.00
Actinium C (AcC)	$^{211}_{83}$Bi	α,β	2.14 m	5.40×10^{-3}	α:6.623 β:0.59
Actinium C' (AcC')	$^{211}_{84}$Po	α	0.53 s	1.31	7.448
Actinium C'' (AcC'')	$^{207}_{81}$Tl	β	4.77 m	2.42×10^{-3}	1.43
Actinium D (AcD)	$^{207}_{82}$Pb	stable			

Table 13.3 The Thorium Series (4n)

Historical Name	Nuclide	Type of Disintegration	Half-life	Disintegration Constant (sec^{-1})	Particle Energy (MeV)
Thorium (Th)	$^{232}_{90}Th$	α	1.40×10^{10} y	1.57×10^{-18}	4.007
Mesothorium1 (MsTh1)	$^{228}_{88}Ra$	β	5.75 y	3.82×10^{-9}	0.048
Mesothorium2 (MsTh2)	$^{228}_{89}Ac$	β	6.13 h	3.14×10^{-5}	2.09
Radiothorium (RdTh)	$^{228}_{90}Th$	α	1.913 y	1.15×10^{-8}	5.423
Thorium X (ThX)	$^{224}_{88}Ra$	α	3.64 d	2.20×10^{-6}	5.686
Th Emanation (Tn)	$^{220}_{86}Rn$	α	55.6 s	1.25×10^{-2}	6.288
Thorium A (ThA)	$^{216}_{84}Po$	α,β	0.15 s	4.62	6.779
Thorium B (ThB)	$^{212}_{82}Pb$	β	10.64 h	1.81×10^{-5}	0.57
Astatine-216 (^{216}AT)	$^{216}_{85}At$	α	3×10^{-4} s	2.3×10^{3}	7.805
Thorium C (ThC)	$^{212}_{83}Bi$	α,β	60.60 m	1.91×10^{-4}	α:6.090 β:2.25
Thorium C' (ThC')	$^{212}_{84}Po$	α	3.0×10^{-7} s	2.31×10^{6}	8.784
Thorium C'' (ThC'')	$^{208}_{81}Tl$	β	3.054 m	3.78×10^{-3}	1.795
Thorium D (ThD)	$^{208}_{82}Pb$	stable			

Table 13.4 The Neptunium Series (4n + 1)

Nuclide	Type of Disintegration	Half-life	Disintegration Constant (sec^{-1})	Particle Energy (MeV)
$^{241}_{94}Pu$	α,β	15 y	1.46×10^{-9}	α:4.896 β:0.0208
$^{241}_{95}Am$	α	433 y	5.07×10^{-11}	5.535
$^{237}_{92}U$	β	6.75 d	1.19×10^{-6}	0.248
$^{237}_{93}Np$	α	2.14×10^{6} y	1.03×10^{-14}	4.787
$^{233}_{91}Pa$	β	27.0 d	2.97×10^{-7}	0.571
$^{233}_{92}U$	α	1.58×10^{5} y	1.39×10^{-13}	4.824
$^{229}_{90}Th$	α	7430 y	2.99×10^{-12}	5.02
$^{225}_{88}Ra$	β	14.8 d	5.42×10^{-7}	0.33
$^{225}_{89}Ac$	α	10.0 d	8.02×10^{-7}	5.83
$^{221}_{87}Fr$	α	4.8 m	2.41×10^{-3}	6.34
$^{217}_{85}At$	α	0.032 s	21.7	7.07
$^{213}_{83}Bi$	α,β	46 m	2.51×10^{-4}	α:5.87 β:1.42
$^{213}_{84}Po$	α	4.26×10^{-6} s	1.65×10^{5}	8.38
$^{209}_{81}Tl$	β	2.2 m	5.25×10^{-3}	1.83
$^{209}_{82}Pb$	β	3.31 h	5.82×10^{-5}	0.635
$^{209}_{83}Bi$	stable			

13.5 Composite Decay

Mixed or **composite decay** results if two or more **genetically unrelated** radionuclides are found in the same sample. In these circumstances the activity of each radionuclide is independent of the other(s). If their half-lives are sufficiently different it may be possible to separate them by means of a straightforward semilog plot of the **composite** activity. For instance, 51.2% of ordinary silver is in the form of ^{107}Ag and the balance, 48.2%, is ^{109}Ag. **Neutron activation,** proceeding via $(n,\ \gamma)$ reactions, produces ^{108}Ag and ^{110}Ag. These isotopes of silver have half-lives of 144 sec (2.4 min) and 24 sec, respectively.

Figure 13.5 gives a composite decay curve for activated silver. By taking data over enough half-lives of the more rapidly decaying component its contribution becomes negligible. The late portion of the composite curve is a **straight line** on a semilog plot. This straight-line

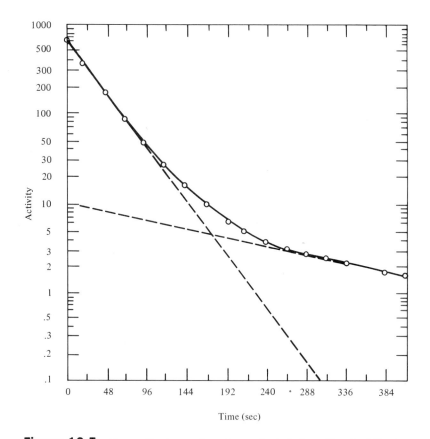

Figure 13.5 Composite decay of neutron activated silver. The short lived component is that due to ^{110}Ag from ^{109}Ag$(n,\ \gamma)$. The long lived component is that due to ^{108}Ag from ^{107}Ag$(n,\ \gamma)$.

portion may be extrapolated back toward zero time, and is shown dashed in Figure 13.5. In this fashion the amount of activity due to the long lived component is determined for the earlier times. Subtracting activities (not the logarithms!) gives early-time activity of the short lived component alone, shown as the dash-dot curve. Mutual adjustment of the separated curves may be needed for self-consistent results in the transition region. With more than two components and/or half-lives closer to each other, the unfolding becomes more difficult and subject to larger errors.

13.6 Alpha-Decay Energetics

The distribution in energy of alpha particles emitted from a radioactive substance forms a **discrete line spectrum.** This experimental fact is consistent with the interpretation that alpha decay is a process in which a parent nucleus in a definite energy state transforms into a new system made up of the alpha particle and a daughter nucleus also in a definite energy state. The difference in mass-energy of the initial and final states of the system appears as kinetic energy of the products. Many alpha emitters produce only a single energy group. Others emit alphas at several different discrete energies. Since the final system consists just of the alpha particle and the daughter nucleus the relationship between the kinetic energy of the alpha, K_α, and the energy of the transformation E_t is straightforward.

To all intents and purposes the parent nucleus is at rest before the transition. The **initial momentum** of the system is zero. Therefore to conserve momentum the daughter nucleus must **recoil** in a direction opposite to that taken by the alpha particle.

$$0 = M_\alpha v_\alpha - M_d v_d \tag{13.27}$$

Conservation of energy specifies that

$$[M_p - (M_\alpha + M_d)]c^2 = E_t = \tfrac{1}{2}M_\alpha v_\alpha^2 + \tfrac{1}{2}M_d v_d^2 \tag{13.28}$$

Equation (13.28) neglects relativistic effects. This is usually satisfactory since most alpha-particle velocities are of the order of 5% the velocity of light. For very precise work, though, relativisticly correct expressions should be used. Combining Eqs. (13.27) and (13.28) gives

$$E_t = K_\alpha \left(\frac{M_d + M_\alpha}{M_d} \right) \tag{13.29}$$

Although it is the alpha-particle energy that is directly observed, it is the transition energy E_t that connects the parent and daughter states, and consequently is of the greater interest.

Examples of radionuclides that produce single-line alpha spectra include $^{208}_{86}\text{Rn}$, $^{218}_{84}\text{Po}$, $^{212}_{84}\text{Po}$, and $^{208}_{84}\text{Po}$. Their alpha-particle energies are 6.141, 5.996, 8.780, 5.108 MeV, respectively. The transition energies may be found with Eq. (13.29). For these and similar simple alpha decays it seems safe to conclude that the transition is from the **ground state** of the parent to the **ground state** of the daughter. In other cases, and there are many such, the spectrum is complex and includes many different alpha-particle energies. Table 13.5 shows alpha-particle groups from the transition $^{227}_{90}\text{Th} \rightarrow {}^{223}_{88}\text{Ra} + \alpha$. The most energetic transition, α_0, is interpreted as leaving the daughter nucleus in its **ground state.** The α_1, α_2, and so on, transitions transfer successively less kinetic energy to the alphas. These transitions leave the daughter nucleus in various **excited states,** the degree of excitation being equal to the "missing" transition energy. These energy differences are listed in the fifth column of Table 13.5.

If this interpretation is correct we should expect the emission of gamma radiation to accompany the deexcitation of the daughter. Gamma emission does indeed accompany alpha decay when the spectrum is complex. Careful measurements of the gamma-ray energies

Table 13.5   Alpha-Particle Groups from the Decay of $^{227}_{90}\text{Th}$

Alpha Group*	K_α (MeV)	E_t (MeV)	Percent of Transitions	ΔE (MeV)
0	6.036	6.144	23	0
1	6.007	6.115	3	0.029
2	5.976	6.083	24	0.061
3	5.959	6.065	3	0.079
4	5.914	6.020	1	0.124
5	5.865	5.970	3	0.174
6	5.805	5.909	1	0.235
7	5.793	5.897	<1	0.247
8	5.761	5.864	<1	0.280
9	5.755	5.858	21	0.286
10	5.712	5.815	5	0.329
11	5.708	5.810	9	0.334
12	5.699	5.801	4	0.343
13	5.692	5.794	2	0.350
14	5.667	5.769	2	0.375

* The groups are numbered away from the most energetic, which is taken to be a ground state to ground state transition. ΔE is the difference in transition energy referenced to the ground state transition. It is equal to the excitation energy of the corresponding level in $^{223}_{88}\text{Ra}$.

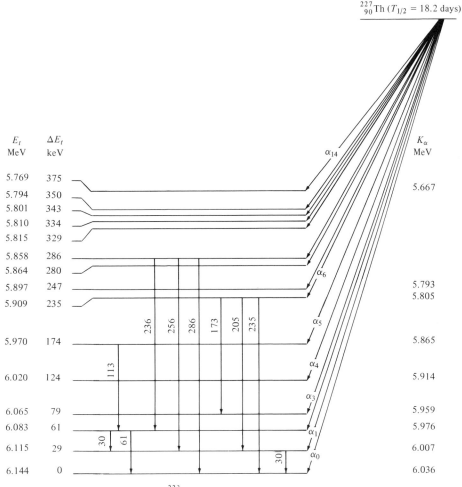

E_t MeV	ΔE_t keV		K_α MeV
5.769	375		
5.794	350	α_{14}	5.667
5.801	343		
5.810	334		
5.815	329		
5.858	286		
5.864	280	α_6	
5.897	247		5.793
5.909	235		5.805
5.970	174	α_5	5.865
6.020	124	α_4	5.914
6.065	79	α_3	5.959
6.083	61	α_1	5.976
6.115	29	α_0	6.007
6.144	0		6.036

$^{223}_{88}$Ra ($T_{1/2}$ = 11.7 days)

Figure 13.6 Energy levels of $^{223}_{88}$Ra. Excitation is by the α decay of $^{227}_{90}$Th. The right hand column tabulates measured energies of α particles in transitions to the ground state and first seven excited states and to the fourteenth state. The measured values for the eighth through thirteenth states are not listed (see Table 13.5 for complete list). The left column lists the transition energies derived from the alpha energies. The next column lists the excitation energies in keV. A few of many γ rays involved in deexcitation are shown by the vertical arrows, with adjacent numbers being the γ energy in kilo electron volts.

reveal correlations with energy differences between various levels. Figure 13.6 gives an energy level diagram for ^{223}Ra based on the alpha transitions shown in Table 13.5. Also shown in Figure 13.6 are ten gamma-ray transitions out of more than 20 known to accompany the alphas. Some levels do not deexcite directly to the ground state but do so only by cascade through intermediate states. For these there is no single E_γ equivalent to the energy of the excited state, but appropriate E_γ summations are found.

13.7 Barrier Penetration and the Decay Constant

Examination of Tables 13.1 through 13.4 reveals some interesting relationships. The ratio of the longest half-life to the shortest is huge, $\sim 10^{24}$. On the other hand, the lowest alpha energy is about 4 MeV and the highest about 8.8 MeV, a rather narrow range. Furthermore, nuclides with high disintegration energies have short lives; those with low disintegration energies have long lives. These facts were apparent to

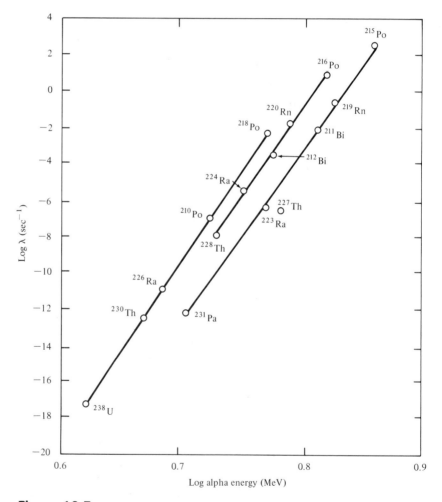

Figure 13.7 The Geiger-Nuttall law.

some of the earliest workers in the field and in 1911 were brought together into the empirical relationship known as the **Geiger-Nuttall law.**

$$\log \lambda = A \log E_\alpha + B \tag{13.30}$$

where A and B are constants. (As originally expressed, the Geiger-Nuttall law used the ranges of alphas in air in lieu of their energies. A and B then take different values.) For best fit B is assigned a different value for each of the radioactive series. Figure 13.7 demonstrates the law. Although this law was entirely **empirical** when set forth, its general features can be shown to derive from quantum mechanical concepts of particle **penetration** through **potential barriers.**

In discussing the energy spectra of alpha decays and the formation of excited states in the daughter nucleus we simply accepted the existence of these transformations. We recognized that there is an energy release E_t in the transition, but we were not concerned about the mechanism of the decay. The treatment, including Eq. (13.29), was quite classical. However, classical physics is unable to reconcile the fact that alpha-particle energies observed in the decay process are considerably less than the height of the Coulomb barrier implied by Rutherford scattering experiments.

Figure 13.8 indicates the general features of the situation. We presume that at sufficiently small r attractive nuclear forces bring about the formation of a potential well of depth U_0. At large r the potential is that of electrostatic repulsion between the alpha particle and the daughter nucleus. Now, Rutherford's experiments included some conducted with alpha particles of 8.8 MeV kinetic energy (supplied

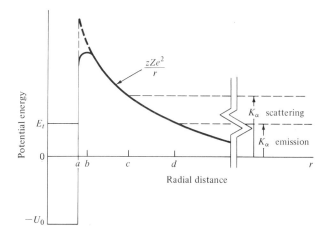

Figure 13.8 The potential barrier for α scattering and emission. Z is the atomic number of the daughter nuclide. $z = 2$ for α particles. Classical turning points are at $r = a$ and $r = d$ for emission energy E_t, and at $r = c$ for the (assumed higher) scattering energy.

by ^{212}Po) incident upon ^{238}U. Since no departure from Coulomb scattering was observed it was concluded that barrier heights in this region of the periodic table must exceed 9 MeV. Classically, on the other hand, any particle that escapes from a potential well must have a total energy, kinetic plus potential, at least as great as the height of the barrier. At large distance of separation the energy of the system is entirely kinetic. Hence E_t, the experimentally determined kinetic energy of decay, is also the total energy. For ^{238}U this is about 4.3 MeV, well below the height of the barrier. It is as though the alpha particle started at point d in Figure 13.8 and slid down the potential slope from there. But the ^{238}U system has a mean-life $\tau \approx 6.5 \times 10^9$ yr. It is inconceivable it would exist in this nearly stable fashion at the separation d. We are forced to believe that prior to its emission the alpha particle is inside the well. Even though classically this is a completely stable situation we see in actuality there is a finite probability of its escape. To do so it must **tunnel** through the barrier. Such tunneling is allowed in quantum mechanical systems and the particle may appear on the other side of the barrier with the same total energy it had inside the well. In this way the paradox is resolved.

Not only does **barrier penetration** explain the spontaneous emission of alpha particles with energies well below the top of the Coulomb barrier, it can also account for the sharp dependence of disintegration constant on emission energy. Let us suppose that the decay probability is the product of two terms

$$\lambda = \nu P \tag{13.31}$$

where ν is the number of times per second that an alpha particle presents itself against the inside of the barrier at a, and P, called the penetrability, is the probability of tunneling through to d. For an estimate of ν we write

$$\nu = \frac{v_\alpha}{2a} = \left[\frac{2(E_t + U_0)}{m}\right]^{1/2} \frac{1}{2a} \tag{13.32}$$

where v_α is the velocity with which the alpha particle rattles around inside the nucleus. This is a rather crude estimate for ν but it turns out that λ is much more sensitive to P than to ν. For ^{238}U $\nu \sim 10^{21}$ sec^{-1}. With $\tau \approx 6 \times 10^9$ yr this implies $\sim 3 \times 10^{38}$ tries for each escape!

Example 13.2   Calculate for ^{238}U the frequency of interior collision with the potential barrier by an alpha particle whose exterior kinetic energy is 4.20 MeV.

Solution.   We take the depth of the potential well to be $U_0 = 45$ MeV and the radius $a = 1.4 \times 10^{-13} \times 238^{1/3} = 8.68 \times 10^{-13}$ cm. Inside the nucleus the kinetic energy of the alpha particle is 49.2 MeV = 7.87×10^{-5} erg. Therefore

$$v_\alpha = \sqrt{\frac{2K}{m}} = \sqrt{\frac{2 \times 7.87 \times 10^{-5}}{4.006 \times 1.660 \times 10^{-24}}} = 4.865 \times 10^9 \text{ cm/sec}$$

The collision frequency $\nu = v_\alpha/2a$

$$\nu = \frac{4.86 \times 10^9}{2 \times 8.68 \times 10^{-13}} = 2.8 \times 10^{21} \text{ sec}^{-1} \quad \bullet$$

In the region between $r = a$ and $r = d$ the kinetic energy of the system is negative. Solutions to the Schrödinger equation in such a region are in the form of exponential decays. Relating the square of the amplitude of the wave function in a given region to the probability of finding the particle in that region gives

$$P = \exp\left\{-\frac{2(2m)^{1/2}}{\hbar} \int_a^d [U(r) - E_t]^{1/2} \, dr\right\} \tag{13.33}$$

The trick now is to select the proper form for $U(r)$ in the region of tunneling. Various forms have been tried that express the rounding at the top of the barrier as well as centrifugal effects that appear if the parent and daughter have angular momentum between them different from zero. By treating the inner radius a as an adjustable parameter, quite good fits to measured values of λ have been obtained. If we examine Eq. (13.33) we see that the exponent becomes more negative, and therefore P smaller, when

1. m increases (i.e., emission of something heavier than an α)

2. a decreases (shell effects may make for a more tightly bound and therefore smaller nucleus)

3. E_t decreases {this increases $[U(r) - E_t]$ and also increases d, the crossover between $U(r)$ and E_t}

4. $U(r)$ increases [for instance, because of the important product zZe^2; the effect of increase in $U(r)$ is much the same as decrease in E_t]

For alpha emission in nuclides at the heavy end of the periodic table, a and Z change but very little and m not at all. E_t is the important factor. The simplest approximation to use for $U(r)$ is to ignore rounding and continue the Coulomb potential down to $r = a$, shown as the dotted potential peak in Figure 13.8. We then have

$$\lambda = \frac{1}{2a}\left[\frac{2(E_t + U_0)}{m}\right]^{1/2} \exp\left\{-\frac{2(2m)^{1/2}}{\hbar} \int_a^d \left[\frac{2Ze^2}{r} - E_t\right]^{1/2} \, dr\right\} \tag{13.34}$$

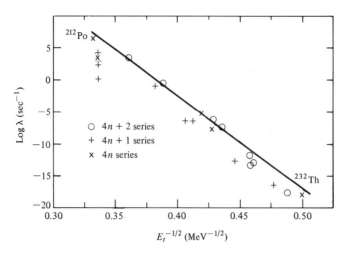

Figure 13.9 Dependence of disintegration constant on α-particle energy.

In Figure 13.9 the solid line is the result of evaluating Eq. (13.34) while the plotted points show experimental values for heavy radionuclides. Even though it was necessary to make a number of approximations to arrive at Eq. (13.34) the fit is quite good. The important feature is the rapid change in λ with E_t. For E_t much greater than 9 MeV ($E^{-1/2} < 0.33$) the disintegration constant becomes so large that such nuclides exist for too short a time to be found. For E_t smaller than 4 MeV ($E^{-1/2} > 0.5$) half-lives become so great the nuclides appear stable even though α emission is energetically possible. In fact, there is evidence that ^{209}Bi, which is the end of the neptunium series and is usually considered stable, is an α emitter of energy ~ 3 MeV and a half-life $\geq 2 \times 10^{18}$ yr.

13.8   Summary

Radioactivity is a statistical process and obeys an exponential decay law.

$$N = N_0\, e^{-\lambda t}$$

The activity of a radioactive species is

$$A = -\frac{dN}{dt} = \lambda N_0 e^{-\lambda t} = \lambda N$$

The mean-life τ is the reciprocal of the decay constant λ. The half-life is more commonly tabulated and is given by

$$T_{1/2} = \frac{0.693}{\lambda} = 0.693\tau$$

Three families, or series, of radioactive nuclides with $Z > 82$ occur in nature.

The uranium series: $A = 4n + 2$

The thorium series: $A = 4n$

The actinium series: $A = 4n + 3$

A fourth series with $Z > 83$ has presumably decayed away. The half-life of its longest-lived member is only 2.2×10^6 years.

The neptunium series: $A = 4n + 1$

Within each series the mass number changes either by four units or by zero in each disintegration, since only alpha or beta emission occurs.

The Bateman equations relate the growth and decay of a daughter product that is itself radioactive.

$$N_2 = \frac{\lambda_1}{\lambda_2 - \lambda_1} N_{1,0} \, (e^{-\lambda_1 t} - e^{-\lambda_2 t})$$

Transient equilibrium is a special case, which occurs when $(T_{1/2})_2 < (T_{1/2}) \approx$ time of observation.

$$N_2 = \frac{\lambda_1}{\lambda_2 - \lambda_1} N_{1,0} \, e^{-\lambda_1 t} \qquad \text{for } t \gg (T_{1/2})_2$$

Secular equilibrium is another special case, which occurs when $(T_{1/2})_2 \ll (T_{1/2})_1 \gg$ time of observation. The Bateman equations lead to

$$N_2 = \frac{\lambda_1}{\lambda_2} N_{1,0} \, (1 - e^{-\lambda_2 t})$$

Alpha decay produces a line spectrum of particle energies. The transition energy is related to the kinetic energy of the alpha by

$$E_t = K_\alpha \left(\frac{M_d + M_\alpha}{M_d} \right)$$

In terms of masses

$$E_t = Q = [M_{\text{parent}} - (M_d + M_\alpha)] \, c^2$$

Alpha decay energies are confined to the range from about 4 MeV to 8.8 MeV. Half-lives range from 1.4×10^{10} years to 3×10^{-7} sec.

Alpha energies are substantially below the height of the Coulomb barrier which characterizes the interaction between the alpha particle and the daughter nuclide. This cannot be understood on the basis of classical mechanics but can be understood on the basis of barrier penetration, a quantum mechanical effect.

Problems

13.1 In a certain half-life measurement, counts were taken at 5-min intervals, the counting time for each observation being 1 min. The following sequence of counts was observed: 1023, 863, 795, 622, 553, 480, 496, 375, 318, 250, 233. Make both linear and semilog plots of these data versus time. To allow for statistical fluctuations in the disintegration rate, draw through each point a vertical "error bar" extending $\pm \sqrt{N}$ above and below the plotted point. From the best-fit curves determine $T_{1/2}$.

13.2 A 100-μg sample of ^{239}Pu is deposited in a spot 2.0 mm diameter on a backing material. A 1-cm diameter detector located 5 cm from the sample registers 43,500 counts/min. Assume the detector is 100% efficient, that is, that it registers a count for each alpha particle that strikes it. Calculate the half-life and decay constant of ^{239}Pu.

13.3 A detector capable of receiving radiation emitted in all directions from a source is said to be one of 4π geometry. Suppose such a detector is used to inspect a sample for the presence of ^{90}Sr, and that because of background counts the sample cannot be detected if its activity is less than 5 counts/sec. What is the minimum detectable mass of ^{90}Sr?

13.4 A sample of ^{222}Rn has an activity of 10^4 counts/min immediately after separation from the other members of the uranium series. At what times do the maximum activities of the daughter, ^{218}Po, and of the combined parent and daughter occur? Calculate these activities.

13.5 A separated source of ^{210}Po has an initial activity of 10^4 counts/min. Calculate the activity of the daughter, ^{206}Tl, at 1, 2, 4, 8, 12, and 16 min. Make a linear plot of the ^{206}Tl activity versus time.

13.6 In a certain accelerator experiment ^{38}Cl is produced by the ^{37}Cl(d, p) ^{38}Cl reaction. If there are 5×10^8 (d, p) reactions/sec, calculate the ^{38}Cl activity at 5, 15, 30, 60, 90, 120 min. Make a linear plot of the ^{38}Cl activity versus time.

13.7 Complete the steps necessary to derive Eq. (13.29).

13.8 (a) Calculate the transition energies for the alpha decay of ^{208}Rn, ^{218}Po, ^{212}Po, and ^{208}Po.
 (b) What are the recoil energies of the daughter nuclei in each of these transitions?

13.9 From alpha-decay energetics determine
 (a) the difference in ground state mass between ^{210}Po and ^{206}Tl.
 (b) the difference in mass between the fourteenth excited state of ^{223}Ra and its ground state.

13.10 Calculate the recoil kinetic energy of ^{223}Ra in the alpha-decay transition that leaves it in the fourteenth excited state.

13.11 Calculate the penetrability factor and the frequency of interior collision with the barrier for the 5.305-MeV alpha decay of ^{210}Po. Refer to Figure 13.8 and assume $U_0 = 45$ MeV, $a = 1.4 A^{1/3}$ fermis, and a Coulomb potential for $r > a$. Determine the decay constant.

13.12 Derive Eq. (13.19).

13.13 Derive Eq. (13.21).

13.14 The half-life of ^{60}Co is 5.26 yrs. If the initial activity of a given sample of ^{60}Co is 4000 disintegrations/sec, calculate the activity at the end of 1 month, 1 yr, 5 yrs, 10 yrs. Determine the amount of ^{60}Ni present at these times.

13.15 The half-life of ^{137}Cs is 30.0 yrs. A 25-μg sample of ^{137}Cs is observed to produce 100 counts/sec in a certain counter. What is the overall counting efficiency?

13.16 ^{238}U constitutes 99.27% of natural uranium and ^{235}U constitutes 0.72%. It is reasonable to assume that these nuclides were present in equal amounts at the time of formation of the elements and that the current difference in abundance arises from their differing rates of decay. On this basis calculate the age of the uranium found in the earth's crust.

13.17 The half-life of ^{14}C is 5730 yrs. Determine the age of an archaeological specimen in which the ^{14}C to ^{12}C ratio is $(10 \pm 1)\%$ of that in a modern specimen. What is the maximum age for which ^{14}C dating is useful if the limit of detection is 0.1%?

Suggestions for Further Reading

Brown, H. "The Age of the Solar System," *Scientific American* (April 1957).

Evans, R. D. *The Atomic Nucleus.* New York: McGraw-Hill Book Company, Inc., 1955. Chapter 15.

Halliday, D. *Introductory Nuclear Physics,* 2nd ed. New York: John Wiley & Sons, Inc., 1955. Chapter 4.

Kaplan, I. *Nuclear Physics,* 2nd ed. Reading, Mass.: Addison-Wesley Publishing Co., Inc., 1962. Chapters 10 and 13.

Beta decay and gamma decay

14.1   Introduction

Emission of beta particles is often, but not always, accompanied by emission of gamma rays. Emission of gamma rays from a radioactive source is always preceded by decay of a parent nucleus. We have seen in the previous chapter that this parent may be an alpha emitter. However, gamma emission is more commonly associated with radionuclides that are beta active than with those that are alpha active.

Furthermore, nearly all artificial radionuclides are beta emitters. Because artificial radionuclides can be produced in nuclear reactors, their number exceeds the number of natural ones by a substantial margin. Ready availability of a wide variety of nuclides has invigorated the study of basic processes involved in radioactive decay.

Because of the close relationship between beta and gamma decay we treat both in this chapter. We shall consider the **energetics** of transitions involved in these decays and the construction of beta-gamma decay schemes, the involvement of **neutrinos** in beta decay, the violation of **parity conservation,** and the **lifetimes** of nuclear states. Although we shall not treat medical and industrial applications, it is interesting to note that radioactive sources for such applications are almost universally beta and gamma emitters.

14.2   Beta Spectrum

The general character of a beta spectrum differs most remarkably from that of an alpha spectrum. Whereas α-particle emission occurs at one, or at most several, discrete and definite energies characteristic of the emitter, beta particles are emitted with a **distribution** in energy which is continuous from essentially zero to some **maximum energy.** Figure 14.1 shows a typical spectrum, that of ^{210}Bi (old name: RaE). The spectrum is **continuous,** but by no means uniform. The most

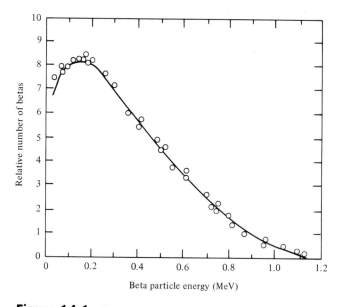

Figure 14.1   The β spectrum of ^{210}Bi. [G. J. Neary: The β-ray spectrum of radium E. *Proc. Roy. Soc.* (*London*), A175:71 (1940).]

probable energy is rather low. Thereafter the distribution falls gradually and finally trails off at an **endpoint,** or maximum, energy of 1.161 MeV. We shall say more later about other ways of plotting that display and locate the endpoint with greater exactness than is achieved in Figure 14.1.

As we shall see, the endpoint energy is equivalent to the **energy of transition.** Furthermore, just as in alpha decay the transition probability is related to this energy. However, the range of half-lives for beta decay is much less than that for alpha decay, whereas the range in transition energy is much larger. The ratio of the longest beta-decay half-life to the shortest among the nuclides of Tables 13.1 through 13.4 is $\sim 10^7$; the ratio of the highest to lowest transition energy is 190. A plot of log λ versus log K_m where K_m is the endpoint energy, was introduced by E. W. Sargent in 1933 and is shown in Figure 14.2. The diagram is analogous to the Geiger-Nuttall rule for alpha decay. There are two apparent groups, designated by lines drawn through the points, but unlike the Geiger-Nuttall plot they do not correlate with the different natural radioactive series. For a given energy the upper group has a decay probability about 100 times as great as the lower group and is designated **allowed decay,** while the lower, less probable group is referred to as **forbidden.**

In a beta-decay process the **atomic number** of the nucleus changes by one unit but the **mass number** remains unchanged. With **negatron**

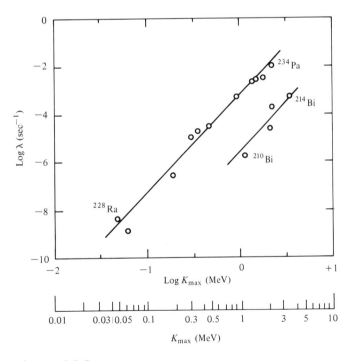

Figure 14.2 The Sargent diagram for heavy radionuclides.

(β^-) emission the atomic number increases, that is, $Z \to Z + 1$; one of the neutrons in the nucleus becomes a proton. With **positron** (β^+) emission a proton becomes a neutron and $Z \to Z - 1$. In a related process an orbital electron of the atom may be **captured** by the nucleus, resulting thereby in the conversion of one of the protons into a neutron. Here, also, $Z \to Z - 1$. These processes may be indicated in the following fashion.

$$n \to p + \beta^- + \bar{\nu} \quad \text{(negatron emission)} \tag{14.1a}$$

$$p \to n + \beta^+ + \nu \quad \text{(positron emission)} \tag{14.1b}$$

$$p + e^- \to n + \nu \quad \text{(electron capture)} \tag{14.1c}$$

where ν and $\bar{\nu}$ are the neutrino and antineutrino, respectively. We shall discuss a little bit later the roles they play.

14.3 Beta-Decay Energetics

In the processes indicated by Eqs. (14.1) the beta particle, or electron, is **created** (processes 14.1a and 14.1b) or destroyed (process 14.1c) at the instant the transition takes place. The alternative hypothesis, that a neutron is simply a proton and electron in close association is unsatisfactory since that would predict integer spin for the neutron instead of the observed half-integer spin. A similar argument applies to any nuclide with odd neutron number. Furthermore, the **electron wavelength** for energies characteristic of beta decay exceeds the size of the nucleus. It is not possible to locate the electron within a region so small.

Nuclear binding energies determine which, if any, of the decay processes is **energetically possible.** The free neutron is unstable against beta decay (process 14.1a) and has a half-life of $\sim$11 min and transition energy of 0.78 MeV. In many nuclides, however, this process is energetically forbidden. They are stable against negatron decay. On the other hand, processes (14.1b) and (14.1c) are energetically forbidden for the free proton but can occur as transitions connecting certain parent and daughter nuclides. Writing mass-energy equations in terms of neutral atomic masses we have

$$^A_Z M \to \,_{Z+}M + Q/c^2 \quad (\beta^- \text{ decay}) \tag{14.2a}$$

$$^A_Z M \to \,_{Z-1}^A M + 2m_0 + Q/c^2 \quad (\beta^+ \text{ decay}) \tag{14.2b}$$

$$^A_Z M \to \,_{Z-1}^A M^* + Q/c^2 \quad \text{(electron capture)} \tag{14.2c}$$

In Eq. (14.2a) the electron mass does not appear on the right side. Although a negative electron is released from the nucleus an extra electron is needed for the $(Z + 1)$ atom; there is no electron left over in the mass-energy balance. Two electron masses appear on the right side of Eq. (14.2b). One is the positive electron from the nucleus, the other is the electron released from the new atom whose atomic number is $(Z - 1)$. In electron capture, Eq. (14.2c), an electron is captured from one of the inner orbits. The resulting atom, of atomic number $(Z - 1)$, is thus neutral but **excited.** We use the asterisk on M^* to indicate excitation.

Energy releases Q can be written

$$Q = (_Z^A M - _{Z+1}^A M) c^2 \qquad (\beta^- \text{ decay}) \qquad \textbf{(14.3a)}$$

$$Q = (_Z^A M - _{Z-1}^A M - 2m_0) c^2 \qquad (\beta^+ \text{ decay}) \qquad \textbf{(14.3b)}$$

$$Q = (_Z^A M - _{Z-1}^A M^*) c^2$$
$$= (_Z^A M - _{Z-1}^A M) c^2 - E_x \qquad (\text{electron capture}) \qquad \textbf{(14.3c)}$$

where E_x is the **excitation energy** of the daughter atom. This excitation energy is quickly released in the form of x rays **characteristic of the daughter.** Process (14.3a) can occur whenever the atomic mass of the daughter is less than that of the parent. For process (14.3b) to occur, the atomic mass of the parent must exceed that of the daughter by at least two electron rest masses. Process (14.3c), which connects the same parent and daughter and therefore is in competition with process (14.3b), demands a mass difference of only E_x. Since this is always much less than 1.02 MeV $(= 2m_0 c^2)$ electron capture may occur even though positron emission cannot.

Example 14.1 The mass of $_{57}^{137}\text{La}$ is 136.906040 u. The masses of neighboring isobars $_{56}^{137}\text{Ba}$ and $_{58}^{137}\text{Ce}$ are 136.905500 u and 136.90733 u, respectively. What types of beta decay are energetically possible? Calculate energies of the emitted particles.

Solution

$_{58}^{137}\text{Ce}$ is heavier than $_{57}^{137}\text{La}$, so β^- decay is not possible.

$_{57}^{137}\text{La}$ is heavier than $_{56}^{137}\text{Ba}$ by 5.4×10^{-4} u.

$\Delta M c^2 = 0.503$ MeV

Since $\Delta M c^2 < 2m_0 c^2$, β^+ decay cannot occur.

Electron capture is possible.

The K-shell binding energy in $_{56}^{137}\text{Ba}$ is 37.4 keV.

Therefore 0.466-MeV neutrinos are emitted, accompanied by one or more x rays of barium with total energy 0.037 MeV.

14.4 The Neutrino Hypothesis

If Eq. (14.3) is valid, every beta decay between a pair of parent and daughter nuclides should result in the same energy release. We certainly expect this on the basis of conservation of energy. The continuous spectrum of beta-particle energies seems to contradict this. To avoid the contradiction, Pauli, in 1931, postulated an **unobserved** particle which could carry away the energy that appears to be missing in those events that produce beta particles of less than endpoint energy. Any particle that is very difficult to detect must be so by virtue of an extremely small probability of interacting with the materials of which detectors are made. As a consequence, the postulated particle is assumed to have **no electric charge** and to have a **rest mass** that is very small, perhaps **zero.** In recognition of these properties it is named the **neutrino.** Whenever a disintegration results in three particles (beta particle, neutrino, daughter atom), the available energy can be shared among the particles in a variety of ways while still conserving linear momentum. Thus betas of low energy are associated with neutrinos of high energy, whereas β's whose energy is near the endpoint are associated with neutrinos whose energy is nearly zero. In each disintegration the daughter atom **recoils** in such a fashion that the **total linear momentum** of the system is the same, usually zero, as that of the parent atom prior to the transition.

If the endpoint of the beta spectrum is associated with neutrinos of zero energy, it follows that the available energy should appear in the form of the kinetic energy of the beta particle and the recoiling daughter. The recoil energy is quite small and can usually be neglected. To test the hypothesis that the beta-spectrum **endpoint energy** is the same as the **transition energy,** it is necessary to measure the transition energy by means independent of the beta spectrum. This has been done for many parent-daughter pairs and the hypothesis is now accepted as thoroughly proved. As an example let us consider the positron decay of ^{13}N, which has a half-life of 10 min and an endpoint energy of 1.20 MeV.

$$^{13}\text{N} \xrightarrow{\beta^+} {}^{13}\text{C} + 1.20 \text{ MeV} \tag{14.4}$$

From Eq. (14.3b) the transition energy is given by

$$\begin{aligned} Q &= [^{13}\text{N} - {}^{13}\text{C}]c^2 - 2m_0c^2 \\ &= [^{13}\text{N} - {}^{13}\text{C}]c^2 - 1.02 \text{ MeV} \end{aligned} \tag{14.5}$$

But ^{13}N and ^{13}C can also be related through the endoergic reaction

$$^{13}\text{C}(p, n)^{13}\text{N} - 2.98 \text{ MeV} \tag{14.6}$$

which can also be written in the form

$$^{13}\text{C} + {}^{1}\text{H} = {}^{13}\text{N} + n - \frac{2.98 \text{ MeV}}{c^2} \qquad (14.7)$$

Rearranging gives

$$[^{13}\text{N} - {}^{13}\text{C}]c^2 = 2.98 - [n - {}^{1}\text{H}]c^2 \qquad (14.8)$$
$$= 2.98 - 0.78$$
$$= 2.20 \text{ MeV}$$

Substituting for $[^{13}\text{N} - {}^{13}\text{C}]c^2$ in Eq. (14.5) we find $Q = 1.18$ MeV, in satisfactory agreement with the observed endpoint energy of 1.20 MeV.

Not only would beta decay violate conservation of **energy** if there were no "extra" particle such as the neutrino, it would also violate conservation of **angular momentum** and of **statistics**. If the transition is between even-A isobars, both parent and daughter have integer spin, J. The emitted beta particle, whether positron or negatron, has half-integer intrinsic spin s. It is not possible to combine J_{final} (an integer) with s (half-integer) to equal J_{initial} (an integer). If there is mutual angular momentum, L, between the β particle and the daughter atom it will be an integer and therefore cannot help achieve balance. If, however, the neutrino is postulated to have **half-integer spin,** this can be combined with the resultant half-integer angular momentum of the other products to produce the requisite integer angular momentum. Similarly, if the parent and daughter are odd-A, their J's are half-integer and s (half-integer) as well as s_β (half-integer) are needed to produce a half-integer resultant. Furthermore, to conserve **statistics** the evenness or oddness of the system must be the same before and after the transition. This requires that the neutrino be a **fermion.**

In connection with Eqs. (14.1) we drew a distinction between the **neutrino** and the **antineutrino.** None of the conservation laws we have discussed thus far would require such a distinction. However, akin to the conservation of nucleon number is another conservation law, the conservation of the number of particles belonging to the **electron family.** All experimental evidence bearing on the subject is in support of the conclusion that the negatron, the β^-, is the same as the electron: The **charge to mass ratio** is the same, the **charge** is the same, the **mass** is the same, both have the same **intrinsic spin,** both **interact with matter** in the same way. Without doubt the β^- is an electron. Except for its opposite electric charge, the properties of the β^+ are the same as those of the electron; the magnitude of the **charge** is the same, the **mass** is the same, and so is the **intrinsic spin.** If an electron and a positron combine, not only do the charges cancel but the masses themselves **annihilate.** Their total mass is converted into energy, $2m_0c^2$, which appears in the form of gamma-ray photons. The positron is clearly the **antiparticle** of the electron. As for the neutrino, it is considered a member of the electron family since it appears only in processes that involve the creation or disappearance of an electron.

To see how the law of **conservation of family number** works we assign the number (+1) to a member of a given family if the member is of ordinary matter, (−1) if it is of antimatter. We assign the number (0) if the particle does not belong to the family at all. Applying this rule to the different types of beta decay we find:

$$(\beta^- \text{ decay}) \qquad n \rightarrow p + \beta^- + \bar{\nu}$$
nucleon family number: $\quad 1 = 1 + 0 + 0$ (14.9a)
electron family number: $\quad 0 = 0 + 1 + (-1)$

$$(\beta^+ \text{ decay}) \qquad p \rightarrow n + \beta^+ + \nu$$
nucleon family number: $\quad 1 = 1 + 0 + 0$ (14.9b)
electron family number: $\quad 0 = 0 + (-1) + 1$

$$(\text{electron capture}) \qquad p + e^- \rightarrow n + \nu$$
nucleon family number: $\quad 1 + 0 = 1 + 0$ (14.9c)
electron family number: $\quad 0 + 1 = 0 + 1$

Only by including the neutrino and antineutrino as members of the electron family and by considering the neutrino associated with β^- decay to be antimatter is it possible to preserve electron family number before and after each transition.

14.5 Neutrinos — Experimental Confirmation

Although it saved the conservation laws, the **neutrino hypothesis** was open to the challenge that it was contrived in an effort to avoid facing a possibly legitimate failure of those laws. The unease thus engendered was put to rest by a series of experiments reported from 1953 to 1960 by Cowan, Reines, and coworkers. Figure 14.3 sche-

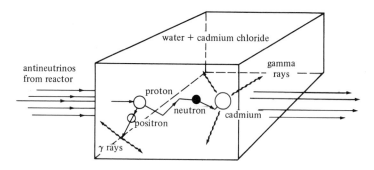

Figure 14.3 Schematic diagram of the antineutrino experiment by Reines and Cowan.

matically diagrams their approach. In their experiments they allowed antineutrinos from beta decays occurring in large nuclear reactors to impinge upon ingenious detection systems. The systems were designed to respond to a particular sequence of events that would serve as the **signature** of the capture of an antineutrino. If the capture is by a proton the reaction may be written

$$\bar{\nu} + p \rightarrow n + \beta^+ \tag{14.10}$$

and is called inverse beta decay. [Compare Eq. (14.1a) with the reaction arrow reversed and the beta particle moved to the other side of the equation and its sign reversed.] To supply the protons a 200-liter tank of water was used. Within about 10^{-9} sec after it is created by process (14.10), the positron annihilates with an electron in the water, resulting in the release of two gamma rays of 0.51 MeV each. To detect the neutron, cadmium salts are dissolved in the water, thereby taking advantage of cadmium's very high cross section for the (n, γ) reaction with slow neutrons. The gamma rays from the **positron annihilation** and from the **neutron capture** are detected by scintillation counters using a large number of photomultiplier tubes. The neutron requires about 10^{-6} sec to be slowed in the water before it can efficiently produce the (n, γ) reaction. Therefore the signature for the inverse beta decay is a pair of 0.51-MeV gamma rays followed by other gamma rays a few microseconds later. In a flux of about 5×10^{13} antineutrinos cm^{-2} sec^{-1} (calculated from the beta-decay rate in the reactor) about three events per hour could be assigned to inverse beta decay in the detector. The difficulty of the experiment may be appreciated upon noting the extremely low probability for neutrino interaction with material through which they pass. This probability is so low that neutrinos can pass through the sun with negligible attentuation. In fact, to attenuate the antineutrino beam of this experiment by a factor of $1/e = 0.37$ would require a layer of water ~3600 light-years thick.

Conservation of electron family number, Eq. (14.9), calls for a neutrino to be emitted in positron decay and an antineutrino in negatron decay. Now, with uncharged particles it sometimes occurs that particle and antiparticle are identical. It is of interest then to test whether this is true of neutrinos. A reaction is needed that ostensibly requires neutrinos. If it will "go" with antineutrinos, then the particles are identical. In 1955 R. Davis, Jr., reported negative results for the reaction

$$\bar{\nu} + {}^{37}\text{Cl} \rightarrow {}^{37}\text{Ar} + \beta^- \tag{14.11}$$

3000 gal of pure carbon tetrachloride were irradiated with antineutrinos from a reactor and were then swept with 12,000 liters of helium to entrain any argon produced. A liquid nitrogen cooled charcoal trap was used to concentrate the argon. Since ${}^{37}\text{Ar}$ is radioactive, decaying by electron capture with a 35-day half-life, its presence could be detected with a Geiger counter responding to the resulting x rays. It was concluded that the upper limit to the cross section for process (14.11) is at least 600 times smaller than that for process (14.10). It is apparent then that the neutrino and antineutrino are different particles.

14.6 Violation of Parity Conservation

Among the eight properties we mentioned in Chapter 10 as being conserved in nuclear reactions is that of **parity.** Until the mid-1950s the law of conservation of parity was believed to apply to all physical processes. About that time, however, a puzzle arose about the nature of some particles observed in experiments conducted at energies in the neighborhood of several hundred megaelectron volts. It appeared that a particular particle could decay sometimes by the emission of **two pions** and other times by the emission of **three pions.** The surprising feature was not that two decay modes could exist but rather that in one mode the end product was a system of **even parity** and in the other mode one of **odd parity.** Believing in parity conservation, one must then say that the parent particle for the first mode has even parity and that for the second mode odd parity. The particles were named **theta** and **tau,** respectively. However, all other properties of the parents seemed to be the same. Simplicity would be served were there but one parent species involved. Since the situation seemed contradictory, it came to be known as the "τ-θ" puzzle.

One of the implications of parity conservation is the principle that the laws of nature are symmetric upon mirror reflection, that it is possible to perform the mirror image of any experiment, and that the result will be the mirror image of the original. This principle is so appealing on intuitive grounds and was so well established by experiments involving **atomic spectroscopy** and **nuclear reactions** that its validity for beta decay was simply taken for granted. In studying the implications of the τ-θ puzzle, however, T. D. Lee and C. N. Yang concluded that definitive experiments had not been performed regarding parity conservation in **weak** reactions. Both τ-θ decays and nuclear beta decays are examples of weak interactions.

Almost immediately after the 1956 suggestion of Lee and Yang, C. S. Wu and her colleagues at Columbia University set about to test the situation experimentally. They chose the beta decay of ^{60}Co as suitable to their purposes in a number of ways. It is a common and familiar radionuclide. It decays by beta emission with an endpoint energy of 0.312 MeV. It has a **nuclear spin** of 5 and a substantial **magnetic moment.** These last features were of particular importance to the parity experiment since they were conducive to achieving substantial

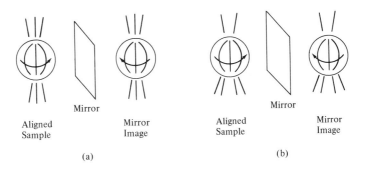

Figure 14.4 Possible outcomes of β-decay measurements on spin aligned samples. (a) Parity conservation is obeyed. (b) Parity conservation is violated.

spin alignment. This alignment was achieved by placing the ^{60}Co, in a matrix of cerium magnesium nitrate, in an external magnetic field at temperatures below 0.1°K.

Figure 14.4 diagrams the essential concept of the experiment. Parts (a) and (a′) schematically represent the experiment, along with its mirror image, under conditions of parity conservation. In (a) the ^{60}Co is shown with its spin vector upward. Its mirror image (a′) has its spin vector down. The short lines drawn outward from the "nucleus" represent beta emission. The density of the lines is used to represent a (hypothetical) angular distribution of beta particles. In this hypothetical case the **mirror image result** is the mirror image of the original experiment, namely, an angular distribution symmetric above and below the equator. Parts (b) and (b′) show a contrasting situation in which parity is not conserved. In the experiment itself the greater beta intensity is in a direction **opposite** to the spin vector, whereas in the mirror image the greater intensity is in the **same direction** as the spin vector. With asymmetric emission of this sort the mirror image result is not just the mirror image of the original experiment since the relationship between intensity distribution and spin direction differs in the two cases. This is an example of a situation in which parity is not conserved.

Figure 14.5 is taken from the results of Wu et al. The cobalt sample was placed in a cryostat and, from a beginning temperature of 0.01°K, it was allowed to warm slowly over a period of some 18 min. The cryostat and sample were placed in a magnetic field to establish spin alignment of the ^{60}Co. The beta emission was counted parallel and antiparallel to the direction of the field. As the sample warmed, the spin alignment of the ^{60}Co was lost due to thermal agitation. The asymmetry in the results for times below 6 min indicates that the beta decay behaves as depicted in Figure 14.4(b). From this it was concluded that parity is not conserved in the beta decay of ^{60}Co and therefore in general need not be conserved in weak interactions. Another way of expressing the situation is to note that had the results been

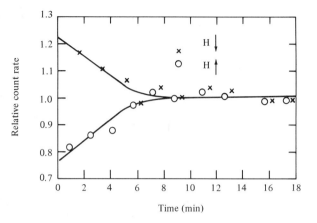

Figure 14.5 Results of measurements by C. S. Wu et al. of the asymmetry in beta emission from ^{60}Co. [C. S. Wu, E. Ambler, R. W. Hayward, D. D. Hoppes, and R. P. Hudson, *Phys. Rev., 105*:1413 (1957).]

symmetric, Figure 14.4(a') could just as well have been the experiment and 14.4(a) its mirror image. The two situations could not be distinguished. In the real world, however, Figure 14.4(b') does not express the behavior of ^{60}Co. Given a beta distribution related to spin direction as in (b') we would immediately know it to be drawn for the mirror image of the sample, not the real sample itself. The **overthrow of parity** has profound implications regarding natural laws in the universe. Even 20 years after the crucial experiments these implications are still being assessed.

14.7   The Kurie Plot

In Figure 14.1 we saw a typical continuous beta spectrum. During the early 1930s Enrico Fermi (1901–1954) developed a theory of beta decay that had considerable success in predicting the shape of the spectrum. We shall not attempt to trace the details of the theory but shall simply give its results. It is usually given in the form of a function describing the probability of finding beta particles in the momentum region between p and $p + dp$, in this fashion:

$$P(p)\, dp = \frac{G^2 |M_{if}|^2}{2\pi^3 \hbar^7 c^3}\, F(Z, E)\, p^2 (E_0 - E)[(E_0 - E)^2$$

$$- (m_\nu c^2)^2]^{1/2}\, dp \qquad \textbf{(14.12)}$$

Here p is the measured **momentum** of the electron and m_ν the **mass** of the neutrino. G is a **constant** introduced by Fermi to express the strength of the interaction, which converts a neutron into a proton or a proton into a neutron. M_{if} is the **matrix element** that connects the initial and final (i.e., the parent and daughter) nuclear states. $F(Z, E)$, often called the **Fermi factor**, expresses the way in which the energy of the beta particle is modified by the Coulomb field surrounding the nucleus. For betas of nonrelativistic energies it is given by

$$F(Z, E) = 2\pi\eta(1 - e^{-2\pi\eta})^{-1} \qquad (14.13)$$

where $\eta = \pm Ze^2/\hbar v$. The plus sign is used in negatron decay, the minus sign in position decay. Although not altogether correct for relativistic energies, Eq. (14.13) is a useful approximation which has less than 5% error if $Z < 30$ and $E \lesssim 5\, m_0c^2$.

In Eq. (14.12) E and E_0 are relativistic beta-particle energies, such that

$$E^2 = (m_0c^2)^2 + (pc)^2$$

and $E_0 = K_0 + m_0c^2$ where K_0 is the endpoint energy of the spectrum. The group of terms

$$p^2(E_0 - E)[(E_0 - E)^2 - (m_\nu c^2)^2]^{1/2}$$

is derived from considerations of phase space available for the beta particle and the neutrino, and expresses the way in which the transition energy is shared.

If we set $m_\nu = 0$ and convert Eq. (14.12) into a proportionality by suppressing the constants the appearance of the expression is considerably simplified. (M_{if} is independent of the electron energy since it involves only the initial and final nuclear states.) We then obtain

$$P(p) \propto F(Z, E)p^2(E_0 - E)^2 \qquad (14.14)$$

which, upon rearrangement, becomes

$$\sqrt{\frac{P(p)}{p^2 F}} \propto (E_0 - E) = (K_0 - K) \qquad (14.15)$$

We see that if the left side of Eq. (14.15) is plotted against electron kinetic energy K, a straight line should result. The reduction of beta-decay data in this fashion is known as a **Kurie plot**. Figure 14.6 is such a plot for the β decay of tritium. Clearly, linear extrapolation determines the endpoint energy with much greater precision than could be obtained from a direct spectrum such as that in Figure 14.1.

In addition to endpoint energy, a Kurie plot gives other information about the beta-decay process. If the neutrino rest mass is not set equal to zero, the shape of the plot is changed, especially near its end-

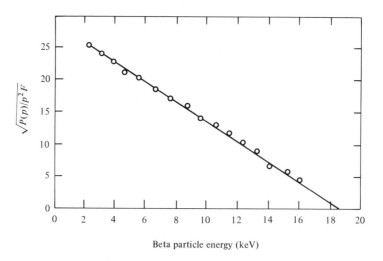

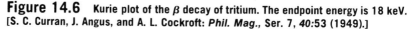

Figure 14.6 Kurie plot of the β decay of tritium. The endpoint energy is 18 keV. [S. C. Curran, J. Angus, and A. L. Cockroft: *Phil. Mag.*, Ser. 7, *40*:53 (1949).]

point. This fact can be used to set an upper limit to the mass of the neutrino. The work of Curran et al., which is the basis for Figure 14.6, established an upper limit of 250 eV for this mass. More recent work has set the upper limit at 60 eV. Included in the assumptions leading to Eq. (14.12) were the restrictions that there be no difference in parity between the parent and daughter nuclear states and that the spin difference ΔJ between these states be $0, \pm 1$. Beta transitions that obey these restrictions are called **allowed,** and are ones that produce linear plots using Eq. (14.15). From Figure 14.6 we see that the β decay of tritium is an allowed transition. If $|\Delta J| > 1$, the transition is called **forbidden.** The probability of transition, and hence the decay constant, λ, becomes very much smaller as ΔJ becomes larger. Data from such transitions do not produce straight lines when plotted according to Eq. (14.15). Linear plots can be obtained from forbidden transitions by modifications in the theoretical analysis, thereby modifying the Kurie expression, Eq. (14.15). By searching out the proper expression to fit the data, the degree of forbiddenness can be determined.

14.8 Gamma-Decay Schemes

The emission of gamma rays is a prominent feature in the decay of many radionuclides. Any process that leaves a daughter in an excited state can lead to gamma emission. Among the natural radionuclides alpha emission and beta emission are the processes that precede gamma decay. Positron emission which is found with certain artificial radionuclides can also be accompanied by gamma emission. Furthermore, many of the nuclear reactions that we discussed in Chapter 11 produce nuclei in excited states, and these states can then deexcite by gamma emission. Figure 14.7(a) shows the decay scheme of ^{60}Co, a negatron emitter, and Figure 14.7(b) shows the scheme for ^{22}Na, a positron emitter. In each of these examples there is a large difference in nuclear spin between the ground states of the parent and daughter nuclides. The beta transition directly to the ground state of the daughter is a forbidden transition and consequently most of the transitions leave the daughter in an excited state.

On the other hand, there are parent radionuclides with multiple decay modes. An example is ^{38}Cl shown in Figure 14.7(c). Three different beta-decay modes are shown; 53% of the beta decays leave

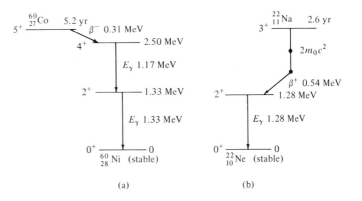

(a)

(b)

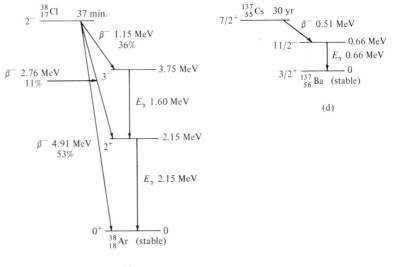

(c)

Figure 14.7 Typical β-γ decay schemes.

^{38}Ar in the ground state, 11% in the first excited state at 2.15 MeV, and 36% in the second excited state at 3.75 MeV. However, in this section our main interest is in the gamma radiation from excited states, no matter what mechanisms formed the states.

Gamma radiation is much more penetrating than alpha or beta radiation. Frequently the container surrounding a radioactive sample, or the walls of a detector, will stop alphas and betas but nevertheless be nearly transparent to gamma radiation. Furthermore, the half-life of the gamma emitting state in the daughter is usually very short compared to the half-life of the parent. Under these circumstances the gamma decay is in **transient equilibrium** with the decay of the parent. Therefore the gamma-ray intensity falls off with the same half-life as the parent. As a consequence it is customary to name the **parent** as the gamma-ray source. Thus in a tabulation of radionuclides we find ^{60}Co listed as the source of gamma rays of 1.17 MeV and 1.33 MeV energies. In a sense this practice is misleading, because it directs our attention away from the fact that these gamma rays are characteristic of transitions in ^{60}Ni. Nevertheless, the practice is thoroughly established and it does at least relate directly to the half-life of the source.

14.9   Lifetimes of Excited States

Considerable information about the characteristic of nuclear states can be obtained from analysis of gamma-ray transitions. For instance, the energy difference between initial and final states can be determined directly by an energy measurement of the gamma rays which link the two states. We shall see in Chapter 15 a number of techniques for making such measurements. Other characteristics include the total angular momentum quantum numbers of the states as well as the lifetimes of those excited states from which transitions originate.

Let us now turn our attention to these lifetimes. Some are relatively long, others may be so short that it becomes a challenge to measure them. All transitions, whether of long or short lifetime, obey the exponential decay laws discussed in Chapter 13. For longer lived states direct electronic counting of activity versus time may be possible. For shorter lived states less direct methods may be required. For instance, by the use of **delayed coincidence** techniques lifetimes as short as 10^{-11} sec can be measured. In this method two detectors are used, one sensitive to the initial event, such as a beta transition or a gamma transition from a higher lever, which excites the level of interest, and the other sensitive to the gamma ray emitted by that level. The signal from the first detector is delayed electronically to produce coincidences with the signal from the second detector. This delay is a measure of the lifetime for a single decay event. The lifetime of the state itself is the **mean** of all decay times observed for that state.

By even less direct methods lifetimes in the range of 10^{-13} to 10^{-15} sec have been measured. This is done by determining the energy **width** of the emitting level and invoking the uncertainty principle.

$$\Delta E\, \Delta t \approx \hbar \tag{14.16}$$

We associate ΔE with the width of the level and Δt with its mean-life. The shorter the lifetime, the broader the level. Even with a lifetime as short as 10^{-15} sec, however, the width of the level is only 0.6 eV. This is much too small to be resolved directly by even the best gamma-ray detectors. Fortunately, though, a technique known as **nuclear resonance absorption** can be used. An absorber that contains the same nuclide as that which emits the gamma rays is introduced between the gamma-ray source and detector. Although the level structure is the same in both the emitter and absorber, resonance absorption ordinarily does not occur. This is because the energy of recoil of the emitter and of the absorber is usually much greater than the level width so there is no energy overlap. To compensate for these recoil energy losses, the emitter is given a large relative velocity toward the absorber, **Doppler-shifting** the gamma radiation to shorter wave lengths. The width of the resonance is obtained by scanning the Doppler shift across it.

14.10 Multipole Radiation

The inverse of the mean-life of a state is the same as its decay probability λ. Although the derivation of decay probabilities is beyond the scope of this book, it will be useful for us to treat some of the considerations involved and to quote results. Gamma rays are **electromagnetic field** radiations that occur when certain rearrangements of nuclear structure take place. The general description of an electromagnetic field can be presented as a series expansion, in terms of **multipole** components: **dipole, quadrupole, octupole,** and so on. The field may be electric or magnetic, according to the nature of the source. In the case of quantized radiation from a quantum mechanical system the terms in the expansion contain the expression $(R/\lambda)^{2l}$ where R is the radius of the radiating system, λ is the reduced wave length, that is, wave length divided by 2π, of the emitted radiation, and l is the index number of the term. l equals 1 for dipole radiation, 2 for quadrupole radiation, 3 for octupole radiation, and so on. The relative number of quanta in each decay mode is given by the corresponding term in the expansion. Thus if R/λ is much less than 1, the series is dominated by the first nonvanishing term. For radiation from atoms

R/λ approximates 10^{-3}. Because this term is small, atomic radiation is essentially **all dipole** in character. Any level that is prohibited from emitting dipole radiation has such a small decay probability and consequently long lifetime that it finds other ways of losing excitation energy, such as by collision with neighboring atoms. In contrast, for nuclei R/λ typically approximates 0.05. Again each term in the expansion is considerably smaller than the preceding one. However, collisions between nuclei are so rare that deexcitation by this mode is unavailable. If radiation of low multipolarity is prohibited, the nucleus will simply remain in its excited state until it finally does emit radiation. Thus in addition to those of dipole character, gamma rays of quadrupole, octupole, and higher multipolarity character can be found.

The selection rules which govern the **multipolarity** character of the radiation arise from conservation of **angular momentum** in the transition. Each photon carries angular momentum $\hbar \sqrt{l(l+1)}$, where l, an integer greater than 0, is the same as the index number of the multipole term in the series expansion. Another way of putting it is that l is the angular momentum **quantum number** of the photon. If J_i is the total angular momentum quantum number of the initial state (upper level) and J_f that of the final state (lower level) connected by the gamma transition, then the rules for combining quantum numbers limit l to integer values such that

$$|J_i - J_f| \leq l \leq J_i + J_f \tag{14.17}$$

For instance, in the 1.60 MeV transition from the second to the first excited states in ^{38}Ar, shown in Figure 14.7(c), we find the allowed values to be $l = 1, 2, 3, 4, 5$. Five multipolarities are allowed for this transition. However, because of the $(R/\lambda)^{2l}$ dependence the transition is predominantly **dipole** in character. In the transition from the 2^+ first excited state to the 0^+ ground state the only possibility is $l = 2$ (**quadrupole**) radiation. Another example is shown in Figure 14.7(d), the decay scheme for ^{137}Cs. The gamma-ray transition in ^{137}Ba is from an $11^-/2$ state to a $3^+/2$ state. The possible values of l are $l = 4, 5, 6, 7$. The radiation is predominantly $l = 4$ (16-pole) in character.

The **electric** or **magnetic** character of the radiation is governed by both the l value and the difference in **parity** between the initial and final nuclear states. If l is even and there is no difference in parity between the states, the emitted radiation is electric multipole. If the states are of opposite parity, the radiation is magnetic multipole. If l is odd and there is difference in parity, the radiation is electric multipole, and is magnetic multipole if there is no difference in parity. In any case the total parity of the final system, photon plus nucleus, is the same as the initial parity of the excited nucleus. This terminology is summarized in the first four columns of Table 14.1.

In addition to $(R/\lambda)^{2l}$, there are other factors that also play a role in determining the **decay probability** for the different types of multipole

Table 14.1 Selection Rules for Gamma Radiation, and Half-lives Based on Weisskopf Estimates for a Nucleus of $A \approx 125$.

Type of Radiation	l	Symbol	Parity Change	Half-life in Seconds — Gamma-Ray Energy		
				10 MeV	1 MeV	0.1 MeV
Elec. dipole	1	$E1$	Yes	2×10^{-19}	2×10^{-16}	2×10^{-13}
Mag. dipole	1	$M1$	No	2×10^{-17}	2×10^{-14}	2×10^{-11}
Elec. quadrupole	2	$E2$	No	8×10^{-17}	8×10^{-12}	8×10^{-7}
Mag. quadrupole	2	$M2$	Yes	9×10^{-15}	9×10^{-10}	9×10^{-5}
Elec. octupole	3	$E3$	Yes	5×10^{-14}	5×10^{-7}	5
Mag. octupole	3	$M3$	No	5×10^{-12}	5×10^{-5}	5×10^{2}
Elec. sixteenpole	4	$E4$	No	5×10^{-11}	5×10^{-2}	5×10^{7}
Mag. sixteenpole	4	$M4$	Yes	5×10^{-9}	5	5×10^{9}

radiation. The form these factors take depends upon the choice of nuclear model. Using the **independent particle model**, V. F. Weisskopf developed the following expressions for the decay probabilities of electric and magnetic radiation.

$$\lambda_E(l) = \frac{2(l+1)}{l[(2l+1)!!]^2} \left(\frac{3}{l+3}\right)^2 \frac{e^2}{\hbar \hbar} \left(\frac{R}{\hbar}\right)^{2l} \tag{14.18a}$$

$$\lambda_M(l) = \frac{20(l+1)}{l[(2l+1)!!]^2} \left(\frac{3}{l+1}\right)^2 \frac{e^2}{\hbar \hbar} \left(\frac{\hbar}{McR}\right)^2 \left(\frac{R}{\hbar}\right)^{2l} \tag{14.18b}$$

where $\lambda_E(l)$ and $\lambda_M(l)$ are the **decay probabilities,** respectively, for **electric** and **magnetic** $2l$-pole radiation, not to be confused with the wavelength of the gamma radiation; e is the **charge** on the electron in esu; M is the **nucleon mass;** $(2l+1)!! = 1 \times 3 \times 5 \dots (2l+1)$. All quantities are in cgs units. Now

$$\lambda = \frac{c}{2\pi\nu} = \frac{\hbar c}{E_\gamma} \quad \text{(cgs units)} \tag{14.19}$$

Substitution of numerical values leads to

$$\lambda = \frac{197}{E_\gamma} \quad (\lambda \text{ in fermis}, E_\gamma \text{ in MeV}) \tag{14.20}$$

Substitution of this and other numerical parameters in Eq. (14.18) leads to

$$\lambda_E(l) = \frac{4.37(l+1)}{l[(2l+1)!!]^2} \left(\frac{3}{l+3}\right)^2 \left(\frac{E_\gamma}{197}\right)^{2l+1} \times R^{2l} \, 10^{21} \text{ sec}^{-1} \tag{14.21a}$$

$$\lambda_M(l) = \frac{1.93(l+1)}{l[(2l+1)!!]^2} \left(\frac{3}{l+3}\right)^2 \left(\frac{E_\gamma}{197}\right)^{2l+1} \times R^{2l-2} \, 10^{21} \; \text{sec}^{-1} \qquad \textbf{(14.21b)}$$

where R is in fermis and E_γ is in MeV. The last three columns of Table 14.1 show half-lives in seconds derived from these equations for a nucleus of $A \approx 125$, $R = 7$ fermis.

Upon comparing experimentally determined half-lives with predictions of Eq. (14.21) it is somewhat disturbing to discover disagreement ranging over several orders of magnitude. We must not be too disappointed however. The derivation of Eq. (14.21) was based on the **extreme independent particle** shell model. Disagreement with experiment simply indicates that the assumptions of the model are too simple, that the deexcitation of a bound nuclear state by gamma emission is likely to involve more than just one nucleon. Nevertheless, the general trends are as predicted. Half-lives become increasingly long as the multipolarity of the radiation increases and as the energy of the transition decreases. Some levels, known as isomeric levels, have lifetimes long enough to be directly measurable. An example is the 0.66 MeV level in ^{137}Ba, shown in Figure 14.7(d). The lowest l value for the transition between this state and the ground state is 4 and there is parity change between the states. The gamma radiation is $M4$ and the observed half-life is 2.6 min. The prediction based on Eq. (14.21b) is 3.1 min, one of the better fits between shell model prediction and experiment.

Example 14.2 $^{110}_{47}$Ag decays by negatron emission to $^{110}_{48}$Cd. Of these transitions, 4.5% populate the first excited state in ^{110}Cd at 0.658 MeV. This is a 2+ (spin and parity) state, while the ground state is 0+. Determine multipolarity characteristics of the gamma transition between these states. Use the Weisskopf expressions to estimate the half-life of the excited state.

Solution

$$|J_i - J_f| \le l \le J_i + J_f$$

Since $J_f = 0$, there is only one choice: $l = 2$. Both states have even parity. Reference to Table 14.1 identifies the radiation ($l = 2$, *no* change in parity), as $E2$, that is, electric quadrupole.

Since the radiation is electric multipole, Eq. (14.21a) is applicable. For use in this expression we note that $R = 1.4 \times 110^{1/3} = 6.708$ fermi

$$(2l + 1)!! = 1 \times 3 \times 5 = 15$$

Substitution in Eq. (14.21a) leads to

$$\lambda_E(2) = \frac{4.37(2+1)}{2 \times 15^2} \left(\frac{3}{2+3}\right)^2 \left(\frac{0.658}{197}\right)^5 (6.708)^4 \times 10^{21} \text{ sec}^{-1}$$

$$= 8.82 \times 10^9 \text{ sec}^{-1}$$

$$T_{1/2} = \frac{0.693}{\lambda} = 7.8 \times 10^{-11} \text{ sec}$$

Experimental studies of this state indicate a half-life of 5×10^{-12} sec.

14.11   Internal Conversion

Internal conversion is an alternate process by which a nucleus in an excited state can lose energy. In this process the energy of transition is delivered directly to an **atomic electron.** The electron is ejected from its shell with a kinetic energy that is the difference between the transition energy and the energy with which the electron is bound in the atom

$$K_e = E_\gamma - BE \tag{14.22}$$

where E_γ is the transition energy otherwise given off in gamma rays and BE is the binding energy of the electron shell. The name internal conversion derives from the early, and **now discredited,** view of the process as involving two steps, the emission of a gamma ray from the nucleus and its **subsequent** conversion to electron energy by photo-electric absorption within the same atom. Instead the process is now viewed as involving only a single step, the **direct transfer** of energy during an encounter between an atomic electron and the nucleus. The process is of considerable interest in determining the parameters of a decay scheme. Electron energies can be measured with great precision, and, before the advent of the Ge(Li) semiconductor detector, internal conversion experiments gave some of the best measurements of transition energies. Furthermore, the way in which a level de-excites, that is, the relative frequency with which conversion electrons are produced in comparison to gamma-ray emission, is a sensitive measure of the multipolarity of the transition. This ratio is called the **internal conversion coefficient.** A separate coefficient is found for each atomic shell. Thus

$$\alpha_K = \frac{N_K}{N_\gamma}$$

$$\alpha_L = \frac{N_L}{N_\gamma} \tag{14.23}$$

$$\alpha_M = \frac{N_M}{N_\gamma}$$

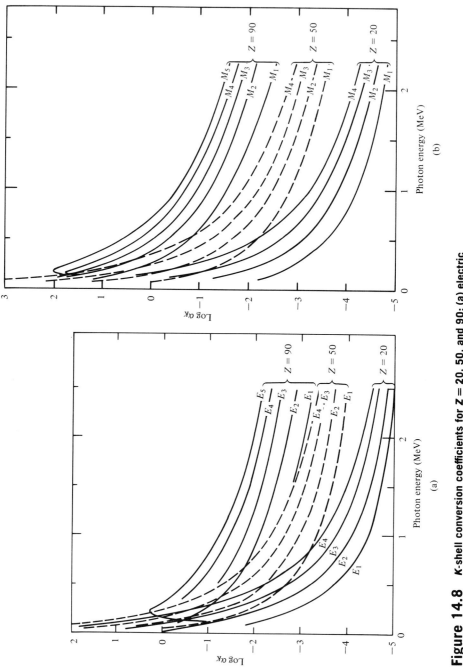

Figure 14.8 *K*-shell conversion coefficients for **Z = 20, 50, and 90**: (a) electric multipole transitions, (b) magnetic multipole transitions. [After M. A. Preston: *Physics of the Nucleus*. Reading, Mass., Addison-Wesley Publishing Co., 1966. Used by permission of the publisher.

where the alphas are the coefficients, N_K, N_L, N_M are the number of electrons from the K, L, and M shells, respectively, and N_γ is the number of gamma rays. As might be expected, electrons in the more tightly bound shells have greater probability of interacting with the nucleus and it follows that

$$\alpha_K > \alpha_L > \alpha_M \tag{14.24}$$

Figures 14.8(a) and 14.8(b) show K-shell conversion coefficients for electric and magnetic transitions as a function of multipolarity and atomic number. Small values of E_γ, large values of l, and large values of Z are conducive to large conversion coefficients. The assignment of l values can be made with much greater confidence on the basis of experimentally determined conversion coefficients than on the basis of gamma-decay probabilities, Eq. (14.21).

14.12 Summary

A beta spectrum exhibits a continuous distribution in energy. However, the transition connects two states of discrete energy and, therefore, must itself be discrete. The energy of the transition is associated with the endpoint, or maximum, energy of the spectrum. The difference in energy is carried away by the neutrino which has zero rest mass and zero charge.

When plotted against energy, the Kurie function produces a straight line. This is a great help in extrapolating to find the endpoint energy. For allowed transitions ($\Delta J = 0, \pm 1$; no difference in parity between the parent and daughter nuclides) the function is of the form

$$\sqrt{\frac{P(p)}{p^2 F(Z, E)}} \propto (E_0 - E) = (K_0 - K)$$

Although emitted from the nucleus, beta particles do not exist within the nucleus. They are created at the time of emission. Energy releases Q can be written

$$Q = ({}_Z^A M - {}_{z+1}^A M)c^2 \qquad (\beta^- \text{ decay})$$
$$Q = ({}_Z^A M - {}_{z-1}^A M)c^2 - 2m_0 c^2 \qquad (\beta^+ \text{ decay})$$
$$Q = ({}_Z^A M - {}_{z-1}^A M)c^2 - BE_e \qquad (\text{electron capture})$$

Neutrinos have very low probability of interaction with matter. A representative attenuation length is 3600 light-years of water! Nevertheless, neutrinos have been detected and their existence firmly established.

Negative beta decay is accompanied by emission of antineutrinos, positron decay by neutrinos. Although both have zero rest mass and zero charge, the neutrino and antineutrino are not identical. Beta decay is a "weak" interaction in which parity conservation is violated.

Gamma rays are photons emitted in transitions between energy levels of a nucleus. Various processes can produce an excited level that subsequently decays by gamma emission. These processes include nuclear reactions induced by particle bombardment or prior alpha or β^- or β^+ decay from a radioactive parent. It is customary to tabulate the radioactive parent as the gamma-ray source.

The decay probability for a gamma transition is very dependent on differences in angular momentum and parity between the two states involved. These differences determine the multipolarity and electric or magnetic character of the radiation. The greater the multipolarity, the smaller the decay constant. For the same multipolarity, magnetic radiation is two orders of magnitude less probable than electric radiation. High-energy transitions are more probable than low-energy ones. The Weisskopf expressions provide useful estimates of decay probabilities.

Internal conversion competes with gamma-ray emission. Conversion coefficients express the ratio of the probability of internal conversion to gamma emission in a transition between levels. Experimental determination of these coefficients is straightforward and is useful in assigning l values to the transition.

Problems

14.1 The endpoint energy of negatrons emitted by ^{12}B is 13.370 MeV. Determine the exact mass of ^{12}B.

14.2 (a) What is the momentum carried by the maximum energy electrons in the beta decay of ^{12}B and of the neutrinos associated with these electrons?

(b) What is the momentum and kinetic energy of the ^{12}C recoil?

14.3 Suppose in the beta decay of ^{12}B an electron is emitted with an energy of 6.685 MeV (50% of the endpoint energy).

(a) What is the energy of the associated neutrino?

(b) Calculate the momenta of the electron and the neutrino.

(c) If the electron and neutrino are emitted in opposite directions, find the momentum and kinetic energy of the ^{12}C recoil.

14.4 A magnetic spectrometer (see Section 15.7) measures momentum p of a particle of charge q according to the expression $BR = pc/q$, where B is the magnetic flux density and R is the radius of the path. The following data come from a spectrometer measurement of the decay of ^{32}P:

BR (gauss-cm)	7000	6000	5000	4000	3000	2000	1000
Relative count	32	330	723	968	968	617	287

Make a Kurie plot from these data and determine the extrapolated endpoint energy.

14.5 The endpoint energy of positrons emitted by ^{27}Si is 3.786 MeV. Determine the difference in mass between ^{27}Si and ^{27}Al.

14.6 The beta decay of ^{27}Mg $\xrightarrow{\beta-}$ ^{27}Al is complex with two endpoint energies of 1.772 MeV and 1.601 MeV. Accompanying the betas are gammas of energy 1.013, 0.842, and 0.171 MeV. Devise a decay scheme for this decay similar to those in Figure 14.7.

14.7 Scintillation spectrometer measurements of the ^{22}Na $\xrightarrow{\beta+}$ ^{22}Ne decay show gammas of 0.51 MeV energy as well as 1.28 MeV, although Figure 14.16b shows only the one at 1.28 MeV. Account for the observation of gammas at 0.51 MeV.

14.8 For the 2.15-MeV transition from the first excited state to the ground state of ^{38}Ar, calculate
(a) the momentum carried by the gamma.
(b) the momentum and kinetic energy carried by the ^{38}Ar recoil.

14.9 The relativistic Doppler effect is given by the expression

$$f' = f(1 + \frac{v}{c} \cos \theta) \left[1 - \left(\frac{v}{c}\right)^2 \right]^{-1/2}$$

where f' is the frequency perceived by an observer due to a source whose velocity is v along a direction at angle θ with the line between source and observer. f is the frequency when the source is stationary. What velocity is required of a source emitting 3.00-MeV gammas to produce an energy shift of 150-eV, if the velocity is
(a) directly toward the gamma detector?
(b) at an angle of 45° with respect to the line between source and detector?

14.10 The first excited state of ^{28}Si is 1.78 MeV above the ground state. For the gamma transition between these states calculate
(a) the momentum carried by the gamma.
(b) the momentum and kinetic energy carried by the ^{28}Si recoil, and hence the amount by which the gamma energy is less than the transition energy.
(c) the $(v/c) \cos \theta$ term in the Doppler effect required to produce resonance absorption of these gammas in a stationary ^{28}Si absorber. (Note that two recoils are involved, one when the source emits the gamma and another when the gamma is captured in the absorber.)
  Suppose the excited ^{28}Si is created by the proton capture reaction, ^{27}Al (p, γ) ^{28}Si, using 5-MeV protons. Calculate
(d) the velocity of ^{28}Si immediately after the proton capture.
(e) the angle θ required for resonance absorption if the ^{28}Si does not slow down before emission of the 1.78-MeV gamma.

14.11 The 1.78-MeV state in ^{28}Si has a half-life of 0.5×10^{-12} sec. By invoking the uncertainty principle, determine the energy width of this level.

14.12 Determine the possible and the most probable multipolarities and the electric or magnetic character of each of the gamma transitions shown in Figure 14.7.

14.13 A beta emitter with endpoint energy of 0.30 MeV is found to have a half-life of 96 hrs. Is this an allowed or forbidden beta decay?

14.14 The free neutron beta decays with a 12 min half-life and an endpoint energy of 0.78 MeV. Determine the difference in mass between the neutron and the proton.

14.15 The mass of $^{108}_{47}$Ag is 107.90595 u, that of $^{108}_{46}$Pd is 107.90389 u, and that of $^{108}_{48}$Cd is 107.90419 u. What types of beta decay are energetically possible? Calculate the endpoint kinetic energies of the emitted particles?

14.16 From the statement that it requires a layer of water 3600 light-years thick to attenuate an antineutrino beam to $1/e$ of its initial value, determine the fraction of neutrinos which interact in a water layer 100 cm thick. (Hint. You may want to use a series expansion for the exponential when it is close to 1.)

14.17 ^{57}Co has a half-life of 267 days. A certain fraction of the decay transitions leave ^{57}Fe in an excited state at 0.014 MeV. What is the recoil energy transferred to the ^{57}Fe nucleus when it deexcites by emission of a 0.014 MeV gamma ray? How many ^{57}Fe masses would have to be involved in a single recoil to reduce the recoil energy to 10^{-8} eV? (If the ^{57}Fe is "locked" in a crystal lattice, a huge number of atoms in the aggregate may act together to take up the recoil momentum. This is the principle of "recoilless" emission and absorption of gamma radiation, otherwise known as the Mössbauer effect.)

Suggestions for Further Reading

Bahcall, J. N. "Neutrinos from the Sun," *Scientific American* (July 1969).

Halliday, D. *Introductory Nuclear Physics,* 2nd ed. New York: John Wiley & Sons, Inc., 1955. Chapters 5 and 6.

Kaplan, I. *Nuclear Physics,* 2nd ed. Reading, Mass.: Addison-Wesley Publishing Co., Inc., 1962. Chapters 14 and 15.

Lederman, L. M. "The Two-Neutrino Experiment," *Scientific American* (March 1963).

Morrison, P. "The Overthrow of Parity," *Scientific American* (April 1957).

Detectors and accelerators

15.1 Introduction

In this chapter we shall consider various devices and apparatus used in nuclear physics experiments. The design and construction of some of these devices, particularly particle accelerators, involve major engineering considerations. In general, however, we shall not try to take up these engineering aspects in any serious fashion but shall devote attention to the basic physics involved and to the ways in which these devices have been used in the study of nuclear physics.

We shall begin by describing devices used for detection of nuclear radiation. Some of these devices merely **count** the radiation events as they occur. Other devices give additional information, such as the **energies** of the events and the **spatial relationships** of the nuclear particles involved. In every case detection and measurement depend upon **electric interaction** of charged particles with the material of the detector. If the primary radiation is electrically neutral (for instance, neutrons or gamma rays), it is not detected until it produces **secondary** charged particle radiation, such as knock-on protons or electrons.

Most of our information about nuclear reactions has come from experiments in which nuclear projectiles bombard target nuclides. Pioneering experiments were done using projectiles found in nature, that is, alpha particles from radioactive material and, at much higher energy, particles found in cosmic rays. However, the development of various charged-particle accelerators has made available a far wider choice of projectiles at controllable energies with much greater intensities than those found in nature. In the last four sections of this chapter we discuss a number of these accelerators.

15.2 Gas-Filled Counters

The ionization chamber, the proportional counter, and the Geiger-Müller counter are all gas-filled counters; each has its own characteristic and applicability.

The **ionization chamber** is often operated with air as the filling gas. The other counters usually use special gas mixtures. Figure 15.1 is a schematic diagram of an ionization chamber. It consists of two electrodes, often in the form of parallel plates or coaxial cylinders. A potential difference, V, of a few hundred volts is maintained between

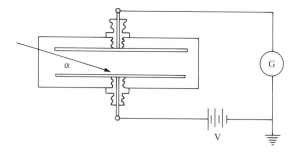

Figure 15.1 Ionization chamber and associated circuit. G is a sensitive indicator of current.

the electrodes by a battery or power supply. G is a sensitive galvanometer or other device for measuring current or charge. When a swift charged particle, such as a proton, alpha, or electron, enters the region between the two plates, it produces **ion pairs** in the gas. In so doing it loses energy and is finally brought to rest. A 5-MeV alpha, for instance, produces approximately 200,000 ion pairs. The electric field sweeps the ions to the electrodes, thereby producing a momentary current in G. In principle these pulses can be used to count individual events. However, other detectors are better suited for this and nowadays the ionization chamber finds its main application as an integrating device.

In an intense radiation field the time between ionizing events may be shorter than the response time of the indicating device, G. The reading at G then is proportional to the rate at which ionization is produced in the chamber. For example, if the chamber is exposed to x rays or gamma rays, electrons may be ejected from the walls and from the atoms of the filling gas by the **photoelectric** or **Compton** interaction. These electrons produce ionization within the chamber. By suitable choice of gas and composition of the chamber walls, the reading of G can be made closely proportional to the rate of biological damage to be expected from the radiation. When used in this fashion, the ionization chamber, with its associated circuitry, becomes a radiation **survey meter.**

In another form, shown schematically in Figure 15.2, the indicating device, E, is an electrometer or electroscope. C is the capacitance of the chamber and associated circuitry. If the voltage supply, V, is connected momentarily and then removed, subsequent ionization occurring within the chamber discharges C, and the reading of the electroscope falls from its initial value. The reading of E then gives the time-integrated ionization within the chamber. With careful attention to insulation such devices can be made to hold charge in the absence of radiation for months at a time. Such chambers, electroscope and all, can be built about the size of a fountain pen and find wide application as **dosimeters.** When used in this fashion, they are carried as pocket instruments by individuals working in the neighborhood of radiation sources. The electroscope may be read at any time and indicates the radiation dose accumulated since the last charging of the chamber.

The **proportional counter** uses gas multiplication to increase the

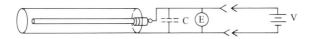

Figure 15.2 An ionization chamber can be used as a radiation dosimeter. V is a battery or power supply which is temporarily connected to the chamber assembly, thus charging the capacitance C. Ionization causes loss of charge, and this is indicated by the reading of a quartz fiber electroscope E which is a permanent part of the assembly.

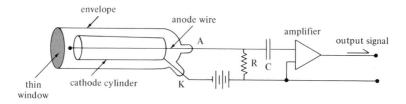

Figure 15.3 Proportional counter and associated circuit. K is the cathode lead attached to a thin-walled metal cylinder. The anode A is a thin wire along the axis of the cylinder.

total ionization collected by the electrodes. Gas multiplication is the result of the production of secondary, tertiary, and so on, ionization within the gas. This extra ionization will arise if electrons from primary ionization gain from the electric field sufficient energy between collisions to ionize the atoms with which they collide. The high fields required may be readily achieved by using coaxial cylinder geometry. Figure 15.3 diagrams such an arrangement. The anode, A, is a small diameter central wire typically 0.2 mm diameter. The cathode, K, is a thin-walled metal cylinder of 2.24 cm diameter. The typical operating potential, V, is 1000 to 3000 V. Gases such as hydrogen or methane or argon at pressures of a few hundred torr may be used to fill the detector. As electrons from primary ionization drift in toward the anode, they encounter higher and higher electric field strengths, given by

$$\mathbf{E} = \frac{V}{r \ \ln(b/a)} \tag{15.1}$$

where r is the distance from the anode wire, a is the radius of that wire, and b the inside radius of the cathode. Within a critical region around the anode the electric field is sufficiently strong to cause the onset of **secondary ionization.** The ionization produced by this avalanche may be several hundred times larger than the primary ionization. The signal across the resistor R is increased by the same factor.

If the secondary ionization is produced entirely by electron collision processes, such as we have described, then the total ionization will be closely proportional to the kinetic energy lost within the gas by the initial particle. If the detector is large enough so that the incident particle is stopped within its volume, the size of the output pulse becomes a measure of the incident kinetic energy. This restriction regarding dimensions limits proportional counters to primary radiation of suitably short range in the gas. Alpha particles of several megaelectron volts are readily handled; protons must be of lower energy since they lose energy in the gas at a much smaller rate than do alphas; and electrons lose energy at such a small rate that from a practical standpoint they must be limited to only a few tens of kiloelectron volts.

Figure 15.3 shows a detector with a thin foil end window to permit the entrance of alpha particles or protons. If more penetrating radiation, such as gamma rays or neutrons, is used, it is not necessary that the end be thin, and the detector, therefore, can be more rugged. To detect neutrons, boron trifluoride (BF_3) is used as the gas for filling the detector, this choice being made for its boron content. The reaction $^{10}B(n, \alpha)^7Li$ has a high cross section for slow neutrons. The Q of the reaction is about 2.8 MeV. The alpha particle receives about 64 percent of this energy, or 1.8 MeV, and is readily detected. The BF_3-filled proportional counter has good efficiency for the detection of slow neutrons, but the reaction cross section is low for fast neutrons. It is possible to slow down fast neutrons before they enter the detector by surrounding it with a paraffin **neutron moderator.** This enhances efficiency of neutron detection, but removes any relationship between neutron energy and pulse height.

The **Geiger counter** construction is very similar to that of the proportional counter. In fact, Figure 15.3 will serve equally well in our discussion of the Geiger counter. The electrical behavior of the detector tube is quite different, however. In Geiger operation the size of the pulse appearing across the resistor, R, is substantially independent of the primary ionization. Response of this type is caused by the presence of considerably higher electric field in the neighborhood of the anode wire than is used for the proportional counter. This higher field may be achieved by increasing the potential difference, V applied to the tube or by using a smaller diameter anode wire, or both.

Let us now examine the sequence of events when the tube is operating in the Geiger region. Because of the higher field the ionization **avalanche** becomes more dense. There is significant opportunity for recombination of electrons with the positive ions formed within this region. Emission of photons in the ultraviolet portion of the spectrum accompanies this recombination. These photons produce **photoionization** throughout the volume of the tube, thus spreading the avalanche along the entire length of the anode wire. Electrons in the avalanche region move rapidly toward the anode, leaving a cylindrical sheath of positive ions moving slowly outward. The presence of this sheath reduces the electric field around the anode below that required to produce additional ionization by collision. In this manner each avalanche becomes of standard size independent of the primary ionization. All of this takes place in about 1 μsec. However, it takes 100 to 200 μsec for the positive ions to reach the cathode. During this time the tube is insensitive to subsequent radiation events.

A typical gas for a Geiger tube is argon. When the positive argon ions reach the cathode, many of them have sufficient energy to eject electrons there. These electrons drift toward the anode and initiate another avalanche, since by now the positive ion sheath is in the neighborhood of the cathode. To avoid producing a continuous train of pulses, a quenching gas such as alcohol vapor or Cl_2 is added to the argon. As the argon ions move outward, they undergo charge-exchange collisions with the molecules of the quenching gas. This neutralizes the argon ion and produces a molecular ion of the other gas component. When these molecular ions reach the cathode, the available energy dissociates the molecule rather than ejecting a free electron. If the quenching gas is an alcohol, the dissociation limits the life of the tube to about 10^{10} counts. With halogens, however, the life of the tube is much greater since the process $Cl + Cl \rightarrow Cl_2$ reforms the diatomic molecule.

Figure 15.4 diagrams the response as a function of applied voltage in a circuit such as that of Figure 15.3. The initial ionization for the upper branch, curve 1, is 10^5 that of the lower branch, curve 2. In region A the tube operates as an **ionization chamber.** In that region the only ionization being collected is the primary ionization, so the pulse height is independent of voltage. Region B is the **proportional counter** region. Pulse-height increases with applied voltage, but the ratio of curve 1 to curve 2 remains the same (10^5 in this instance). Note that the ordinate scale is logarithmic. C is a transition region and D is the **Geiger** region, where pulse height is independent of the initial ionization. Since this pulse height is large, the associated amplifier can be much simpler than that for the proportional counter.

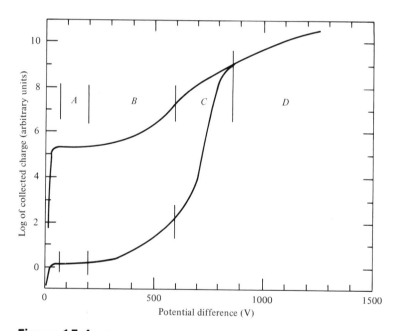

Figure 15.4 Ionization versus voltage characteristics for a gas-filled cylindrical counter. The primary ionization in branch (1) is 10^5 that in branch (2). In region A the counter acts as an ionization chamber. Region B is the proportional region. Region D is the Geiger region in which pulse height is independent of primary ionization.

Gas-filled detectors are relatively simple and cheap. However, if they are to produce output pulses whose sizes are proportional to the incident kinetic energy, they may become quite large and bulky. This follows from the fact that the active material, being a gas, is of relatively low density. Long pathlengths within it are required to stop the primary particles. Also, if "heavy" incident particles are to be counted, a thin entrance window is required. This entrance window separates the filling gas from the atmosphere around it. It is usually very fragile and prone to breakage. The efficiency for detection of gamma rays is low because of the requirement that conversion electrons be produced by the photoelectric or Compton process. The density of atoms within the gas is small so very few conversion electrons are produced in it. The cathode cylinder is solid and therefore has a relatively high density of atoms. However, only those conversion electrons produced close enough to its inner surface to be projected into the sensitive gas volume of the detector are effective. Increasing the wall thickness of the cathode beyond this value does not increase counting efficiency. We shall see in the next sections that there are other types of detectors that overcome these deficiencies.

15.3 Scintillation Counters

The **scintillation counter** was one of the very earliest detectors of nuclear radiation. However, its original form was considerably different from detectors used today. To detect alpha particles scattered by thin gold and other foils, Geiger and Marsden used a screen made of a thin layer of ZnS crystals. When an alpha particle strikes ZnS, a tiny flash of light is given off. This flash is known as a scintillation and can be seen by a dark-adapted eye aided by a simple magnifier. In modern practice a variety of scintillation materials are used in conjunction with **photomultiplier tubes** which gather and convert the light flashes to electrical pulses. These pulses are recorded and stored by electronic circuits. The widest use of scintillation counters today is in detection and measurement of gamma radiation.

Figure 15.5 is a schematic representation of a scintillator crystal optically coupled to a photomultiplier tube. We begin our discussion by considering the interaction of radiation with the scintillator itself. Sodium iodide (NaI) grown in the form of a single crystal is an efficient scintillator when optically activated by admixture of a small quantity of thallium. The abbreviation for thallium activated sodium iodide is NaI(Tl). When a charged particle passes through the crystal, it ionizes and excites atoms of the crystal. These atoms quickly deexcite by emission of photons, primarily in the visible spectrum. The total amount of light given off is proportional to the energy that the charged

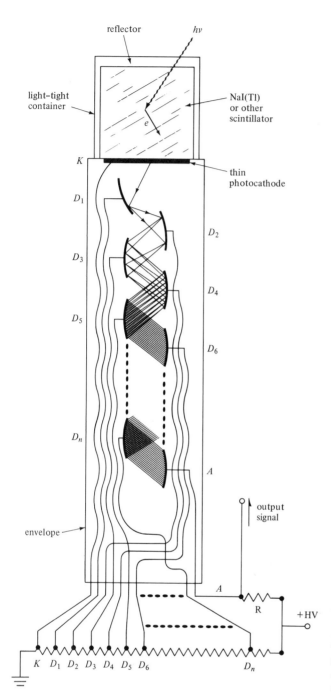

reflector

$h\nu$

light–tight
container

NaI(Tl)
or other
scintillator

e

K

D_1

thin
photocathode

D_2

D_3

D_4

D_5

D_6

D_n

A

output
signal

envelope

A

R

$+HV$

K D_1 D_2 D_3 D_4 D_5 D_6

D_n

Figure 15.5 Section view of scintillator and photomultiplier tube assembly.
The scintillator, at the top of the assembly,
is usually cylindrical and is surrounded by
a reflector and light-tight container. Light
from the scintillator ejects photoelectrons
from the cathode **K**. A string of dynodes
D_1, D_2, ..., D_n is connected to tap points
on a resistive voltage divider, creating an
accelerating voltage across each stage of
the tube. Electrons from the photocathode
strike the first dynode and produce secondary electrons. These are accelerated to the
second dynode, producing additional secondary electrons, and so on along the
string of dynodes. If the secondary emission ratio is ρ at each stage, the overall
gain in the tube is ρ^n. (In this sketch $\rho = 2$.)
Electrons from the last dynode are col-
lected at anode **A** and produce a negative-
going pulse. This signal is ordinarily
coupled to the input of an amplifier for
further electronic processing.

particle delivers to the crystal. If the incident radiation consists of gamma ray photons, then the charged particles to which the scintillator responds consist of conversion electrons produced within the scintillator itself. There are three main conversion processes that may occur. If the interaction is **photoelectric absorption** of the gamma ray, the kinetic energy of the ejected electron equals the photon energy, less the ionization energy of the atom from which the electron was ejected. Most of these electrons will be stopped within the crystal, thereby delivering all their kinetic energy to it. Furthermore, as the ionized atom deexcites, it gives off photons which add to those produced by the passage of the electron. The result is a pulse of light made up of many photons within the visible spectrum, a pulse whose magnitude is proportional to the energy of the original gamma ray.

Compton scattering of the incident gamma ray may also occur. In this interaction the gamma photon behaves as a particle making a "billiard ball" collision with an electron. The collision scatters the photon through an angle that may have any value between $0°$ and $180°$. The energy of the scattered photon is reduced by an amount which depends on the angle. The electron recoils with an energy equal to the original photon energy less the energy carried away by the scattered photon. If the scattered photon escapes from the crystal, the resulting light pulse is due only to the energy carried by the electron and is thus smaller than the full-energy pulses produced by the photoelectric interaction. The kinetic energy of the recoil electron is given by

$$K_e = h\nu_0 - h\nu \tag{15.2}$$
$$= \frac{(1 - \cos \phi)\,(h\nu_0/m_0 c^2)}{1 + (h\nu_0/m_0 c^2)\,(1 - \cos \phi)}\,h\nu_0$$

where $h\nu_0$ is the energy of the incident photon, $h\nu$ that of the scattered photon, and ϕ the angle of photon scattering. The kinetic energy delivered to the electron varies from 0, when $\phi = 0°$, to a maximum given by

$$K_{e_{max}} = \frac{h\nu_0}{1 + (m_0 c^2/2h\nu_0)} \tag{15.3}$$

when $\phi = 180°$. A continuous distribution of electron energies and, consequently, magnitude of the light pulses from the scintillator results. The drop-off in the distribution which occurs at the energy given by Eq. (15.3) is referred to as the **Compton shoulder**. Figure 15.6 shows the pulse-height distribution from a 2-in. diameter $\times$ 2 in. high NaI(Tl) irradiated with ^{137}Cs gamma rays (0.661 MeV). The full-energy peak is located at A, the Compton shoulder at B. The peak at C is called the **back-scatter peak**. It is due to gamma rays Compton-scattered through large angles by nonscintillating material in the neighborhood of the crystal, that is, the photomultiplier structure, the

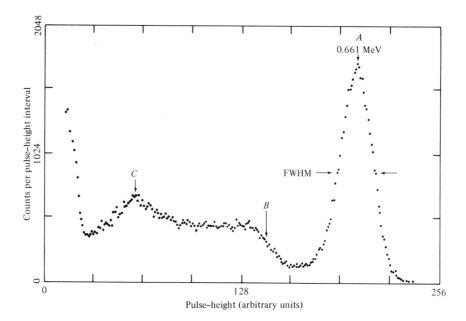

Figure 15.6 Pulse-height distribution from a 2 in. × 2 in. NaI(Tl) crystal irradiated with 0.661 MeV gamma rays.

reflector, the light-tight container, the container for the radioactive sample, and so on. Its energy is equal to the energy difference between the full-energy peak and the Compton shoulder.

Not all the events recorded in the full-energy peak arise exclusively from primary photoelectric interactions. There is a finite chance that photons scattered by the Compton process within the scintillator may undergo photoelectric interactions before escaping from the scintillator. The energy of the photoelectrons so produced is added to that of the original Compton electron. These events happen so close together in time that the photomultiplier registers them as a single event, whose total energy is that of the original gamma photon. This process adds extra events to the full-energy peak and reduces the number recorded as Compton events. If the crystal is small, the probability of escape of the Compton-scattered photon is high. For a large crystal the probability of escape is reduced, thus increasing the relative number of events recorded in the full-energy peak.

If the gamma photons have sufficient energy, a third type of interaction known as **pair production** is possible. As its name implies, this process results in the production of an electron-positron pair. The conservation of mass-energy dictates that

$$h\nu_0 = 2m_0c^2 + K_{e-} + K_{e+} \tag{15.4}$$

We see that there is a threshold for pair production given by

$$h\nu_{min} = 2m_0c^2 = 1.02 \text{ MeV} \qquad \textbf{(15.5)}$$

As photon energy increases above the threshold, the probability for pair production increases gradually, the excess energy appearing as the combined kinetic energies of the electron and positron. Although mass-energy is conserved readily enough according to Eq. (15.4), the simultaneous conservation of linear momentum cannot be satisfied unless another particle is involved. Pair production is usually accomplished by the interaction of an energetic photon with a heavy nucleus. By recoiling, the nucleus takes up the required momentum, but since it is very massive, it acquires negligible kinetic energy. The probability of pair production increases rapidly with atomic number of the absorbing material.

If pair production occurs within a scintillator, the kinetic energy of the electron-positron pair is converted to visible light in a fashion identical to that of a photoelectron or a Compton electron. The energy available, however, is 1.02 MeV less than that of the original photon. Figure 15.7 shows a pulse distribution which results when a NaI(Tl) crystal is irradiated with gamma rays from natural thorium in equilibrium with all its daughter products. The last radioactive member of the chain is $^{208}_{81}$Tl. Its decay produces gamma rays at 2.615 MeV. These gamma rays produce the three peaks in the right half of the figure. The left half of the figure, plotted on a smaller scale, is due to lower energy gamma rays coming from other members of the radioactive chain.

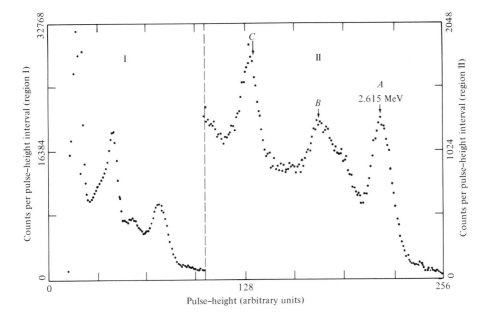

Figure 15.7   Pulse-height distribution from a 2 in. × 2 in. NaI(Tl) crystal irradiated with 2.615 MeV gamma rays.

Peak A is the full-energy peak and represents 2.615 MeV. Peak C shows the combined kinetic energy of the electron and positron which result from a pair production interaction. It has 1.022 MeV less energy than the full-energy peak. After the positron comes to rest, it combines with one of the electrons in the crystal, producing annihilation radiation consisting of two gamma photons of 0.511 MeV each. The escape of both of these photons without further interaction in the crystal produces events in peak C, the **double escape peak.** If one of the photons produced by positron annihilation is absorbed in the crystal, say by photoelectric interaction, whereas the other photon escapes, then 0.511 MeV is added to the pulse produced by the electron-positron pair. This sequence leads to events in peak B, the **single escape peak.** If both annihilation gammas are absorbed in the crystal, their energy is delivered to the crystal along with that of the electron-positron pair, and the total is registered in the full-energy peak. The larger the crystal the greater the probability that the annihilation radiation will be absorbed, thus increasing the relative number of events registered in peak A. The presence of three peaks with 0.511 MeV spacing is characteristic of the pair production interaction in a moderate sized scintillator.

We have focused our discussion on sodium iodide as a scintillator. Scintillations are also produced by other materials, such as potassium iodide, cadmium tungstate, calcium tungstate, anthracene, naphthalene, *trans*-stilbene, and *p*-terphenyl, in liquid solution and in plastics. In general, the higher the atomic number of the atoms contained in the scintillator, the greater the efficiency for the detection of gamma rays. With heavy inorganic materials the efficiency may approach 100%, a great advantage over the 1% or 2% counting efficiency exhibited by Geiger counters. Scintillators are also much faster than gas-filled counters in recovering after they have recorded an event. In NaI(Tl) the decay time of the scintillation is approximately 0.25 μsec. Although the organic scintillators have lower efficiency for the detection of gamma rays, their scintillation decay times are very short. For *trans*-stilbene it is of the order of 0.006 μsec. Short recovery times make possible corresponding accuracy in determining time relationships between coincident or successive nuclear radiations.

We have seen in Figures 15.6 and 15.7 that the response of a scintillator to monoergic gamma rays is fairly complicated. The **Compton interaction** produces a continuum of pulse sizes, **pair production** produces three separate peaks, and even within the **full-energy** peak there is a range of pulse sizes. This range results from a number of factors. The efficiency of the reflector which surrounds the scintillator may not be completely uniform. Scintillations occurring in different parts of the crystal may result in different amounts of light reaching the photocathode of the photomultiplier tube. There are statistical variations in the conversion of electron energy to light quanta. If the photocathode is not completely uniform, light striking one part of it may produce a different number of photoelectrons than light striking a different part. Similarly, variations in the efficiency with which the dynodes produce secondary electrons may change the number of electrons which finally reach the anode. In addition to all of these, there are **dark current** pulses within the photomultiplier itself, pulses that occur even though there is no light falling on the photocathode. Although these pulses are small, they add in a random fashion to the true pulses. All of these effects contribute to the spread, such that in the full-energy peak the full width at half maximum height (FWHM) is typically 10% to 15% of the pulse size of the peak itself.

Example 15.1 The $^7\text{Li}(p, \gamma)^8\text{Be}$ reaction produces gamma rays whose energy is 17.64 MeV. What is the energy of the Compton shoulder and of the single and double escape peaks in pair production?

Solution

$$K_{e\,\text{max}} = \frac{h\nu_0}{1 + m_0 c^2/(2h\nu_0)} = \frac{17.64}{1 + 0.511/(2 \times 17.64)} = 17.39 \text{ MeV}$$

The Compton shoulder is 0.25 MeV below the full-energy (photo) peak.

The single escape peak is 0.51 MeV below the full-energy peak, that is, at 17.13 MeV.

The double escape peak is at 16.62 MeV.

15.4 Semiconductor Detectors

A semiconductor detector is a large *pn* junction operated as a reverse-biased diode. Some semiconductor detectors are designed primarily for the detection of charged particles, such as protons, deuterons, alphas, and so on. Others are designed primarily for the detection of gamma rays. Charged particle detectors are usually made of silicon wafers approximately 1 mm thick and a few millimeters to a few centimeters in diameter. For detection of x rays and very-low-energy gamma rays, those below 40 keV, silicon crystals 3 to 5 mm thick may be used. For higher energy gamma rays, which may range up to 10 MeV, large germanium crystals approximately 5 cm in diameter × 5 cm high can be used. In general, semiconductor detectors exhibit far better energy resolution than gas-filled and scintillation detectors.

In Chapter 9 we discussed the properties and behavior of semiconductor diodes. We saw that in the region of contact between *n*-type and *p*-type material the majority carriers of each diffuse into the other. This depletes the carrier concentration in the neighborhood of the

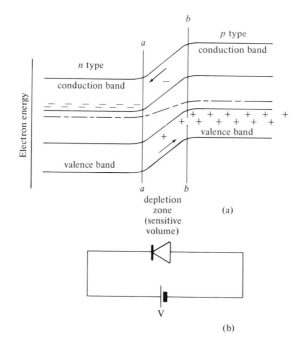

n type
conduction band

b
p type
conduction band

a

valence band

+ + + + + + +
+ + + + + + +
valence band

+

a   b
depletion
zone   (a)
(sensitive
volume)

V

(b)

Figure 15.8 Biased semiconductor detector: (a) the electron band structure, (b) schematic circuit diagram. An electron-hole pair is shown in the depletion zone. Many such pairs are produced when ionizing radiation passes through the zone.

junction, creating what is known as a **depletion zone.** Figure 15.8 is a repetition of Figure 9.17. The region between the lines *a-a* and *b-b* is the depletion zone. In a reverse-biased diode the thickness of this zone depends upon the applied voltage and is found to be proportional to $V^{1/2}$. We have seen that when a charged particle passes through any material, it loses energy by ionizing collisions with the atoms. When this happens within the depletion zone, electron-hole pairs are produced and the electric field sweeps them to the opposite faces of the junction. When used as a detector, the depletion zone forms a sensitive volume for collection of charge.

In Figure 15.8 we show a charge pair formed midway within the depletion zone. The electron "slides down" its potential gradient while the hole "floats up" along its gradient. In silicon 3.6 eV are required on the average to produce an electron-hole pair. A 5-MeV alpha thus produces about 1.4×10^6 electron-hole pairs. As these are swept through the junction, a charge pulse appears at the electrodes connected to the diode. The behavior is entirely similar to that of an ionization chamber and, in fact, semiconductor detectors are sometimes alluded to as solid-state ionization chambers. External electrical connections are diagramed in Figure 15.9. Note the similarity to the connections shown in Figure 15.3.

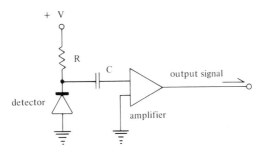

Figure 15.9 Circuit diagram for a semiconductor detector.

The energy required to produce an electron-hole pair in silicon is about one eighth that required to produce an electron-ion pair in a gas. Correspondingly greater primary ionization is produced in a semiconductor than in a gas counter. Although it might appear possible to operate a semiconductor detector in the **proportional region,** attempts to do so result in destruction of the detector. Although we must forego charge amplification within the detector, excellent external amplifiers are available with adequate gain. Furthermore, the large number of primary pairs produced in the semiconductor detector results in a relatively small fractional spread in pulse size. This spread may be as low as 12 to 15 keV FWHM for alpha particles of energy 5.5 MeV.

In addition to its superiority in converting particle energy to electrical pulses of uniform size, the silicon semiconductor detector is considerably more convenient to use than either a gas-filled detector or a scintillator-photomultiplier combination. Since it is solid, the density of the semiconductor detector is several thousand times that of the gas in a gas-filled detector. Consequently, a much shorter pathlength is required to extract fully the kinetic energy of a charged particle. The detector can be very compact. Also, since it is solid, it is not necessary to have a foil window between it and the atmosphere or vacuum around it. To be sure, the *p*-type or *n*-type material that constitutes the front face of the detector behaves in much the same manner as a thin window, since electron-hole pairs created there do not contribute to the electric signal. However, this "window" can be made as thin as 200 Å, far thinner than any gas-retaining foil. The energy precision of these detectors, their ruggedness, compactness, and their ability to operate directly in the vacuum of a scattering chamber have brought them into wide use.

In general, gamma-ray cross sections for production of conversion electrons increase with increasing atomic number of absorbing materials. It is clear, therefore, that a gamma-ray detector should be constructed of materials having as high an atomic number as practicable. For semiconductors this means that germanium is superior to silicon. Gallium-doped germanium forms a *p*-type semiconductor. When lithium is diffused into the surface of such a crystal and then **drifted** deeper by application of an electric field, a sensitive volume approximately 2 cm thick can be achieved. Gamma-ray interaction with the crystal follows the same three processes, photoelectric, Compton, and pair-production, discussed in connection with the NaI(Tl) scintillator. The conversion electrons produce a large number of electron-hole pairs in the sensitive volume of the Ge(Li) detector. Charge carried by these pairs is made available directly at the detector electrodes. This bypasses the remaining sequence of events (conversion to light, reflection of light to the photocathode, production of a photoelectron at that cathode, and electron multiplication by the dynode string of the photomultiplier) required when a scintillator is used. Some 4×10^5 electron-hole pairs are produced when there is a full-energy interaction with a 1.3 MeV gamma. The pulses produced by direct collection of this charge are much more uniform in size than those produced by a scintillator-photomultiplier combination. The peaks that result are narrower by a factor of 50 to 75.

Narrow peaks, of course, make it possible to resolve gamma rays that differ only slightly in energy. Furthermore, since the area under a peak is equivalent to the number of events registered in the peak, the product of peak height and width is approximately constant for a given number of events. It follows that if a full-energy peak with a Ge(Li) detector is $\frac{1}{50}$ as wide as one with a scintillator, the height of the peak will be 50 times greater. Because of this considerable increase in the heights of individual peaks, a Ge(Li) detector can detect low energy gamma rays superimposed upon Compton plateaus due to gamma rays of higher energy. The heights of the plateaus are not increased, since they inherently consist of continuous distributions of pulse sizes. Increasing the precision with which the kinetic energies of Compton electrons are translated into detector output pulses does not change the continuous nature of the distributions.

The germanium crystal itself is not bulky, but it must continuously be maintained at liquid nitrogen temperature, not only while it is in use but 24 hr a day, 7 days a week, throughout its entire life. This requirement is occasioned by the fact that the semiconductor properties of lithium-drifted germanium are not stable at room temperature. A dewar flask of appreciable size, sufficient to retain liquid nitrogen for several days between fillings, becomes an inherent part of the detector assembly. The entire assembly is quite bulky and more awkward to use than a silicon semiconductor detector, or for that matter a scintillator with photomultiplier. However, the peak width for a 1.3 MeV gamma ray can be as narrow as 2 keV FWHM when a Ge(Li) detector is used in conjunction with a good solid-state amplifier. Most users of these detectors conclude that the bulkiness of the dewar flask and the inconvenience of continuous maintenance are small prices to pay for the precision and resolution which this detector brings to gamma ray spectroscopy.

15.5 Detectors That Record Tracks

With proper arrangement of apparatus it is possible to make the actual path followed by charged particles in passing through gas, liquid, or solid visible for inspection. Also, it may be possible to arrange things so that the gas or liquid or solid used to record tracks contains a material in which nuclear reactions of interest take place. If so, the tracks of all charged bombarding particles involved in the reaction can be recorded. A graphic display of the event is obtained. If one (or more) of the particles involved is neutral, it does not leave a track. Its presence can be inferred, though, from momentum calculations involving observed particles. The picture formed by a track-recording device can be of considerable help in understanding many reactions.

The **cloud chamber** was the earliest of the track-recording devices. In 1895, C. T. R. Wilson observed that the adiabatic expansion of moist air in a chamber would result in a supersaturated vapor. If dust or other nucleation centers are present, water droplets form upon them in preference to other locations in the vapor. Whether a droplet disappears by evaporation or grows by condensation from the vapor around it is determined by whether its own vapor pressure at its surface is greater or less than that of the vapor. If it is greater, the drop will evaporate; if it is less, the drop will grow. The ratio of the vapor pressure p of a drop of radius r to the vapor pressure p_0 of a plane surface of a liquid whose density is ρ is given by

$$\log \frac{p}{p_0} = \frac{2\tau M}{\rho r R T} \tag{15.6}$$

where τ is surface tension, M is molecular weight, R is the universal gas constant, and T is the absolute temperature. We see that if r is small, p is large, and the drop tends to evaporate. If r is sufficiently large, p becomes small so there is condensation onto the drop and it grows even larger.

When a swift charged particle passes through the vapor, it leaves a trail of ions in its wake. Many of these ions become attached to small drops, which therefore become charged. This considerably changes the condensation versus evaporation conditions because electrostatic energy introduces an additional term into Eq. (15.6).

$$\log \frac{p}{p_0} = \frac{M}{\rho r R T} 2\tau - \frac{q^2}{8\pi r^3} \frac{\epsilon - 1}{\epsilon} \tag{15.7}$$

where q is the charge on the drop and ϵ is the dielectric constant of the liquid. For very small r, the logarithm is negative and p is smaller than p_0. Condensation occurs and the drop grows. As the drop grows, the electrostatic term becomes smaller. The drop pressure rises,

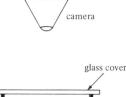

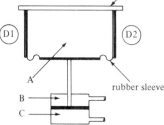

Figure 15.10 Section view of an expansion cloud chamber. Compressed air, admitted alternately to chambers **B** and **C**, causes rapid expansion of chamber **A**, followed by recompression. During expansion vapor in chamber **A** becomes supersaturated and sensitive to droplet formation along particle tracks.

reaches a maximum, and then falls off. For drops of large radius the electrostatic term is negligible and the behavior is like that of the uncharged drop. Growth of the drop ceases when its pressure reaches that of the vapor around it. In a supersaturated vapor the vapor pressure is larger than p_0 and larger drops are produced. Inspection of Eq. (15.7) shows that p does not increase without limit; it reaches a maximum and then falls off as the electrostatic term becomes less significant. If the degree of supersaturation is such that the pressure of the vapor exceeds the maximum droplet pressure, the drops grow to large size. With proper illumination these drops become visible and can be photographed, thus providing a permanent record of the passage of the charged particle.

Figure 15.10 is a schematic representation of an expansion cloud chamber. It consists of a cylindrical chamber A, closed at the top by a glass plate, through which the tracks can be observed, and at the bottom by a movable piston, which is sealed to the cylinder by a rubber sleeve. Water vapor in air is often used as the working medium in chamber A. A small pool of water at the bottom of the cylinder assures that the atmosphere inside is normally saturated. Admitting compressed air to chamber B and exhausting it from chamber C drives the piston assembly rapidly downward and expands the gas in chamber A. During a fast expansion, heat flow into chamber A is negligible. In this adiabatic expansion the temperature of the gas drops. A supersaturated vapor results. The degree of supersaturation can be controlled by the expansion ratio. This ratio should be about 1.3:1 for an air-water mixture initially at 20°C and 1.5 atm pressure. Immediately after expansion, tracks in the chamber are photographed. The chamber is then recompressed to its original condition. To improve the probability that tracks are actually present in the chamber at the time of expansion, detectors D1 and D2 may be used to trigger the expansion.

Another version of the cloud chamber produces supersaturation by temperature stratification of the vapor. Such a device is called a

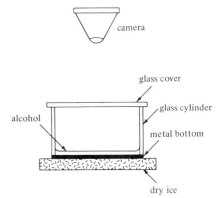

Figure 15.11 Section view of a
diffusion cloud chamber.

diffusion cloud chamber and is shown in Figure 15.11. The chamber
consists of a glass cylinder closed at the top by a glass plate and at the
bottom by a metal pan which rests on a block of dry ice. There is a pool
of alcohol in the bottom pan. Blotter paper lines the inside wall of
the glass cylinder and acts as a wick to draw liquid alcohol to the top
of the cylinder. Since the temperature there is relatively warm, alcohol
evaporating from the blotter maintains a vapor pressure higher than
that characteristic of temperatures at lower levels in the cylinder. As
the vapor drifts downward under the influence of gravity, it becomes
supersaturated. About halfway down conditions are suitable for the
formation of tracks. Such a chamber is continuously sensitive and does
not require the complicated mechanisms needed to cycle the expansion
chamber.

Both versions of the cloud chamber suffer from the limitation
inherent in all gas-filled detectors, namely the low density of gas. Energetic particles may pass entirely through the chamber without depositing all of their kinetic energy. Furthermore, the probability of an
interesting event, such as a nuclear reaction or a scattering, occurring
within the chamber is small. Because of its simplicity the diffusion
cloud chamber is still used, but largely as a piece of demonstration
apparatus. The expansion cloud chamber is of interest primarily from
a historical standpoint.

The **bubble chamber** behaves in a fashion analogous to that of the
expansion cloud chamber. However, it is filled with a liquid instead of
a vapor so there is a much greater probability of a reaction occurring
within its volume. The liquid is held at conditions of pressure and
temperature close to the boiling point. If the pressure is suddenly reduced, the boiling point is lowered below that of the liquid. For a short
time the liquid exists in a superheated condition before boiling occurs.
If there are ions within the liquid, they serve as nucleation centers for
the formation of vapor bubbles. A track of bubbles will be formed if an
energetic charged particle passes through the liquid during a critical
time following expansion and before boiling occurs throughout the
bulk of the liquid.

The timing of events is much more critical for the bubble chamber than it is for the cloud chamber. As a consequence, bubble chambers find use almost exclusively with high-energy accelerators. These accelerators provide beams of particles in very short bursts at precisely timed intervals. The bubble chamber is expanded just prior to a burst. After the bubbles have had time to grow to a suitable size, they are photographed in the illumination of a high-power flash lamp. Liquid hydrogen is a common choice for the working medium. Chambers several meters in diameter have been built. The construction of such chambers is a major engineering effort, and many precautions must be observed to achieve safe operation.

The **spark chamber** is usually made up of a stack of electrically insulated plates, the spacing between them being a few millimeters. Alternate plates are connected to ground; the other plates are connected to a potential of approximately 20 kV. If a charged particle sufficiently energetic to penetrate the plates passes through the stack, it leaves a trail of ions within the air gaps. Sparking is initiated within the region of ionization; the track is recorded as a series of sparks between plates. These sparks may be photographed with a stereoscopic camera looking into the stack from the edges of the plates. A spark chamber may be operated in a continuous mode or in a pulsed mode. When operated in the continuous mode high voltage is applied to the plates continuously. A series resistor or other device lowers the voltage on the plates after each spark to prevent continuous breakdown. However, operation in the pulsed mode is more stable and reliable. In this mode high voltage is applied to the plates only after an external detector has indicated the occurrence of an interesting event. Figure 15.12 is a simplified diagram of such a chamber. A nonionizing particle has entered the chamber from the right and has initiated in the third plate a reaction that has produced two charged particles, which pass

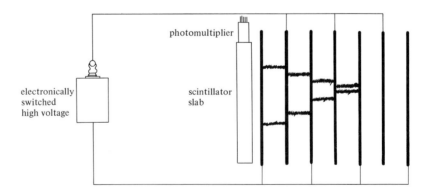

Figure 15.12   Schematic diagram of a seven-plate spark chamber. A neutral particle has entered the chamber from the right and produced a pair of charged particles in the third plate. Passage of these particles is recorded by a series of sparks between the remaining plates.

out of the chamber at its left side. As they pass through the scintillator, they initiate a signal that triggers the electronic switching circuit to apply high voltage to the plates. Spark chambers are particularly useful when a very massive detector is required. This would be the case when the probability of the primary interaction is low, as in neutrino studies.

Nuclear emulsions are similar to photographic emulsions, but the composition is modified to enhance the recording of individual tracks produced by ionizing particles. The silver bromide content is much greater in the nuclear emulsion. Individual grains are very small and separated from their neighbors, and the emulsion may be as much as 2 mm thick. When an ionizing particle passes through a silver bromide grain, it renders it developable. Thus the passage of a charged particle through the emulsion leaves a line of developable grains. Figure 15.13 is a photomicrograph of a nuclear emulsion taken with dark field

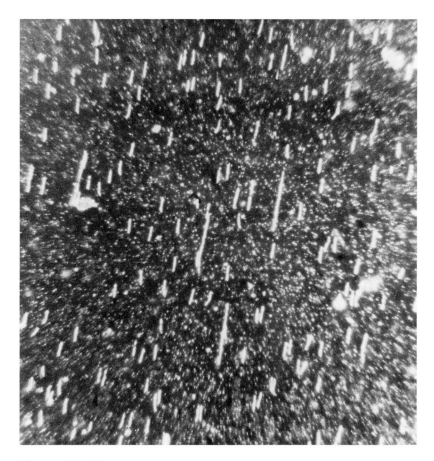

Figure 15.13 Photomicrograph of proton and deuteron tracks in a nuclear emulsion. The longer tracks are due to 3-MeV protons, which have a range of 75μ in the emulsion. Collimation of the incident particles causes the tracks to be nearly parallel.

Table 15.1 Ranges of Protons and Alphas in Nuclear Emulsion

Energy (MeV)	Proton Range (microns)	Alpha Range (microns)
10	545	58
5	175	21
1	15	3

illumination. The long tracks are due to 3-MeV protons and the short ones due to 1.5-MeV deuterons.

The rate at which a charged particle loses energy by ionization depends not only upon the material through which it passes but upon the nature of the particle itself and upon its velocity. Different particles, having the same kinetic energy, may have considerably different ranges in the emulsion. Table 15.1 shows ranges for protons and alpha particles at three different energies. The grain density along the path is proportional to the rate of ionization unless that rate is so high that every grain becomes developable. Combined measurements of total range and of grain density in the track can be used to identify the particle and its energy. Since the tracks are relatively short and the grains very small (approximately 0.5 μm diameter), very good microscopes of high resolving power must be used to measure them. If measurements are to be made when the particle range exceeds the thickness of the emulsion, two stratagems are available. The simplest is to use collimated incident particles at grazing incidence. If this is not possible, perhaps because the particles of interest arise from reactions occurring within the emulsion itself, it is possible to produce the equivalent of a very thick emulsion by using a multilayer stack. This, of course, requires careful indexing of the layers so that a portion of a track in one layer can be properly correlated with its continuation in the next layer.

Thin emulsions are usually mounted on glass plates for dimensional stability and ease of handling. Their processing through developer, fixer, and washing cycles is not much different from that of photographic plates. Emulsions thicker than 100 μm require rather special developing and fixing procedures to assure that the entire thickness of the emulsion receives the same development time.

The hydrogen content of emulsions makes them useful for measuring neutrons of medium and high energy. When a neutron makes an elastic collision with a stationary proton, the proton recoils and leaves a track in the emulsion. If the neutrons are collimated in a single direction, the analysis of the proton tracks is simple, since those protons that recoil in the forward direction have the same kinetic energies as the incident neutrons. Emulsions may also be "loaded" with a variety of additional elements for special purposes. For instance, the $^6Li(n, \alpha)^3H$

reaction has a high cross section for slow neutrons. Its Q value of 4.8 MeV provides sufficient energy to the alpha particle and triton in the reaction to assure that they produce readily discernible tracks. This is of considerable advantage when working with low-energy neutrons, since "knock-on" protons have correspondingly low energy and produce tracks too short to be found in the emulsion.

Emulsions are lightweight, compact, continuously sensitive, highly reliable, and free of spurious effects associated with electronic noise. The main drawback to their use is the labor required in extracting information from them. It may require several days to many weeks of scanning and measurement under a high-power microscope.

Solid dielectric track recorders are used in much the same fashion and serve many of the same purposes as nuclear emulsions. In many dielectrics the ionization damage caused by the passage of an energetic charged particle causes permanent structural changes at the molecular level. For some materials it is possible to choose a suitable chemical etchant such that the etching rate is far greater in the region of ionization damage than in the bulk material. If the ionizing particles reach the surface of the dielectric, either upon entrance or exit, the region of damage becomes accessible to the etchant. Etch pits are produced of depth equal to the track length in the material; the diameter is a measure of the degree of damage, that is, the density of ionization produced by the particle. For a given ionization density this diameter also depends upon the choice of track-recording material and etching conditions.

Measurements of etch pit diameters can be exploited in a number of ways. Density of ionization depends upon the atomic number of the incident particle and also upon its velocity. With careful control of the etching conditions in a track recorder of known characteristics, a good estimate of the atomic number of the incident particle can be obtained. Also, it turns out that when ionization density is below a certain threshold the damage produced is inadequate for production of etch pits at all. This threshold can be controlled over rather wide limits by the choice of track-recording material and etchant conditions (time, temperature and concentration). Thus it is possible to discriminate against the recording of light particles (for instance, alphas) while still recording heavier particles, such as fission fragments. Although techniques have been developed recently by which particles as light as protons can be recorded, most track-recording materials are insensitive to particles lighter than alphas. Even with these the energy must be below 5 MeV since at higher energies ionization density is too small for etch pit formation. The complete lack of response which such track-recording foils exhibit toward protons and deuterons has been exploited in measurements of (d, α) reactions having very small cross section. The recorder ignores altogether protons from competing (d, p) reactions, as well as the large number of deuterons elastically scattered by the target. Discrimination of this sort is not attained with nuclear emulsions.

Like emulsions, track recorders are free of electronic noise. They are insensitive to light and can be handled under conditions of room illumination. Even in foil form they are more rugged and less subject to damage than emulsions. The process of etching is less complex than the development and fixing process required for thick emulsions. The inability to register the passage of lightly ionizing particles is an obvious limitation, but it can be turned to good use for some experiments.

15.6   Pulse-Height Measurements

We have seen in Sections 15.2 through 15.4 that many types of detectors produce electric pulses that can be used to measure the energy of ionizing radiation. In Figures 15.6 and 15.7 we showed pulse-height distributions from a NaI(Tl) scintillator. We did not at that point undertake a discussion of the electronic techniques for measuring these distributions. A detailed treatment of these electronic techniques is beyond the scope of this book, but it is possible to give a general idea of the way in which pulse-height measurements are made.

Figures 15.5 and 15.9 show electrical connections for representative detectors. A bias voltage V is applied to the top end of resistor R. A pulse in the detector rapidly lowers the voltage at the bottom end of R. This voltage pulse is coupled to the input of the amplifier through capacitor C. Such amplifiers are highly specialized and the design of a good one is a real art. Input characteristics must properly match the output of the detector. Great care must be taken to minimize introduction of electronic noise, and the design must be such that over a wide range of pulse sizes the height of the output pulse is linearly related to that of the detector. Although the amount of amplification required depends upon the detector and upon the radiation being measured, a voltage gain of 1000 is fairly typical. This gain is usually achieved by the use of two amplifiers, a preamplifier close to the detector and a main amplifier with a variety of gain controls and pulse shaping.

Figure 15.14 shows a cathode ray oscilloscope display of pulses used in the production of Figure 15.6. This is a time exposure and many pulses are superimposed on the oscilloscope screen. Although the pulses are randomly distributed in time, a new oscilloscope sweep is generated for each pulse so all pulses appear to start at the origin. It can be seen that there is a high density of pulses with heights between 9 and 10 V. These are full-energy pulses. Below 7 V there are numerous pulses that arise from Compton events. Between 7 and 8 V there are few pulses. This is the valley between the full-energy peak and the Compton shoulder. With a photograph of this sort it is possible

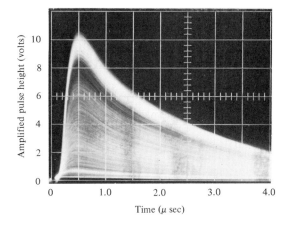

Figure 15.14 Cathode ray tube display of photomultiplier pulses due to interaction of gamma rays with a NaI(TI) crystal. The source was ^{137}Cs, which produces gamma rays of 0.661 MeV. Full-energy pulses rise to heights between 9 and 10 V. Pulses that do not rise above 7 V are due to Compton events. There are very few pulses between the Compton shoulder and the full-energy region.

to get a distribution in pulse height by measuring the photographic density along a vertical line through the crests of the pulses, that is, at 0.5 μs in Figure 15.14. This technique was used before the development of digital electronic techniques for pulse-height analysis. Clearly, though, it is awkward and leaves much to be desired.

Figure 15.15 is a schematic diagram of the digitizer section of a multichannel pulse-height analyzer. At the right side of the figure are three voltage-versus-time displays, which would be seen with an oscilloscope measuring the voltages at points *a, b,* and *d*. The output of the detector amplifier is applied at the input *a*. Two pulses of differing heights are shown in the corresponding voltage display. Although shown one above the other, they actually arrive at different times.

Let us begin by considering the analysis of the smaller pulse. It passes through the diode D and is used to charge the capacitor C. The diode prevents discharge of the capacitor after the input pulse dies away. The block circuit T is a trigger circuit that detects the fact that the capacitor is no longer at zero voltage. After a time delay sufficient to allow the incoming pulse to reach its full height, the trigger circuit closes the electronic switches S1 and S2. Circuit I is a constant current generator; closing switch S1 allows a constant negative current to flow into the capacitor, thus discharging it linearly with time. The linear discharge is called a **ramp** and is shown in the display of V_b between t_0 and t_1. Switch S2 closes simultaneously with S1, thus permitting pulses from a square wave oscillator (OSC) operating at a frequency of tens of megahertz to be counted by an address scaler (AS). At time t_1 the capacitor has been discharged back

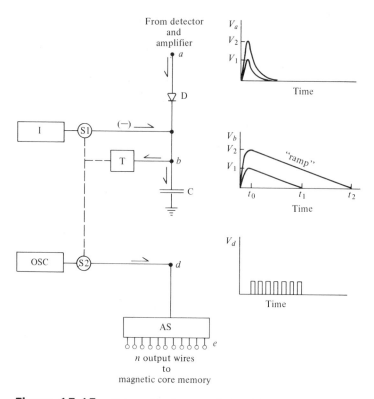

Figure 15.15 Schematic diagram of the digitizer section of a multichannel pulse-height analyzer.

to zero voltage. The trigger circuit detects this condition and immediately opens switches S1 and S2. The address scaler registers a binary number, which is a measure of the time interval $(t_1 - t_0)$. Since I is a constant current generator, the ramp is linear and its slope is independent of the height of the pulse being measured. Consequently, $(t_2 - t_0)/(t_1 - t_0) = V_2/V_1$, and therefore, the number registered by the address scaler is proportional to pulse height. With an address scaler having n binary stages, the total pulse-height region is divided into $2^n - 1$ pulse-height addresses or channels. The binary number determined by the address scaler (output e) is presented to a magnetic core memory. A memory cycle is initiated in which the number of previous events stored at that memory address is read out, increased by one, and restored to the memory. During digitizing and memory cycle times, the analyzer is busy and is forced to ignore new events. Upon completion of the memory cycle the analyzer is ready to accept new pulses. Each plotted point in Figures 15.6 and 15.7 represents information stored at one of 256 memory channels. The ordinate is proportional to the number of counts stored in each channel and the abscissa is proportional to channel number.

15.7   The Magnetic Spectrometer/Spectrograph

For some experiments it is necessary to achieve energy resolution superior to that provided by semiconductor detectors. Improvement by an order of magnitude or more can be achieved with **magnetic spectrometers** and **spectrographs,** in which magnetic fields deflect and focus charged particles. If a single narrow detector is used at the focus, thus registering one energy at a time, the device is a spectrometer; if a photographic plate or detector array registers simultaneously many energies, the device is a spectrograph. Many versions of these instruments have been built in various laboratories throughout the world. They exist in a wide variety of styles and in sizes ranging from those which can be mounted on a laboratory bench to some which weigh hundreds of tons. We shall give scant attention to the major engineering considerations involved in the design of these instruments. Our main concern will be with basic physics principles. We shall consider styles particularly suitable for work with ions of low atomic mass in the MeV region, that is, protons, tritons, alphas, and so on.

The motion of a charged particle in a magnetic field is governed by the Lorentz force

$$F = \frac{1}{c} q\mathbf{v} \times \mathbf{B} \tag{15.8}$$

expressed here in gaussian units (q is in electrostatic charge units, esu, $\mathbf{B}$ is in gauss, and force and velocity terms are in customary cgs units). Since at all times the force acts normal to the velocity, its direction alone changes, not its magnitude. If velocity is initially normal to the magnetic field, it will remain so and the path will be an arc of a circle, with the Lorentz force providing the required centripetal force. Thus

$$\frac{1}{c} qvB = \frac{mv^2}{R} \tag{15.9}$$

from which

$$BR = \frac{cmv}{q} = \frac{cp}{q} \tag{15.10}$$

where R is the radius of the arc and p is the momentum of the particle. From this we see that determination of the radius of curvature of a particle of known charge in a magnetic field of known strength can be used to determine its momentum. Equations 15.9 and 15.10 are correct for relativistic as well as classical velocities. To calculate kinetic energy we use

$$p^2c^2 + (m_0c^2)^2 = E^2 = K^2 + 2m_0c^2K + (m_0c^2)^2 \qquad \text{(15.11)}$$

so that

$$K^2 + 2m_0c^2K - p^2c^2 = 0 \qquad \text{(15.12)}$$

the solution for which is

$$K = \sqrt{(m_0c^2)^2 + p^2c^2} - m_0c^2 \qquad \text{(15.13)}$$
$$= \sqrt{(m_0c^2)^2 + B^2R^2q^2} - m_0c^2$$

In the classical limit this becomes

$$K = \frac{1}{2}\frac{B^2R^2q^2}{m_0c^2} \qquad \text{(15.14)}$$

If one constructs several arcs of equal radius diverging in slightly different directions from a common starting point, an interesting geometrical property is discovered. During the first 90° of path the arcs diverge, but during the second 90° of path they converge, arriving at an approximate focus when they have completed a full semicircle. Figure 15.16 illustrates the application of this focusing property. There is a uniform magnetic field normal to the plane of the diagram and directed outward. Charged particles originate at the source S, pass through the slit S1, which defines the angular divergence, and come to a focus at slit S2. Detector D counts particles passing through S2. All particles entering the detector have a well defined BR product and thus well defined momentum and energy. If the magnetic field

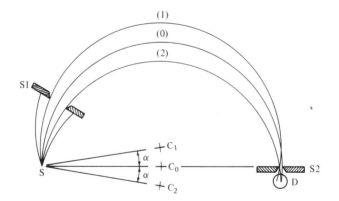

Figure 15.16   The 180° focusing properties of a uniform magnetic field.

strength is increased in small increments and the number of counts registered by the detector is recorded for each setting of the field, a momentum spectrum is produced that can be converted to an energy spectrum by application of Eq. (15.13) or Eq. (15.14). With care, energy resolution better than 0.1% can be obtained. Used in this fashion the instrument is a **magnetic spectrometer.**

An energy resolution of 0.1% implies 1000 settings of the magnetic field to cover the energy region, for instance, between 1 MeV and 2 MeV particle energy. This is a very tedious and time-consuming procedure, but fortunately there is a way around it. When particles with a given BR product focus on slit S2, other particles with lower product focus along the diameter of the spectrometer closer to source S, and those with larger product farther out. A plane through S and S2, normal to the plane of the diagram, is a focal surface of the instrument. If a nuclear emulsion several centimeters long is situated in the focal surface as a substitute for slit S2 and detector D, a substantial range of particle energies can be recorded with a single setting of the magnet. For example, a 5-cm strip of emulsion used in conjunction with a magnet of 1 m diameter spans a momentum interval of 5%. This corresponds to 10% in energy. A hundredfold increase in gathering power is achieved. Moreover, no sacrifice in resolution is required since the position where each charged particle strikes the emulsion is recorded microscopically by the track it produces. When used in this fashion, the instrument is designated a **magnetic spectrograph.**

The 180° magnetic spectrograph suffers two main limitations. In Figure 15.16 three rays, all with the same radius, are drawn. Careful construction of the three semicircular arcs shows that both the inner and outer rays (designated 1 and 2) strike the focal surface slightly closer to the source than the position where the central ray strikes. This focusing aberration limits the resolving power of the instrument. The aberration becomes rapidly worse as the angle of acceptance defined by slit S1 is increased. The other limitation arises from the requirement that the source and detection systems be between the poles of the magnet. This makes for very cramped working conditions which are especially irksome if the source is inside the target chamber of an accelerator.

Figure 15.17 shows the geometry of a magnetic spectrograph that alleviates the limitations of the 180° spectrograph. Focusing aberrations are smaller and the source and detection systems are outside the magnetic field. The magnetic pole face has a circular boundary of radius R, the source is at a distance R from the point at which the particles enter the magnetic field, and prime focus F_0 is at a distance R from the point at which the particles exit from the field. The magnetic field is adjusted so that the radius of curvature of the paths of the particles within the field is also R. Thus the central ray undergoes a 90° deflection. Slit S1 defines the angle of acceptance, which may be

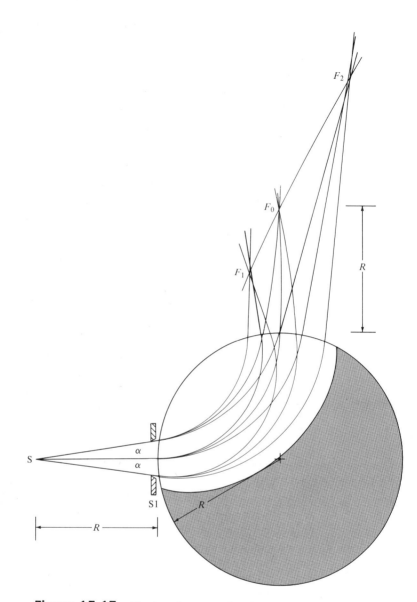

Figure 15.17 The focusing properties of a broad-range magnetic spectrograph.
[C. P. Browne and W. W. Buechner *Rev. Sci. Instr.*, **27**:899 (1956).]

much larger than that in Figure 15.16. Although best focus is obtained for deflections of 90° (that is, at F_0), acceptable focus is obtained along a tilted focal surface F_1-F_0-F_2. In practice the region of acceptable focus can cover an energy ratio of approximately 2:1; that is, the energy of particles at F_2 is twice that of particles at F_1. Since there are no particle trajectories in the lower and right-hand portions of the magnet pole face, shown shaded in Figure 15.17, it is wasteful to provide magnetic field there. For large spectrographs the poles are not complete cylinders but are shaped to provide magnetic flux only in the region of actual use, thus achieving considerable saving in the amount of power required to operate the magnet.

Example 15.2 An electromagnet with a 10-in. diameter pole face is available. Its maximum field is 12,000 gauss. If it is used as the main element of a spectrograph of the type shown in Figure 15.17, calculate the maximum energy of beta particles and of alpha particles that can be brought to the prime focus.

Solution

$$(BR)_{max} = 12 \times 10^3 \times 5 \times 2.54 = 1.52 \times 10^5 \text{ gauss-cm}$$

For betas we use Eq. (15.13).

$$K_e = \sqrt{(.511 \times 1.6 \times 10^{-6})^2 + (1.52 \times 10^5 \times 4.8 \times 10^{-10})^2}$$
$$- 0.511 \times 1.6 \times 10^{-6}$$
$$= 7.21 \times 10^{-5} \text{ erg} = 45 \text{ MeV}$$

We estimate the alphas to be nonrelativistic and therefore use Eq. (15.14).

$$K_\alpha = \frac{1}{2} \frac{(1.52 \times 10^5 \times 9.6 \times 10^{-10})^2}{4.006 \times 931 \times 1.6 \times 10^{-6}} = 1.78 \times 10^{-7} \text{ erg} = 1.11 \text{ MeV}$$

The alphas are indeed nonrelativistic, since $M_\alpha c^2 \gg 1.11$ MeV.

15.8   DC Accelerators

The **Cockcroft-Walton** generator maintains a high dc potential by means of the voltage multiplying scheme shown on the left side of Figure 15.18. The right side of the figure diagrammatically represents a high-voltage terminal that surrounds an ion gun. The ion gun consists of an ionization region in which low pressure gas is used, for example, hydrogen to produce protons, deuterium to produce deuterons, and so on, which, in conjunction with extractor and focusing electrodes, directs a well-formed beam of ions into an evacuated acceleration tube. The acceleration tube extends from the high-voltage terminal to ground potential. After reaching ground potential, the ions pass through a field-free drift tube and are focused on a target.

Suppose, in Figure 15.18, the secondary of a transformer T operates at a peak voltage V. Diode D_1 and capacitor C_1 form a half-wave rectifier. The peak inverse voltage which appears across D_1 is $2V$.

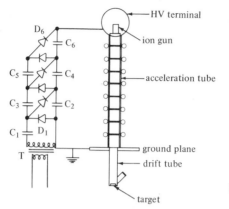

Figure 15.18 Schematic diagram of a Cockcroft-Walton accelerator.

Through diode D_2 this is transferred to capacitor C_2, charging it to a potential of $2V$. Diode D_3 transfers charge to capacitor C_3, charging it also to a potential of $2V$. Charge transfer continues up the ladder, each capacitor except C_1 being charged to a potential difference of $2V$. The number of stages is arbitrary and may be greater than that shown in our illustration. However, as more stages are added the problem of voltage ripple becomes increasingly severe. Also, the current that can be drawn decreases as the number of stages increases. This must be true since the power available at the high voltage end of the ladder cannot exceed that provided by the transformer at the bottom. Most Cockcroft-Walton generators operate at potentials below 10^6 V. They survive today primarily for production of neutrons through the ^{2}H$(d, n)^3$He and ^{3}H$(d, n)^4$He reactions, and for injection of ions into other accelerators which boost the energy to much higher values.

The **Van de Graaff** generator is a dc machine that can achieve much higher voltages than the Cockcroft-Walton generator. Figure 15.19 shows such a generator. Like the Cockcroft-Walton it utilizes an ion gun, acceleration tube, and drift tube. Its generation of voltage, however, utilizes electrostatic principles. An insulating conveyer belt runs between two pulleys P driven by the motor M. A power supply PS operating at 30 to 60 kV sprays charge onto the belt by means of corona needles that face it. The belt, being an insulator, conveys the charge to the high voltage terminal where it is removed from the belt by another set of corona needles. Charge is transferred efficiently from belt to terminal, since the latter acts as a Faraday pail. As more charge collects on the terminal its potential rises. When charge transfer down a resistor string (not shown in the figure) alongside the acceleration tube, and by corona current between the high voltage terminal and ground, equals the rate of transfer by the belt, equilibrium is achieved and the voltage becomes constant. If the entire apparatus is enclosed within a pressure tank, higher voltages can be achieved before the onset of sparking. Air is a much better in-

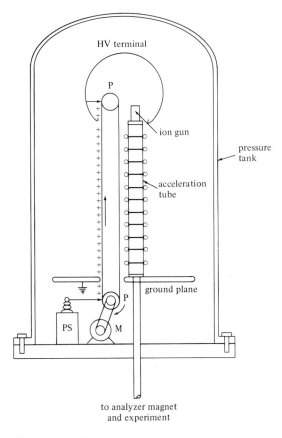

Figure 15.19   Schematic diagram of a Van de Graaff accelerator.

sulator at elevated pressures than at 1 atm. Also, with the use of a tank, the air can be kept dry, thus improving the electrostatic behavior of the generator. Further improvement is achieved by adding special insulating gases, such as freon (CCl_2F_2) or sulphur hexafluoride (SF_6). Total pressures of 10 to 20 atm are common. Although compressed air is a fairly good insulator, its oxygen concentration at elevated pressures greatly increases the hazard of fire. To eliminate this hazard, a mixture of nitrogen and carbon dioxide may be substituted.

An outstanding feature of the Van de Graaff generator is the steadiness of its voltage and freedom from ripple. Ion energies can be controlled to approximately 1 part in 10^4 by use of circuits which sense the position of the ion beam after magnetic deflection and feed back information to control the potential of the high voltage terminal. A few machines have achieved terminal potentials in the neighborhood of 10^7 V, but machines designed for about half this voltage are much more common.

The **tandem Van de Graaff** accelerator is illustrated in Figure

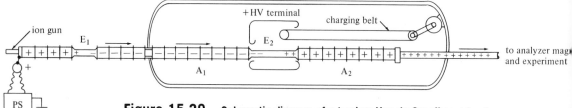

Figure 15.20 Schematic diagram of a tandem Van de Graaff accelerator.

15.20. At the left an ion gun produces positive ions that are accelerated by the power supply PS. In region E_1 a small flow of gas is introduced in order to transfer electrons to the moving ions, some of which are thereby converted to negative ions. In the first acceleration region A_1, negative ions acquire kinetic energy enroute to the positive high voltage terminal. Inside the high voltage terminal they pass through a thin foil or gas canal, where electrons are stripped from them. They emerge as positive ions and proceed through the second acceleration region A_2, acquiring additional kinetic energy. Hydrogen ions acquire twice the kinetic energy they would acquire in a single-stage Van de Graaff. For ions of higher atomic number the kinetic energy is given by

$$K = N_1 eV + N_2 eV \qquad\qquad (15.15)$$

where N_1 is the ionization state during the first stage of acceleration, N_2 that during the second stage of acceleration, e the electronic charge, and V the potential of the high-voltage terminal. Almost always N_1 is unity but with certain ions N_2 may be quite large, thus giving a very substantial gain in kinetic energy.

Example 15.3 A tandem Van de Graaff is used to accelerate oxygen ions. What is the kinetic energy achieved if the high-voltage terminal operates at +10 MV and the oxygen is an O^- ion for the first stage of acceleration and an O^{++} ion for the second stage?

Solution

$$K = 1e \times 10 \text{ MV} + 4e \times 10 \text{ MV}$$
$$= 50 \text{ MeV}$$

15.9 The Cyclotron

High voltage breakdown between terminal and ground limits the maximum ion energy that can be obtained with dc machines. The

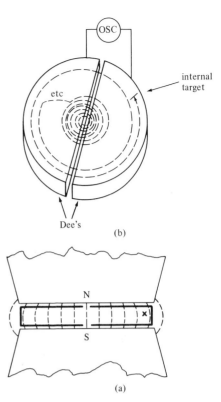

Figure 15.21 The magnetic field and dee electrodes of a cyclotron: (a) section view of magnet poles and the dee's, (b) isometric view of the dee's with spiral path of particles shown by dashed line. The drawing omits the vacuum housing for the dee's and the remainder of the massive magnet and electric coils used to produce the magnetic field.

cyclotron sidesteps this problem by causing ions to be accelerated over and over again through a modest and more manageable voltage drop. This is achieved by placing a pair of hollow electrodes, called dees, into a magnetic field that guides the ion trajectory into a spiral. Figure 15.21 is a schematic illustration of these parts. Figure 15.21(a) is a side view showing the dees placed between magnet poles. The shape of the magnetic field is shown by dashed lines between the N and S poles. Figure 15.21(b) shows a view of the dees removed from the magnet. The dashed spiral is the path taken by the ions undergoing accleration. (To visualize the shape of the dees, imagine a flat pill box separated into two halves by cutting it along a diameter.) The dees are connected to an oscillator so that the electric field in the gap between them is alternately to the right and to the left.

Ions are produced in an ion source placed at the center of the gap between the dees. When the dee on the left is positive, ions are accelerated to the right and enter the interior of the right-hand dee. Since the dees are conductors this region is free of electric field. However, the magnetic field penetrates the dees and the ions follow a circular path as developed in Eqs. (15.8) through (15.14). It is instructive to solve for the velocity. We find

$$v = \frac{BRq}{cM} \tag{15.16}$$

If we divide by the circumference of the circular path, we obtain the number of revolutions per second.

$$f = \frac{v}{2\pi R} = \frac{Bq}{2\pi cM} \qquad (15.17)$$

We see that in a constant magnetic field the frequency of revolution depends upon the charge-to-mass ratio, but does not depend upon the radius of the orbit. If the frequency of the oscillator is set to that given by Eq. (15.17), an ion of constant mass (that is, nonrelativistic) and charge travels a half circle of path during a half cycle of the oscillator. At each gap crossing the voltage between the dees is in the direction to give further acceleration. After each acceleration the radius of the path increases. The entire path is made up of a series of semicircles of increasing radius and, thus, approximates a spiral. Ions that cross the gap at the peak of the oscillator voltage receive maximum increments of energy and produce a spiral with coarser pitch than those for ions which arrive at less favorable phases of the oscillator. In any case, however, the ions remain in step with the oscillator and are successfully accelerated. Near the outer rim of the dees they may be focused upon an internal target or may be deflected into a path which escapes from the magnetic field, thus producing an external beam. The final kinetic energy is

$$K = \frac{1}{2}Mv^2 = \frac{1}{2}\frac{B^2R_0^2}{c^2}\frac{q^2}{M} \qquad (15.18)$$

where R_0 is the outer radius of the path.

For a 20-MeV proton, in a magnetic field of 12,000 gauss, R_0 is a little over 0.5 m. If the peak voltage of the oscillator is 100 kV, the path length in the spiral is several hundred meters. Because of this long path we must be concerned about focusing the ion stream; will the ions actually follow the predicted spiral path or will small perturbations deflect them so they do not reach the target? The 180° focusing properties of the uniform magnetic field (see Section 15.7) provide focusing against perturbation in the radial direction. Focusing in the z direction is provided by shaping the magnetic field so that it is convex outward, as sketched in Figure 15.21(a). With this field shape there is an outward radial component of magnetic field above the median plane and an inward component below it. A positive ion that circulates counterclockwise experiences a downward force when above the median plane and an upward force when below. These ions execute small amplitude oscillations centered on the median plane. The more convex the field lines, the stronger becomes the focusing in the z direction. However, a price is paid, since shaping the field this way inherently requires the z component of magnetic field to become weaker as we move outward from the center of the pole face. This results in weaker focusing in the radial direction. Appropriate compromise, however, can produce simultaneous focusing in both directions.

The application of Eq. (15.17) to provide constant oscillator frequency assumes ion energies that are nonrelativistic. For this reason the classical cyclotron experiences an energy limitation for protons at about 20 MeV. Above this energy the increase in mass, although slight, results in loss of resonance.

If we recast Eq. (15.17) we see that

$$f = \frac{Bq \sqrt{1 - \beta^2}}{2\pi c M_0} \tag{15.19}$$

where $\beta = v/c$. When β^2 is not negligible with regard to unity, either f must be adjusted downward or B upward. At first sight one might consider making the magnetic field stronger at the outer radii where ion energies are high. This, however, results in a field shape concave outward and produces defocusing in the z direction. Under these circumstances the beam is very quickly lost to the upper and lower faces of the dees. The other choice is to program the oscillator frequency to follow a group of ions to very high energies. When this is done, the instrument is known as a frequency-modulated, or **synchro-cyclotron.** Proton energies of 740 MeV have been achieved in this fashion.

Example 15.4 For a 20-MeV proton in a uniform and constant field of 15,000 gauss, compare the classical cyclotron frequency and the relativistically correct frequency. How many complete path revolutions can the proton make before it falls out of phase by 90° with the fixed classical frequency oscillator?

Solution

$$f_{cl} = \frac{Bq}{2\pi M_0 c} = \frac{15 \times 10^3 \times 4.8 \times 10^{-10}}{2\pi \times 1.67 \times 10^{-24} \times 3 \times 10^{10}} = 22.8726 \text{ MHz}$$

$$\sqrt{1 = \beta^2} = \frac{M_0 c^2}{M c^2} = \frac{938.256 \text{ MeV}}{938.256 + 20 \text{ MeV}} = 0.979129$$

$$f_{rel} = 22.3952 \text{ MHz}$$

During n proton revolutions the classical oscillator will complete $(n + \frac{1}{4})$ cycles for a 90° phase difference.

$$\frac{n}{f_{rel}} = \frac{(n + \frac{1}{4})}{f_{cl}}$$
$$n f_{cl} = (n + \frac{1}{4}) f_{rel} = 0.979 \, f_{cl}(n + \frac{1}{4})$$
$$0.021 \, n = 0.245$$
$$n = 11.7 \text{ rev} \quad \bullet$$

Example 15.4 is the extreme case of phase difference, since at 90° no further acceleration is provided by the dee voltage. To achieve even 20-MeV proton energy with a fixed frequency oscillator, it is necessary to use large dee voltages to minimize the required number of revolutions.

15.10   The Linear Accelerator

The linear accelerator is another device for the production of high-energy ions by the repetitive application of a modest accelerating voltage. It consists of a series of hollow cylindrical electrodes mounted along the axis of a long vacuum tank. As shown in Figure 15.22, these electrodes are connected in alternate fashion to the terminals of a high-frequency oscillator. Ions are injected into the assembly from a preacceleration source, such as a Cockcroft-Walton or Van de Graaff generator. While traveling down the axis of each electrode, the ions drift at constant velocity since the interior is field-free. At each gap, however, they encounter an electric field. If the oscillator phase is proper when the ions cross the gaps, they will gain additional energy. The length of each drift tube must be planned so that

$$L = \frac{vT}{2} = \frac{v}{2f} \tag{15.20}$$

where v is the ion velocity within the drift tube and T is the period of the oscillator. As v increases, drift tube length must increase.

The trick, of course, is to keep the ions in phase with the accelerating voltage across the gaps. When this is accomplished, the accelerator is said to exhibit **phase stability.** Figure 15.23 illustrates how this is achieved. The increase in energy, and consequently in velocity, acquired by an ion at each gap crossing depends upon the péak voltage of the oscillator and also upon the phase of the voltage wave at the time of crossing. To achieve phase stability, drift tube lengths are designed to accommodate ions that cross the gaps when the wave is positive and rising. These times are shown in the figure as t_0, $t_0 + T$,

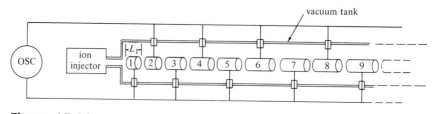

Figure 15.22   Schematic diagram of a linear accelerator.

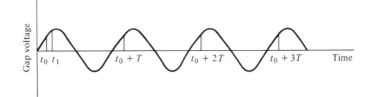

Figure 15.23   Phase stability diagram for the linear accelerator.

$t_0 + 2T$, and so on. An ion that arrives a bit late, as at t_1, will receive extra acceleration and will be less tardy upon its arrival at the next gap. It will continue to receive extra accelerations and, after several cycles, it arrives at a gap just at the time $t_0 + nT$. By then, however, it has a velocity greater than the design velocity, and at the next several gaps it will arrive early. When arriving early, it receives smaller than normal accelerations until it again arrives at the gaps at $t_0 + n'T$, but now with less than the design velocity. The net result is that the phase of the ion oscillates around the design phase. Proton linear accelerators use oscillator frequencies of about 200 megahertz (MHz) and have been built for energies of 30 to 100 MeV.

Linear accelerators can also be used for acceleration of electrons. In some respects the design is simpler than for heavy ions. At 2 MeV the electron velocity is about 98 percent the velocity of light. The drift tubes can be of constant length and the structure excited in a traveling wave mode. Electron linear accelerators have been built to cover a variety of energies. Some can be operated as low as 6 MeV, whereas for high energies there is the 2-mile-long accelerator at the Stanford Linear Accelerator Center (SLAC), which produces a beam of 11 to 23 GeV (10^9 eV) electrons.

15.11   The Synchrotron

The synchrotron, like the linear accelerator, uses high-frequency rf cavities to provide electric fields for ion acceleration. These cavities are located at one or more positions along the circumference of a circle. Ions are guided and focused to follow that circle by a ring of magnets. As the momentum of the particle increases, the magnetic field is increased to hold the radius of the path constant. The ions make many turns around the circle, using the same rf cavities over and over again. As a consequence, cavity length cannot be designed for a specific velocity, as in the case of the proton linear accelerator. Instead, cavity frequency is modulated, increasing as the velocity of the particle increases. For relativistic particles the velocity is essentially equal to the velocity of light; no further increase occurs as energy and momentum increase. When these conditions are achieved cavity frequency is constant.

The electric power required to excite a magnet depends upon the total magnetic flux, that is, upon the product of magnetic field strength and pole-face area. In the synchrotron this area is confined to an annular region of fairly limited radial extent. It is thus practical to design for a much larger path radius than is feasible for synchrocyclotron, whose pole face must cover the entire area enclosed by the path of maximum momentum. At very high energies this distinction in magnet design becomes of great importance. For instance, at the Fermi National Accelerator Laboratory the radius of the main ring accelerator is 1 km. It would be completely out of the question to provide a magnet with a pole face covering the entire area of a circle this size.

Magnet power depends upon the length of the gap between the poles as well as upon the pole-face area. These parameters can be kept to a minimum if the ion beam can be kept well focused with respect to both height and radial position. In the discussion of the cyclotron we pointed out that a magnetic field convex outward provides focusing in both directions. For focusing of this sort the radial field index n, given by $-(R/B)(dB/dR)$, must be between 0 and 1. This focusing is weak, however, and permits rather large excursions around the equilibrium path. As applied to a synchrotron, this would require wide pole faces and large gaps. Under these conditions large quantities of iron and excessive electric power are needed.

Fortunately, better focusing is possible. If $n > 1$ (field lines sharply convex outward) there is strong focusing in the vertical direction, but there is defocusing in the radial direction. Let this apply, however, only to a short sector of the ring magnet. It is followed by another sector in which $n < -1$, thus providing strong focusing in the radial direction, accompanied by defocusing in the vertical direction. The combined effect is strong focusing in both directions.

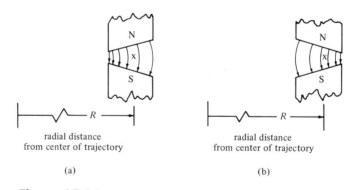

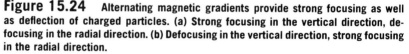

(a)   (b)

Figure 15.24 Alternating magnetic gradients provide strong focusing as well as deflection of charged particles. (a) Strong focusing in the vertical direction, defocusing in the radial direction. (b) Defocusing in the vertical direction, strong focusing in the radial direction.

That strong focusing can be accomplished by combining focusing and defocusing elements of equal strength can readily be understood by considering an analogy with optical lenses. Suppose we give attention to the behavior in the vertical plane. The first magnet sector behaves like a lens of focal length $+F_1$. The second magnet sector behaves like a lens of focal length $-F_1$ at a distance b from the first. The combined focal length is F_1^2/b and is positive. In the horizontal plane the positive and negative lenses are interchanged, but the focal length of the combination is again F_1^2/b and positive.

The desired magnetic field shape is achieved by use of sloping pole faces, as suggested in Figure 15.24, which shows the two types of magnets. The view is along the axis of the ion beam, which is directed into the plane of the drawing. Synchrotrons with this type of focusing are called **alternating gradient** (AG) synchrotrons and may use several hundred magnet sectors.

Another type of strong focusing is provided by a **quadrupole magnet.** As its name implies, this consists of four poles, which alternate N-S-N-S. An end view of such a magnet is shown in Figure 15.25. The

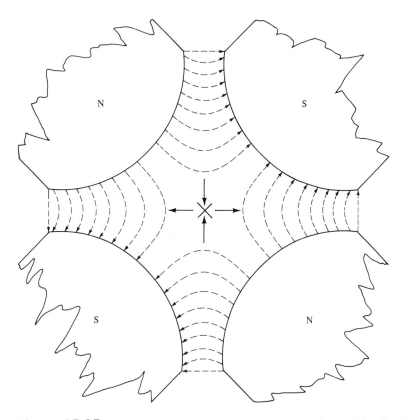

Figure 15.25   An element of a magnetic quadrupole lens. For particles directed into the plane of the drawing, there is focusing in the vertical direction, defocusing in the horizontal direction.

axis of the beam passes down the center between the poles and is shown directed away from the viewer. If the poles are shaped as hyperbolic cylinders, the field increases linearly with distance from the center. Strong focusing is achieved in one plane, but there is defocusing in the other plane. In Figure 15.25 there is focusing in the vertical plane, defocusing in the horizontal plane. If this quadropole is followed by a second one in which the north and south poles are interchanged, the combination exhibits strong focusing in both vertical and horizontal planes. In some synchrotrons only weak or zero focusing is provided by the bending magnets, the major focusing being provided by quadrupole lenses placed between magnet sectors.

Ions to be injected into a synchrotron ring must have an initial kinetic energy. This may be provided by a Cockcroft-Walton or Van de Graaff generator or even by a linear accelerator. The higher this initial energy the better, since then cavity frequency and strength of the mag-

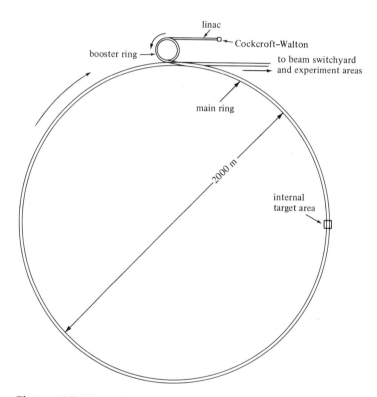

Figure 15.26 Layout of the Fermilab accelerator. Protons are given initial acceleration by a 0.750 MeV Cockcroft-Walton generator. A linear accelerator 175 m long brings them to 200 MeV. A booster synchrotron of 150 m diameter gives further acceleration to 8 GeV. The protons are then injected into the main ring accelerator, which has a diameter of 2 km, where energies of 500 GeV are achieved. Experiments may be performed with targets internal to the main ring, or the beam may be deflected out of the ring and into the magnetic "switchyard." The switchyard steers the beam to one of several different experimental areas.

netic field need not cycle over wide limits. At the Fermilab accelerator the main ring is preceded by three stages of acceleration: a Cockcroft-Walton providing protons at 0.750 MeV, a linear accelerator providing 200 MeV, and a booster synchrotron providing 8 GeV. These are coupled as indicated schematically in Figure 15.26. The final energy of the main ring is 500 GeV. By using superconducting magnets within the same tunnel that houses the present main ring, it is believed that the energy can be raised to 1000 GeV.

15.12 Summary

Charged-particle radiation is detected by the ionization it produces. The ionization can be made manifest by a charge pulse in an electronic detector or by a track in a cloud chamber, bubble chamber, nuclear emulsion, or solid state dielectric. Neutrons are detected by means of charged particles produced by a nuclear reaction or by knock-on protons in hydrogenous material. Gamma rays are detected by the production of conversion electrons through the photoelectric, Compton, and pair-production interactions.

$$K_e = h\nu_0 - BE_e \qquad \text{(photoelectric)}$$

$$K_e = \frac{(1 - \cos\theta)\,(h\nu_0/m_0c^2)}{1 + (h\nu_0/m_0c^2)\,(1 - \cos\theta)}\,h\nu_0 \qquad \text{(Compton)}$$

$$K_{e\,\text{max}} = \frac{h\nu_0}{1 + (m_0c^2/2h\nu_0)} \qquad \text{(Compton)}$$

$$K_{e-} + K_{e+} = h\nu_0 - 2m_0c^2 \qquad \text{(pair production)}$$

The Geiger counter and spark chamber give signals that are independent of the initiating radiation. In other detectors the signal (pulse-height in a counter or ionization density in a track) depends upon the radiation. Scintillation detectors are used primarily for gamma-ray measurements, silicon semiconductor detectors for charged particles or low energy gamma rays, Ge(Li) detectors for higher energy gamma rays.

For the passage of a charged particle through a magnetic field

$$BR = \frac{pc}{q} = \frac{\sqrt{K^2 + 2KM_0c^2}}{q}$$

$$K = \sqrt{(M_0c^2)^2 + (BRq)^2}$$

The magnetic spectrograph exploits these relationships and achieves high accuracy in the measurement of kinetic energies of charged particles. These relationships apply also to the motion of a particle in a cyclotron and a synchrotron.

The cyclotron frequency is given by

$$f = \frac{Bq \sqrt{1 - \beta^2}}{2 M_0 c}$$

A useful conversion factor is

$$1 \text{ MeV} = 1.6 \times 10^{-6} \text{ erg}$$

With the synchrocyclotron, proton energies of ~740 MeV have been achieved, 500 GeV with the synchrotron, and 200 MeV with the linear accelerator. Electron linear accelerators have been built for energies to 23 GeV and electron synchrotrons to 10 GeV.

Direct acceleration machines, the Cockcroft-Walton, the Van de Graaff generator, and the tandem Van de Graaff, are energy limited by sparking from the high-voltage terminal. Below the sparking voltage, however, they are characterized by beam energies that are precisely defined and easily varied.

Problems

15.1 An ionization chamber is used to detect the passage of 5.3-MeV alpha particles. It requires, on the average, 29 eV to produce an ion pair in the gas. The interelectrode capacitance plus that of the associated circuit is 30 pF. Calculate the electrode current in amperes if 10^5 alpha particles per second enter the chamber and stop in the gas.

15.2 The same detector described in Problem (15.1) is operated in the proportional region with a gas gain of 80. If the alpha particle rate is reduced so individual pulses can be resolved, calculate
(a) the magnitude of the charge collected per pulse at each electrode.
(b) the magnitude of the voltage pulse.

15.3 A proportional counter filled with BF_3 is used for the detection of neutrons. Calculate the maximum alpha-particle energy in the detector from the $^{10}B(n, \alpha)^7Li$ reaction for neutrons of energy 0.025 eV, 100 keV, 1.0 MeV, and 5.0 MeV.

15.4 In a certain ten-stage photomultiplier tube the emission ratio of secondary to primary electrons is, on the average, 2:1 at each dynode.
(a) How many electrons reach the anode for each electron ejected from the cathode?
(b) What should be the value of the secondary to primary emission ratio to reduce by a factor of two the number of electrons reaching the anode?

15.5 Calculate the electron energy at the Compton shoulder for gammas of energy 200 keV and for those of 3.0 MeV.

15.6 A silicon semiconductor particle detector is used to measure the energies of a group of 5.3-MeV alphas. It has a depletion zone 100 μm thick and a frontal area of 1.0 cm². The capacitance of the depletion zone is 100 pF and that of the associated circuit is 40 pF. It requires, on the average, 3.6 eV to produce an ion pair in silicon. Calculate
(a) the magnitude of the charge per pulse released in the junction.
(b) the magnitude of the voltage pulse.

15.7 It is desired to momentum analyze in a cloud chamber beta particles of endpoint energy 2.0 MeV. This is to be done by placing the chamber in a magnetic field and photographing the tracks. What should be the strength of the magnetic field if no track is to have a radius of curvature greater than 10 cm?

15.8 A magnetic spectrograph is designed for use with 20-MeV protons. What magnetic field strength is required if the radius of the path to the prime focus is 50 cm?

15.9 What are the energies of deuterons, tritons, and alphas if with the same magnetic field setting in a spectrograph they come to the same focus as 20-MeV protons?

15.10 Make a careful scale drawing of the focusing properties of a broad-range magnetic spectrograph. Note that each ray proceeds in a straight line from the source until it reaches the boundary of the magnetic field. It then continues as the arc of a circle, tangent to the straight ray. Upon reaching the exit face, it proceeds as a straight line, tangent to the arc. Use ~7.5° for the half-angle of acceptance at the entrance slit. (A large scale improves accuracy, but $R = 10$ cm will keep the drawing on a $8\frac{1}{2}'' \times 11''$ sheet. A drafting machine is useful, but not essential. Construction with compass and straightedge works.)

15.11 A 5-MV Van de Graaff generator carries a belt current of 500 μamp. What is the horsepower of the motor required to drive the belt? (Actually a larger motor is used to allow for windage losses in high-pressure gas and for driving an alternator inside the high-voltage terminal.)

15.12 A synchrocyclotron accelerates protons to 700 MeV. If it uses a uniform magnetic field of 19,000 gauss, calculate
(a) the radius of the proton path at maximum energy.
(b) the frequency on the dees at low proton energy.
(c) the frequency at maximum proton energy.

15.13 A 60-MeV proton linear accelerator is operated at a cavity frequency of 201 MHz. It is designed for phase stability at 400 keV per gap. If protons are injected from a preaccelerator with an energy of 500 keV, find
(a) the length of the first drift tube.
(b) the length of the seventy-fifth drift tube.
(c) the length of the final drift tube.

15.14 The main ring of the Fermilab accelerator accelerates protons from 8 GeV at injection to a final energy of 400 GeV. The rf cavities increase the energy by 2.8 MeV in each revolution around the ring. At the final energy the magnetic field is 18,000 gauss. Calculate
(a) the radius of the proton path within the bending magnets.

(b) the magnetic field at time of proton injection.

The perimeter of the ring is 6.28 km, differing from $2\pi R$ because of gaps between magnet sectors and several straight sections which are required for injection of the beam, for the rf cavities, for beam extraction, and for experiment stations internal to the ring.

(c) How many revolutions per second around the ring do the protons make?

(d) How far do they travel to achieve final energy?

(e) How much time in the main ring is required to achieve final energy?

Suggestions for Further Reading

Enge, H. A. *Introduction to Nuclear Physics.* Reading, Mass.: Addison-Wesley Publishing Co., Inc., 1966. Chapter 12.

Howard, R. A. *Nuclear Physics.* Belmont, Calif.: Wadsworth Publishing Co., Inc., 1963. Chapter 4, detectors; Appendix A, accelerators.

Rose, P. H. and A. B. Wittkower. "Tandem Van de Graaff Accelerators," *Scientific American* (August 1970).

The following review articles contain a wealth of information and are not beyond reach of undergraduate students:

Blewett, M. H. "Characteristics of Typical Accelerators," *Annual Review of Nuclear Science,* vol. 26. Palo Alto, Calif.: Annual Reviews, Inc., 1976.

Fleischer, R. L., P. B. Price, and R. M. Walker. "Solid-State Track Detectors: Applications to Nuclear Science and Geophysics," *Annual Review of Nuclear Science,* vol. 15. Palo Alto, Calif.: Annual Reviews, Inc., 1965.

Sanford, J. R. "The Fermi National Accelerator Laboratory," *Annual Review of Nuclear Science,* vol. 26. Palo Alto, Calif.: Annual Reviews, Inc., 1976.

Tavendal, A. J. "Semiconductor Nuclear Radiation Detectors," *Annual Review of Nuclear Science,* vol. 17. Palo Alto, Calif.: Annual Reviews, Inc., 1967.

Elementary particles

16.1 The Positron

The Schrödinger theory is a nonrelativistic theory and thus constitutes an approximate theory. An obvious relativistic version of the Schrödinger equation is the Klein-Gordon equation. The Klein-Gordon equation is, as discussed in Chapter 4, the quantum mechanical operator version of the relativistic energy momentum relation, $E^2 = p^2c^2 + m_0^2c^4$. The Klein-Gordon equation, however, fails to describe the electron. The relativistic quantum theory that successfully describes the electron is the Dirac equation, which P. A. M. Dirac discovered in 1928. The Dirac equation differs from the Klein-Gordon equation in that the derivatives of space and time coordinates in the Dirac equation are both of first order.

The immediate successes of the Dirac theory included the predictions of the electron spin and of the hydrogen spectrum. Dirac's hydrogen spectrum automatically accounted for fine structure and relativistic effects. Concerning the spin of the electron, the theory predicted that if the electron's intrinsic magnetic momentum vector μ is written in the form

$$\mu = g\mu_B S \tag{16.1}$$

where μ_B is the Bohr magneton and S is the electron spin vector, then the g factor must be precisely

$$g = 2 \tag{16.2}$$

This result was, at the time, in agreement with observation. The currently accepted value of the g factor of the electron is

$$g = 2.003192 \tag{16.3}$$

where the slight deviation from Dirac's prediction is accounted for by the theory of quantum electrodynamics. The existence of the intrinsic spin of the electron is a natural feature of Dirac's relativistic quantum theory.

Another consequence of the Dirac theory is that an electron — a free electron, for example — may exist in a negative energy state as well as in a positive energy state. Now the existence of negative energy states is rooted in the relativistic energy-momentum relation

$$E = \pm\sqrt{p^2c^2 + m_0^2c^4}$$

Hence the relativistic classical theory also admits the possibility of negative energy states. However, negative energy states pose no problem in classical theory, because the theory requires changes to take place in a continuous manner. Therefore transitions from positive energy states to negative energy states are forbidden unless there exists a continuum of intervening states. But no states of the electron exist between $E = +m_0c^2$ and $E = -m_0c^2$. It follows that no transitions take place between these states, and as a consequence, in classical theory negative energy states of the electron are discarded as being unphysical.

In the relativistic quantum theory, on the other hand, transitions from a positive energy state to a negative energy state are allowed. A quantum electron, for example, can make a quantum jump over the forbidden energy gap to the "sea" of negative energy states. The problem with such transitions is that an electron in the "Dirac sea" of negative energy states has a negative mass and is therefore unphysical. There is nothing in the Dirac theory, however, that would prevent an ordinary electron from making an initial transition to a negative energy state and then subsequent transitions to still lower negative energy states, radiating photons in the process. In fact, such would indeed happen. Thus if negative energy states were available, an ordinary electron would have a transitory existence.

It is an experimental fact that the lifetime of an ordinary electron is, for all practical purposes, infinite. It is thus obvious that an ordinary electron does not normally make a transition to the Dirac sea of negative energy states. Dirac explained this fact in terms of the Pauli exclusion principle. He assumed that the negative energy states are all filled.

Consider now a filled Dirac sea. There is no reason why an electron in the Dirac sea could not be excited to a positive energy state. As Dirac suggested, such an upward transition could be achieved by an energetic photon. Suppose that an electron initially in a negative energy state is excited, say, by a photon, to a positive energy state. Such a transition would create a pair of particles—an ordinary electron and a "hole" in the otherwise filled Dirac sea. It was proved that a hole in the Dirac sea would have the same mass as an electron. Moreover, a hole would have a charge opposite to the charge of the electron and hence would represent a positively charged particle. Notice that the hole owes its mass to an absence of negative mass and its charge to an absence of negative charge. Dirac called the hole an **antielectron,** which is now also called a positron.

A hole in the Dirac sea is, under normal conditions, likely to have a transitory existence. A hole implies an unfilled negative energy state. An unfilled negative energy state, in turn, implies that an ordinary electron can now make a downward transition to the unoccupied state, radiating energy in the process. The result is the annihilation of an electron-positron pair. Thus the fate of a positron, and of antimatter in general, is annihilation.

The existence of the positron was predicted by Dirac in 1930. Its existence was confirmed by C. Anderson in 1932. It is now an established fact that to every particle there exists an antimatter counterpart. In fact, the existence of antimatter is an essential feature of the relativistic quantum theory.

16.2   The Photon and the Electromagnetic Interaction

The interaction of a photon with an electron and, more generally, of radiation with matter, is referred to as an electromagnetic interaction. The electromagnetic interaction is rigorously described by quantum electrodynamics, which is a theory constructed by imposing quantum conditions on Maxwell's electromagnetic theory. Quantum electrodynamics was proposed by Dirac in 1926 to explain the absorption, the emission, and the scattering of radiation by atoms. The theory was modified independently by R. P. Feynman, J. Schwinger, and S. Tomonaga in the 1940s.

In the quantum electrodynamics of Feynman, the positron is treated as a negative-energy electron moving backward in time. Thus the electron and the positron need not be viewed as separate concepts; rather, they can be viewed as two different aspects of the same thing. The annihilation of an electron-positron pair, for example, is then described as follows: An electron moving forward in time creates a photon at a point in space-time called a vertex, and, in so doing, reverses its direction in time (Figure 16.1). The depiction of an interaction in a space-time graph is referred to as a **Feynman diagram.** The Feynman diagram for the creation of an electron-positron pair is as follows: A photon moving forward in time is absorbed by an electron

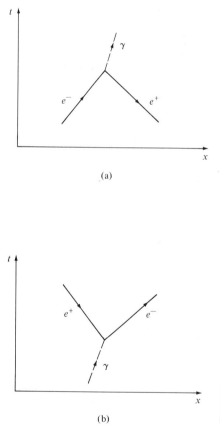

(a)

(b)

Figure 16.1 Feynman diagrams for (a) pair-annihilation and (b) pair-creation.

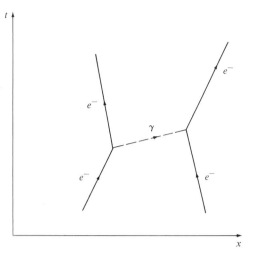

Figure 16.2 Feynman diagram for the exchange of a photon between two electrons.

moving backward in time, and such results in the reversal of the electron's backward motion in time.

A fundamental idea of quantum electrodynamics is that the electromagnetic interaction is mediated by a photon. For example, when an electron interacts electromagnetically with another electron, the interaction occurs in two stages: one electron emits a photon, thus bringing about a change in its state of motion, and the other electron absorbs the photon, resulting in a change in its state of motion (Figure 16.2). Hence the electromagnetic interaction between two electrons consists of at least two vertices, a photon-creating vertex and a photon-annihilating vertex. In general, electromagnetic interactions, at the fundamental level, take place between particles and photons. The necessity of mediation via photons is a consequence of relativity theory, which states that no signal can propagate faster than the speed of light.

The notion of mediation via a photon is actually quite familiar in the case of a static interaction between two charged particles. A photon, or a propagating electromagnetic field V, is described by a wave equation

$$\nabla^2 V - \frac{1}{c^2}\frac{\partial^2 V}{\partial t^2} = 0 \tag{16.4}$$

In the static case the wave equation reduces to

$$\nabla^2 V = 0 \tag{16.5}$$

which, in spherical coordinates, takes the form

$$\frac{1}{r^2}\frac{d}{dr}\left[r^2\frac{dV(r)}{dr}\right] = 0 \tag{16.6}$$

Equation (16.6) is satisfied by a Coulomb potential

$$V(r) = \frac{q_1 q_2}{r} \tag{16.7}$$

Hence, it can be said that the electrostatic force between two charged particles is described by the same equation that describes the propagation of a photon. In general, when an interaction is said to be mediated via a photon, or to arise from the "exchange of a photon," it means that the force and the exchanged particle are described by the same equation and the exchanged photon is viewed as the carrier of the force.

The electromagnetic force is of infinite range. The range of an interaction is, as discussed in the following section, inversely proportional to the rest mass of the exchanged particle. Hence the rest mass of the photon is zero. The charge of a photon is also zero. As a result, the charge of a particle emitting or absorbing a photon remains unchanged. This fact gives rise to the statement that the electromagnetic interaction is an interaction between two **neutral currents.** A moving charged particle may indeed by viewed as an electric current. Finally, the strength of the electromagnetic interaction is the fine structure constant:

$$\alpha = \frac{e^2}{\hbar c} = \frac{1}{137.036} \tag{16.8}$$

16.3   The Pion and the Strong Interaction

The force that holds the nucleus together is called a nuclear, or strong force. Nuclear force has two distinctive features. It is, first of all, a force of great strength. It must be strong enough to overcome the Coulomb repulsion between protons in the nucleus. Secondly, nuclear force is a short range force. It does not extend much beyond 10^{-13} cm from its source.

The first successful attempt at arriving at a quantum description of the nuclear force was made in 1935 by H. Yukawa. He attacked the problem by analogy with the quantum description of the electromagnetic force. Accordingly, he proposed that a quantum field is associated with the nucleon and that the interaction between nucleons arises from the exchange of a quantum of the nucleon field. He then assumed that the field of the exchanged particle is described by the Klein-Gordon equation

$$\nabla^2 \psi(\mathbf{r}, t) - \frac{1}{c^2} \frac{\partial^2 \psi(\mathbf{r}, t)}{\partial t^2} = K^2 \psi(\mathbf{r}, t) \tag{16.9}$$

where

$$K = \frac{mc}{\hbar} \tag{16.10}$$

and m is the rest mass of the exchanged particle.

Now the potential energy $V(\mathbf{r})$ of two stationary nucleons emerges from the static case of the Klein-Gordon equation

$$\nabla^2 V(\mathbf{r}) = K^2 V(\mathbf{r})$$

or, in spherical coordinates,

$$\frac{1}{r^2} \frac{d}{dr} \left[r^2 \frac{dV(r)}{dr} \right] = K^2 V(r) \tag{16.11}$$

A solution of Eq. (16.11) is

$$V(r) = g^2 \frac{e^{-Kr}}{r} \qquad r > 0 \tag{16.12}$$

where g^2 gives the strength of the Yukawa potential. The range of the potential is determined by $R = 1/K$. The observed nuclear potential has a range $R = 1/K = 1.4 \times 10^{-13}$ cm, and hence the mass of the exchanged particle comes out to be

$$\begin{aligned} mc^2 &= \hbar c K \\ &\approx 140 \text{ MeV} \end{aligned} \tag{16.13}$$

Note that the range of the Yukawa potential is inversely proportional to the mass of the exchanged particle and is equal to the Compton wavelength of the particle, namely, $R = \hbar/mc$. The strength of the potential needed to fit observation is

$$\frac{g^2}{\hbar c} \approx 15 \tag{16.14}$$

The quantum of the Yukawa field is called a pion, or pi-meson. Yukawa's model of nuclear force envisions one nucleon emitting a pion to be later absorbed by another nucleon. Such an exchange, however, violates the conservation of energy. During the transit of a pion from one nucleon to another, the total energy of the nucleon system exceeds the amount allowed by energy conservation by at least the rest energy of the pion

$$\Delta E = m_\pi c^2 \tag{16.15}$$

Now no process that violates the principle of energy conservation can be directly observed. Such a process can, nevertheless, exist by virtue of the uncertainty principle. The uncertainty principle permits energy conservation to be violated by an amount ΔE if such violation lasts no longer than a duration of time Δt given by the uncertainty relation

$$\Delta E \Delta t = \hbar \tag{16.16}$$

A process that violates energy conservation in accordance with the uncertainty principle is called a **virtual** process, and a particle that owes its transitory existence to the uncertainty principle is called a virtual particle. Thus the Yukawa theory envisions the strong interaction between nucleons to arise from the exchange of virtual pions (Figure 16.3).

Pions in the nucleon or in the nucleus are virtual particles and are not directly observable. However, if sufficient energy is supplied to a nucleon or a nucleus, pions may attain the status of a real particle and thus become directly observable. Charged pions were first observed by C. F. Powell and collaborators in 1947 in cosmic rays. Pions are now readily available in high-energy nucleon-nucleon collisions in accelerators.

There are three kinds of pions—π^-, π^+, and π^0. The positively charged pion is the antimatter counterpart of the negatively charged pion, and hence they have the same mass. The neutral pion is less massive than the charged pion by 4.61 MeV/c^2. The difference in mass is thought to be related to the electromagnetic interaction in which the charged pions participate but the neutral pions do not.

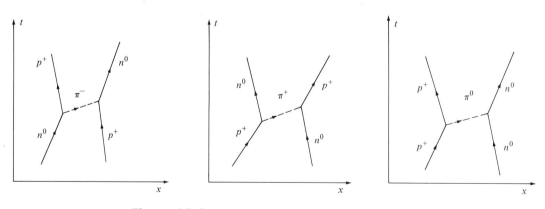

Figure 16.3 Exchange of virtual pions between two nucleons.

Pions are radioactive particles. The lifetime of a neutral pion is 0.8×10^{-16} sec, and the lifetime of a charged pion is 2.6×10^{-8} sec. A neutral pion decays into two gamma rays. The decay of a neutral pion is thus a photon-creating process and is described by the electromagnetic interaction. A charged pion, on the other hand, decays into a muon and a muonic neutrino, and its decay is governed by an interaction that is much weaker than the electromagnetic interaction. The interaction that is responsible for the decay of the charged pion is called a **weak interaction,** because the strength of the weak interaction is weaker than the electromagnetic interaction by a factor of 10^{10}. The weak interaction, nevertheless, is very important in the submicroscopic world of particles, and to its study we now turn.

16.4   The Weak Interaction

The first model of the weak interaction was proposed by E. Fermi in 1934. In his theory of beta decay, Fermi envisioned the emission of an electron-antineutrino pair, for example, to be conceptually similar to the emission of a photon from an atom. The emission of a photon, in the quantum theory of radiation developed initially by Dirac and later by Fermi himself, is essentially a creation process. The creation of a photon by an atom takes place at the moment of a quantum transition in the atom. In formulating his theory of beta decay, Fermi proposed that beta decay be likewise viewed as a creation process, in which the transformation of a nucleon from one charge state to another charge state is achieved by the simultaneous creation of a charged beta particle and a neutral neutrino. The beta particle and the neutrino are then described by the Dirac equation. Because the Dirac equation admits the existence of antimatter, Fermi's theory of beta decay describes the emission of particles as well as antiparticles.

In the Fermi theory, the emission of an electron from a radioactive nucleus arises from the transformation of a neutron into a proton with the creation of an electron-antineutrino pair.

$$n \rightarrow p^+ + e^- + \bar{\nu} \tag{16.17}$$

where the bar over the neutrino denotes an antineutrino. The emission of a positron occurs when a proton inside a nucleus is transformed into a neutron by the creation of a positron-neutrino pair.

$$p^+ \rightarrow n + e^+ + \nu \tag{16.18}$$

In certain nuclei, a proton can capture an innermost atomic electron and thus transform into a neutron with the simultaneous creation of a neutrino.

$$e^- + p^+ \rightarrow n + \nu \qquad \qquad \text{(16.19)}$$

Notice that, in all three types of nucleon transformation in beta decay, a pair of nucleons is involved as well as a pair of a beta particle and a neutrino. The beta particle and the neutrino are classified to be members of the electron family. The neutron and the proton, on the other hand, belong to the so-called baryon family. The proton is the lightest member of the baryon family. The occurrence of these particles in pairs is suggestive of an underlying conservation law. Indeed, it is found that the family number is conserved not only in weak interactions but in all physical processes. According to the principle of conservation of the family number, a positive unity is assigned to a particle member of a family, a negative unity is assigned to an antiparticle member, and a zero is assigned to a particle not belonging to the family.

$$n \rightarrow p^+ + e^- + \bar{\nu}$$

electron family number: $\quad 0 = 0 + (+1) + (-1) \qquad \text{(16.20)}$

baryon family number: $\quad (+1) = (+1) + 0 + 0$

It is interesting to note that it is the conservation of the baryon family number that prevents a proton from decaying into lighter particles.

The conservation of the electron family number is not explicitly assumed by the Fermi theory of beta decay. The simultaneous creation of a beta-neutrino pair in beta decay, however, establishes the neutrinos as bona fide members of the electron family. Once neutrinos are so established, the conservation of the electron family number implies that an electron must be created with an antineutrino and a positron must be created with a neutrino.

The processes of electron emission, positron emission, and electron capture are all beautifully described by the Fermi theory. Fermi then showed that the force responsible for the creation of a beta-neutrino pair is much weaker than the nuclear or the electromagnetic force. In fact, the weak force is some 10^{10} times weaker in strength than the electromagnetic force, so that the beta decay of a nucleus, for example, proceeds some 10^{10} times more slowly than a comparable electromagnetic process.

The question now arises: Is the weak interaction also mediated by a particle? To postulate the existence of such a particle would be in keeping with the observation that both strong and electromagnetic interactions arise from the exchange of particles. Indeed, a particle to mediate the weak interaction has been hypothesized and is called the **intermediate vector boson** or the W particle. It is thought that the W particle, like the photon, carries one unit of spin angular momentum. The beta decay of a neutron, for example, can then be described by the exchange of a charged W particle between the neutron-proton "current" and the electron-neutrino "current." More specifically, the neutron emits a W^- particle and becomes a proton, and a neutrino absorbs the W^- particle and becomes an electron. The neutron-

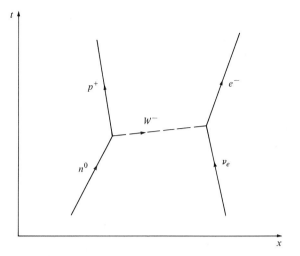

Figure 16.4 Exchange of a **W⁻** particle between two charged currents giving rise to a weak interaction between them.

proton and the electron-neutrino currents are called **charged currents,** because the charge within the current undergoes a change at the emission or absorption of a W particle. Consequently, the weak interaction is said to arise from the exchange of a charged W particle between two charged currents (Figure 16.4).

The electromagnetic interaction, on the other hand, is an interaction between two neutral currents, because the photon which mediates the interaction is neutral in charge. When an electron interacts electromagnetically with a proton, for example, a virtual photon is exchanged between the electron-electron current and the proton-proton current. The interaction does not cause a change in the charge of the current. It is then reasonable that a unified field theory proposed by S. Weinberg, which seeks to give a unified description of the weak and the electromagnetic interactions, predicts the existence of neutral currents for the weak interaction and hence the existence of a neutral vector boson. An example of the neutral current of the weak interaction is a neutrino which, in an interaction with a proton, emits a neutral W particle and thus remains a neutrino.

Table 16.1   The Fundamental Interactions

Interaction	Relative Strength	Mediating Particle
strong	1	mesons
electromagnetic	10^{-2}	photon
weak	10^{-13}	intermediate vector boson
gravitational	10^{-38}	graviton

16.5   Two Types of Neutrinos

Consider now the beta decays of the neutron and the pion. In the case of the neutron an electron is emitted, accompanied by an antineutrino. The pion, on the other hand, decays into a muon and a neutrino. The question then arises: Is the neutrino created with an electron identical to the neutrino created with a muon? As no existing theory provides an answer to that question, the issue must be settled by experiment.

The question is important, because the muon is identical to the electron in all of its interactions except for a difference in mass. The muon is about 200 times as massive as the electron. The origin of the muon's anomalously large mass is a great mystery. It is also puzzling that nature has seen fit to admit it into her scheme of things when the muon seems completely redundant as no more than a merely "heavy" electron.

The conceptual basis for a two-neutrino experiment is simple. The theory of weak interactions predicts that, if a neutrino is absorbed by a neutron, a proton and an electron would be produced.

$$\nu + n \rightarrow p^+ + e^- \qquad\qquad (16.21)$$

If the energy of the neutrino is increased so as to provide enough energy to produce a muon, the reaction would also produce a proton and a muon.

$$\nu + n \rightarrow p^+ + \mu^- \qquad\qquad (16.22)$$

Thus if the muon-type neutrino is identical to the electron-type neutrino, the absorption of high-energy neutrinos by neutrons would result in the creation of equal numbers of electrons and muons.

An experiment, performed by L. Lederman, M. Schwartz, and J. Steinberger in 1962, consisted in bombarding a target with a beam of neutrinos from the decay of a pion beam. An energetic pion beam can readily be produced in a powerful accelerator. As these pions decay in flight, muons and muon-type neutrinos are produced. The muon-type neutrinos move predominantly in the forward direction, thus forming a high-energy neutrino beam. Upon bombarding a target, such a neutrino beam would then produce equal numbers of electrons and muons if these neutrinos were identical to electron-type neutrinos.

The experimental result was that only muons and no electrons were produced by the neutrino beam. The observation was thus interpreted as establishing two types of neutrinos. The neutrino created with a muon, as in the decay of a pion, was simply not capable of producing an electron. Similarly, a neutrino produced with an electron would not be able to create a muon, although an electron-type neutrino beam energetic enough to produce muons is not available. The muon-type and the electron-type neutrinos must, therefore, be viewed as fundamentally different. No physical difference, however, is apparent between them. They both seem to have identical properties: negligible mass, zero charge, and spin $\frac{1}{2}$.

When the muon was first discovered, the muon was thought to belong to the electron family. However, the difference between the muon-type and the electron-type neutrino now separates the muon family from the electron family, and each family number is conserved separately. A muon, decaying into an electron and two neutrinos, actually produces two different types of neutrinos. The family number is then assigned to the neutrinos by invoking the conservation of family numbers:

$$\mu^- \rightarrow \quad e^- + \quad \bar{\nu}_e \quad + \quad \nu_\mu$$

muon number: $\qquad (+1) = \quad 0 \ + \quad 0 \ + (+1)$ **(16.23)**

electron number: $\qquad 0 \ = (+1) + (-1) + \quad 0$

Table 16.2   Leptons

Family Name	Particle Name	Particle Symbol	Mass (MeV)	Spin ($\hbar$)	Charge (e)	Lifetime (sec)
electron family	neutrino	ν_e	0	$\frac{1}{2}$	0	stable
	electron	e^-	0.511	$\frac{1}{2}$	-1	stable
muon family	neutrino	ν_μ	0(?)	$\frac{1}{2}$	0	stable
	muon	μ^-	105.66	$\frac{1}{2}$	-1	2.2×10^{-6}

16.6   Hadrons

Hadrons are strongly interacting particles, and currently comprise all the particles other than the photon, the graviton, and the leptons. Hadrons are divided into two classes of particles: mesons and baryons. Mesons are particles with integral values of spin, and baryons are particles with half-integral values of spin.

The lightest member of the baryon family is the proton. The proton is stable by virtue of the conservation of the baryon family number. As far as its strong interaction is concerned, the neutron is also stable; the neutron decays only via the weak interaction. Moreover, the mass difference between the neutron and the proton is about 1.3 MeV, which is only 0.1% of the proton mass. Hence the proton and the neutron are

Table 16.3   Hadrons

Family Name	Particle Name	Particle Symbol	Mass (MeV)	Charge (e)	Spin ($\hbar$)	Parity P	Isospin I_3	Strangeness S	Lifetime (sec)	Typical Decay Mode
mesons	pion	π^+	139.57	+1	0	odd	+1	0	2.60×10^{-8}	$\pi^+ \rightarrow \mu^+ + \nu_\mu$
		π^-	139.57	−1	0	odd	−1	0	2.60×10^{-8}	$\pi^- \rightarrow \mu^- + \bar{\nu}_\mu$
		π^0	134.96	0	0	odd	0	0	0.8×10^{-16}	$\pi^0 \rightarrow \gamma + \gamma$
	kaon	K^+	493.7	+1	0	odd	$+\frac{1}{2}$	+1	1.24×10^{-8}	$K^+ \rightarrow \pi^+ + \pi^0$
		K^0	497.7	0	0	odd	$-\frac{1}{2}$	+1	0.88×10^{-10}	$K^0 \rightarrow \pi^+ + \pi^-$
									5.2×10^{-8}	$K^0 \rightarrow \pi^+ + e^- + \bar{\nu}_e$
	eta	η	549	0	0	odd	0	0	2.5×10^{-19}	$\eta \rightarrow \gamma + \gamma$
baryon family	nucleon	p^+	938.26	+1	$\frac{1}{2}$	even	$+\frac{1}{2}$	0	stable	$n \rightarrow p^+ + e^- + \bar{\nu}_e$
		n^0	939.55	0	$\frac{1}{2}$	even	$-\frac{1}{2}$	0	920	
	lambda	Λ^0	1115.6	0	$\frac{1}{2}$	even	0	−1	2.5×10^{-10}	$\Lambda^0 \rightarrow p^+ + \pi^-$
	sigma	Σ^+	1189.3	+1	$\frac{1}{2}$	even	+1	−1	0.80×10^{-10}	$\Sigma^+ \rightarrow n^0 + \pi^+$
		Σ^-	1197.3	−1	$\frac{1}{2}$	even	−1	−1	1.5×10^{-10}	$\Sigma^- \rightarrow n^0 + \pi^-$
		Σ^0	1192.5	0	$\frac{1}{2}$	even	0	−1	$<10^{-14}$	$\Sigma^0 \rightarrow \Lambda^0 + \gamma$
	xi or cascade	Ξ^-	1321.3	−1	$\frac{1}{2}$	even	$-\frac{1}{2}$	−2	1.7×10^{-10}	$\Xi^- \rightarrow \Lambda^0 + \pi^-$
		Ξ^0	1315	0	$\frac{1}{2}$	even	$+\frac{1}{2}$	−2	3.0×10^{-10}	$\Xi^0 \rightarrow \Lambda^0 + \pi^0$
	omega	Ω^-	1673	−1	$\frac{3}{2}$	even	0	−3	1.3×10^{-10}	$\Omega^- \rightarrow \Xi^0 + \pi^-$

the same particle in their strong interactions. However, they differ in their electromagnetic interactions, and they constitute two different charge states of the same particle, namely, the nucleon.

The different charge states of a strongly interacting particle are represented by the "z or third components" of an **isotopic spin,** or isospin, vector. The isospin of a particle has nothing to do with its intrinsic angular momentum, but the mathematics of isospin is exactly analogous to the mathematics of ordinary spin. The isospin quantum number of the nucleon, for example, is $I = \frac{1}{2}$, because the charge multiplicity of the nucleon is two, that is, multiplicity $= 2I + 1$. The proton is assigned the isospin-up state, $I_3 = +\frac{1}{2}$, and the neutron is assigned the isospin-down state, $I_3 = -\frac{1}{2}$. The charge multiplicity of the pion is a triplet, and hence the isospin quantum number of the pion is $I = 1$. The pion multiplets are: π^+ with $I_3 = +1$, π^0 with $I_3 = 0$, and π^- with $I_3 = -1$. Note that ΔI_3 equals the difference in charge.

The strong interaction conserves isospin, that is, both the magnitude of the isospin vector and its third components are conserved. The conservation of isotopic spin implies that the strong interactions are independent of electric charge. For example, the probability of interaction between a pion and a nucleon does not depend on the charge state of the interacting particles. The electromagnetic interaction, on the other hand, depends on charge, and therefore does not conserve isotopic spin. It is clear, then, that the conservation of charge that is respected by all interactions and the conservation of isospin that is respected only by strong interactions are two entirely different ideas. The forbidden reaction of $\pi^- + p \rightarrow \Lambda^0$, for example, conserves charge but does not conserve isotopic spin, that is, $-1 + \frac{1}{2} \neq 0$.

The reaction $(\pi^- + p)$ leads via a strong interaction to the so-called **associated production** of a lambda particle and a K-meson.

$$\pi^- + p \rightarrow \Lambda^0 + K^0 \tag{16.24}$$

The conservation of isospin yields $I_3 = -\frac{1}{2}$ for the neutral kaon. It is experimentally established that the isospin quantum number of the kaon is $\frac{1}{2}$, and the other member of the kaon doublet is the K^+ meson with $I_3 = +\frac{1}{2}$.

The neutral kaon does not decay via a strong interaction, although it is a strongly interacting particle. However, this fact is not surprising because, except for a class of hadrons called the resonances, all strongly interacting particles decay via either an electromagnetic interaction or a weak interaction. The neutral pion, for example, decays via an electromagnetic interaction into two gamma rays with an average lifetime of 10^{-16} sec. The neutral kaon, on the other hand, decays only via a weak interaction with a lifetime of 10^{-10} sec. This fact is then suggestive of the possibility that there exists a conservation principle that is respected by the electromagnetic interaction but is violated by the weak interaction.

A new conservation law was proposed independently by M. Gell-Mann and K. Nishijima in 1953. The new conserved quantity is called **strangeness,** and is conserved in all strong and electromagnetic interactions. However, strangeness is not conserved in weak interactions. The neutral kaon must then be a particle with a nonzero strangeness quantum number, or simply a strange particle. The pion and the nucleon, on the other hand, are not strange particles; their strangeness quantum number is $S = 0$. Now the $(\pi^- + p)$ reaction leading to the associated production of $(\Lambda^0 + K^0)$ proceeds via a strong interaction, and hence must conserve strangeness.

$$\pi^- + p \rightarrow \Lambda^0 + K^0$$

strangeness: $0 + 0 = (-1) + (+1)$ $\hspace{3cm}$ **(16.25)**

A similar reaction leads also to the production of a K-meson and a sigma particle:

$$\pi^- \;\; + \;\; p \rightarrow \Sigma^0 \;\; + K^0$$

strangeness: $0 \;\; + \;\; 0 = (-1) + (+1)$ $\hspace{2cm}$ **(16.26)**

I_3: $(-1) + (+\tfrac{1}{2}) = (0) \;\; + (-\tfrac{1}{2})$

However, such a reaction does not produce a cascade particle and a K-meson,

$$\pi^- \;\; + \;\; p \nrightarrow \Xi^0 \;\; + K^0$$

I_3: $(-1) + (+\tfrac{1}{2}) \neq (+\tfrac{1}{2}) + (-\tfrac{1}{2})$ $\hspace{2cm}$ **(16.27)**

because it violates the conservation of isospin. A reaction that conserves isospin is

$$\pi^- + p \rightarrow \Xi^0 + K^0 + K^0 \hspace{3cm} \textbf{(16.28)}$$

whence a strangeness number $S = -2$ must be assigned to the cascade particle. As usual, the strangeness of an antiparticle is opposite in sign to the strangeness of the corresponding particle. The members of an isospin multiplet share the same strangeness.

The existence of the strangeness quantum number was shown by Gell-Mann to lead to the grouping of baryons and mesons in a simple pattern. The patterns of particles appear when the strangeness quantum number S is plotted against the third component I_3 of the isospin vector. The patterns of particles become symmetric about the origin if a new quantum number, namely, **hypercharge,** is defined as $Y = S + B$, where B is the baryon number of the particle, and the hypercharge is then plotted versus I_3 (Figure 16.5).

In 1963, Gell-Mann showed that the groupings of particles according to their patterns in a Y versus I_3 plot can be traced to the existence of hypothetical particles called **quarks.** All the hadrons

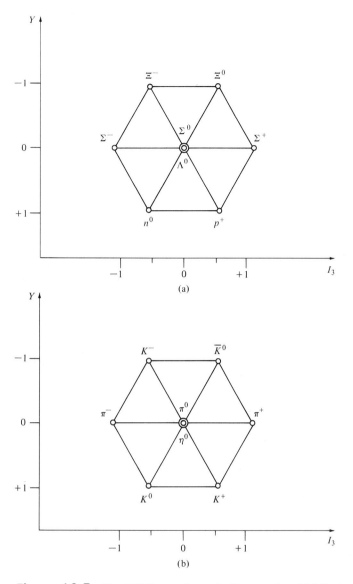

Figure 16.5 The Gell-Mann scheme for the grouping of (a) the baryon octet and (b) the meson octet.

would then be composites of quarks. The number of quarks he needed for his scheme was three, namely, the proton type, the neutron type, and the lambda type, and they were characterized by fractional charges. Baryons required three quarks, and mesons a quark and an antiquark. The proton, for example, is made up of two proton-type quarks and one neutron-type quark, namely $(q_1q_1q_2)$. The quark combination for the lambda particle is $(q_1q_2q_3)$. The π^+ consists of a q_1 and an anti-q_2, and the K^+ consists of a q_1 and an anti-q_3. All observed

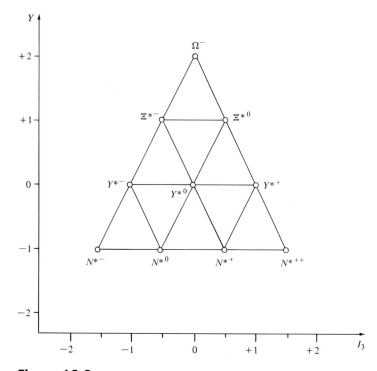

Figure 16.6   Gell-Mann's grouping scheme leading to the prediction of Ω^-.

particle multiplets were accounted for on the basis of three quark combinations for baryons and two quark combinations for mesons. Moreover, the quark model predicted the existence of the Ω^- particle, which has since been amply confirmed (Figures 16.6 and 16.7).

Table 16.4   Quark Quantum Numbers

Quark	Symbol	S	B	q/e	Y	I_3	Spin	Parity
p type	q_1	0	$\frac{1}{3}$	$+\frac{2}{3}$	$\frac{1}{3}$	$+\frac{1}{2}$	$\frac{1}{2}$	even
n type	q_2	0	$\frac{1}{3}$	$-\frac{1}{3}$	$\frac{1}{3}$	$-\frac{1}{2}$	$\frac{1}{2}$	even
Λ type	q_3	-1	$\frac{1}{3}$	$-\frac{1}{3}$	$-\frac{2}{3}$	0	$\frac{1}{2}$	even

The original quark hypothesis has now been extended so as to require more quarks than the original three. The successes of the quark model remain impressive, and if the model turns out to be ultimately correct, particles of higher mass and spin presumably exist corresponding to orbital and radial excitations of various quark systems. As yet, however, free quarks have not been observed. Furthermore, the present theory does not constitute a fundamental theory of elementary particles. It seems that physics remains, as it always has been, an intellectual discipline open to daring and imaginative thought.

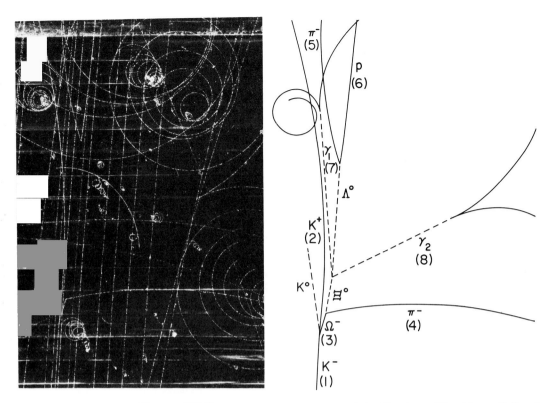

Figure 16.7 Photograph of the decay of Ω^-. [Courtesy of Brookhaven National Laboratory.]

16.7 Summary

Dirac's relativistic quantum theory predicts the existence of the positron and a g factor of 2 for the spin of the electron. The existence of antimatter is an essential feature of all relativistic quantum theories, and the slight deviation of Dirac's g factor from the experimental value is fully accounted for by quantum electrodynamics.

There are four fundamental interactions in nature. In quantum field theories these interactions are mediated by particles: the strong interaction by mesons, the electromagnetic interaction by photons, the weak interaction by intermediate vector bosons, and the gravitational interaction by the graviton.

The muon and the electron are identical in their interactions but have their own neutrinos. Thus they belong to the muon and the electron families, respectively, and their family numbers are separately conserved. Called leptons, they are regarded as truly elementary particles in that they have no internal structure.

All unstable hadrons, except the resonances, decay via an electromagnetic interaction or a weak interaction. The charge multiplicity of a hadron is given by $(2I + 1)$ where I is its isospin quantum number. The strong interaction conserves isotopic spin and thus is independent of electric charge. The electromagnetic interaction, however, does not conserve isotopic spin. Strangeness is conserved in strong and electromagnetic interactions but not in weak interactions. All baryons and mesons can be built up of fractionally charged particles called quarks, but free quarks have not been observed.

Problems

16.1 Discuss the nature of the physical vacuum in the Dirac theory, for example, the mass, the charge, and the polarizability of the vacuum.

16.2 Verify that the Coulomb potential is a solution of Eq. (16.6).

16.3 Verify that the Yukawa potential satisfies Eq. (16.11).

16.4 Calculate the range of a weak force mediated by a charged intermediate vector boson with a mass of the order 40 times the mass of the proton.

16.5 The classical radius of the electron is defined by $r = e^2/mc^2$. Compare the classical electron radius to the range of a strong force mediated by a charged pion and express the ratio in terms of the fine structure constant.

16.6 Calculate the time during which a charged pion can exist in a virtual state.

16.7 Explain why the decay of a muon into an electron and a photon is forbidden.

16.8 Calculate the energy of the muon-type neutrino in the decay of a pion at rest.

16.9 Give the quark constitution of Ω^-.

16.10 If the kinetic energy of a pion is 860 MeV, calculate the kinetic energies of the decay products, namely, the neutrino and the muon, for the case in which the neutrino moves forward.

Suggestions for Further Reading

Chew, G. F., M. Gell-Mann, and A. H. Rosenfeld. "Strongly Interacting Particles," *Scientific American* (February 1964).

Dyson, F. J. "Field Theory," *Scientific American* (April 1953).

Feinberg, G., and M. Goldhaber. "The Conservation Laws of Physics," *Scientific American* (October 1963).

Fowler, W. B., and N. P. Samios. "The Omega-Minus Experiment," *Scientific American* (October 1964).

Lederman, L. M. "The Two-Neutrino Experiment," *Scientific American* (March 1963).

Weinberg, S. "Unified Theories of Elementary-Particle Interactions," *Scientific American* (July 1974).

Perkins, D. H. *Introduction to High Energy Physics*. Reading, Mass.: Addison-Wesley Publishing Co., 1972.

Weisskopf, V. F. *Physics in the Twentieth Century*. Cambridge, Mass.: The MIT Press, 1972.

Appendix

Table of Selected Nuclides

Z	Element Symbol	A	Mass Excess M − A μu	J^π	Half-life Sec	Abundance Percent	Decay Modes and Energies MeV
0	n	1	8665	$1/2^+$	720		β^- 0.78
1	H	1	7825	$1/2^+$		99.99	
		2	14102	1^+		0.015	
		3	16050	$1/2^+$	3.87 E 8		β^- 0.0181
2	He	3	16030	$1/2^+$		0.0001	
		4	2603	0^+		100	
3	Li	6	15125	1^+		7.42	
		7	16004	$3/2^-$		92.58	
		8	22487	2^+	0.85		β^- 13
4	Be	7	16929	$3/2^-$	4.58 E 6		EC; γ 0.48
		8	5309	0^+	~1 E −16		α 0.048
		9	12186	$3/2^-$		100	
		10	13534	0^+	8.52 E 13		β^- 0.56
5	B	10	12939	3^+		19.78	
		11	9305	$3/2^-$		80.22	
		12	14354	1^+	0.020		β^- 13.4, 9.0; γ 4.4
6	C	10	16810	0^+	19.0		β^+ 1.9; γ 7.2, 1.04
		12	0	0^+		98.89	
		13	3354	$1/2^-$		1.11	
		14	3242	0^+	1.81 E 11		β^- 0.156
7	N	13	5738	$1/2^-$	600		β^+ 1.19
		14	3074	1^+		99.63	
		15	108	$1/2^-$		0.37	

Table of Selected Nuclides (Continued)

Z	Element Symbol	A	Mass Excess M − A μu	Jπ	Half-life Sec	Abundance Percent	Decay Modes and Energies MeV
8	O	15	3070	1/2⁻	124		β^+ 1.74
		16	−5085	0⁺		99.76	
		17	−867	5/2⁺		0.037	
		18	−840	0⁺		0.204	
9	F	17	2095	5/2⁺	66.0		β^+ 1.74
		19	−1595	1/2⁺		100	
		20	−13	2⁺	11.0		β^- 5.40; γ 1.63
10	Ne	20	−7560	0⁺		90.92	
		22	−8615	0⁺		8.82	
11	Na	22	−5563	3⁺	8.14 E 3		β^+ 0.54; γ 1.28; EC
		23	−10229	3/2⁺		100	
12	Mg	24	−14958	0⁺		78.70	
		25	−14161	5/2⁺		10.13	
		26	−17407	0⁺		11.17	
13	Al	27	−18461	5/2⁺		100	
		28	−18095	3⁺	138		β^- 2.87; γ 1.78
14	Si	28	−23070	0⁺		92.21	
		29	−23504	1/2⁺		4.70	
15	P	31	−26235	1/2⁺		100	
		32	−26091	1⁺	1.24 E 6		β^- 1.71
16	S	31	−20389	1/2⁺	2.60		β^+ 4.4; γ 1.27
		32	−27926	0⁺		95.0	
		34	−32135	0⁺		4.22	
17	Cl	37	−34102	3/2⁺		24.47	
		38	−31995	2⁻	2.24 E 3		β^- 4.8, 1.1, 2.8; γ 2.2, 1.6
18	Ar	38	−37272	0⁺		0.063	
		40	−37616	0⁺		99.60	
19	K	39	−36290	3/2⁺		93.10	
		40	−36000	4⁻	4.10 E 16	0.0118	β^- 1.32; γ 1.46; EC
		41	−38168	3/2⁺		6.88	
20	Ca	40	−37411	0⁺		96.97	
		43	−41220	7/2⁻		0.145	
21	Sc	41	−30753	7/2⁻	0.550		β^+ 5.61
		45	−44081	7/2⁻		100	
26	Fe	56	−65063	0⁺		91.66	
		57	−64602	1/2⁻		2.19	
27	Co	57	−63704	7/2⁻	2.31 E 7		EC; γ 1.22, 0.014, 0.137
		59	−66810	7/2⁻		100	
		60	−66186	5⁺	1.66 E 8		β^- 0.31; γ 1.33, 1.17
28	Ni	58	−64658	0⁺		67.88	
		59	−65657	3/2⁻	2.53 E 12		EC
		60	−69213	0⁺		26.23	
		64	−72042	0⁺		1.08	

Z	Element Symbol	A	Mass Excess M − A μu	J^π	Half-life Sec	Abundance Percent	Decay Modes and Energies MeV
29	Cu	59	−60504	3/2⁻	81.0		β^+ 3.78; γ 1.30, 0.87, 0.46, 0.34
		63	−70408	3/2⁻		69.09	
		64	−70241	1⁺	4.64 E 4		β^- 0.57; β^+ 0.66; γ 1.34; EC
30	Zn	63	−66794	3/2⁻	2.28 E 3		β^+ 2.35; γ 0.67, 0.97; EC
		64	−70855	0⁺		48.89	
38	Sr	88	−94359	0⁺		82.56	
		90	−92253	0⁺	8.84 E 8		β^- 0.54
40	Zr	90	−95300	0⁺		51.46	
42	Mo	98	−94591	0⁺		23.78	
		99	−92280	1/2⁺	2.38 E 5		β^- 1.23, 0.45; γ 0.74, 0.041−0.78
43	Tc	98	−92890		4.75 E 13		β^- 0.3; γ 0.75, 0.66
46	Pd	108	−96109	0⁺		26.71	
47	Ag	107	−94906	1/2⁻		51.82	
		108	−94051	1⁺	144		β^- 1.65; β^+ 0.7; γ 0.63; EC
		109	−95244	1/2⁻		48.18	
48	Cd	108	−95813	0⁺		0.88	
		114	−96639	0⁺		28.86	
		118	−93030	0⁺	3.00 E 3		β^- 1.8, 2.3; γ 0.42
49	In	115	−96129	9/2⁺	1.58 E 22	95.72	β^- 0.5
		119	−94010		120		β^- 1.6; γ 0.82
50	Sn	120	−97801	0⁺		32.85	
51	Sb	121	−96183	5/2⁺		57.25	
55	Cs	133	−94645	7/2⁺		100	
		137	−93230	7/2⁺	9.47 E 8		β^- 0.51, 1.17; γ 0.662
58	Ce	140	−94608	0⁺		88.48	
74	W	184	−48975	0⁺		30.64	
		186	−45560	0⁺		28.41	
78	Pt	195	−35187	1/2⁻		33.8	
		200	−28570	0⁺	4.14 E 4		β^- 0.70
79	Au	197	−33459	3/2⁺		100	
		201	−28080		1.32 E 3		β^- 1.5; γ 0.53
80	Hg	202	−29358	0⁺		29.80	
		206	−22487	0⁺	450		β^- 1.31
81	Tl	203	−27647	1/2⁺		29.50	
		207	−22550	1/2⁺	287		β^- 1.44; γ 0.89
		208	−17987	5⁺	186		β^- 1.80, 2.38; γ 2.61, 0.58
82	Pb	204	−26956	0⁺	4.42 E 24	1.48	α 2.6
		208	−23350	0⁺		52.3	
		214	−234	0⁺	1.61 E 3		β^- 0.7; γ 0.352, 0.295, 0.050−0.295
83	Bi	209	−19606	9/2⁻		100	
84	Po	208	−18757	0⁺	9.15 E 7		α 5.11; γ 0.6, 0.28
		210	−17124	0⁺	1.20 E 7		α 5.30; γ 0.80
		212	−11134	0⁺	3.0 E −7		α 8.78
		218	8930	0⁺	183		α 6.00; β^- 0.30

Table of Selected Nuclides (Continued)

Z	Element Symbol	A	Mass Excess M − A μu	J^π	Half-life Sec	Abundance Percent	Decay Modes and Energies MeV
86	Rn	208	−10210	0⁺	1.38 E 3		EC; α 6.14
90	Th	232	38124	0⁺	4.45 E 17		α 4.01, 3.95; γ 0.059; SF
91	Pa	233	40132	3/2⁺	2.37 E 6		β⁻ 0.26, 0.15, 0.51; γ 0.31
92	U	233	39522	5/2⁺	5.11 E 12		α 4.82, 4.78, 4.73; γ 0.043, 0.054
		235	43915	7/2⁻	2.25 E 16	0.72	SF; α 4.18−4.56; γ 0.195
		238	50770	0⁺	1.42 E 17	99.27	α 4.19; SF; γ 0.048
93	Np	239	52924	5/2⁺	2.03 E 5		β⁻ 0.72; γ 0.045−0.355
94	Pu	239	52146	1/2⁺	7.69 E 11		α 5.15, 5.13; γ 0.013; SF
		240	53882	0⁺	2.13 E 11		α 5.16, 5.12; γ 0.045; SF
95	Am	240	55280		1.84 E 5		EC; γ 1.00, 0.90, 1.40, 0.043, 0.099
		241	56714	5/2⁻	1.45 E 10		α 5.48, 5.31−5.53; γ 0.060; SF

Notation: Mass excess is given in micro mass units and is the difference between exact mass and mass number, thus $^4_2$He = 4.002603 u and $^{16}_8$O = 15.994915 u. For half-lives computer notation is adopted to express powers of ten, thus the half-life of $^3_1$H is 3.87×10^8 seconds. EC denotes electron capture and SF denotes spontaneous fission. Sources: J. H. E. Mattauch, W. Thiele, and A. H. Wapstra, *Nuclear Physics*, **67**, 1 (1965); General Electric *Chart of the Nuclides*, October, 1972; K. Way et al., *Nuclear Data Sheets*, National Academy of Science-National Research Council, Oak Ridge, Tennessee.

Answers to Numerical Problems

Chapter 1

1.1   (a) $x' = 0$, $y' = v_0't' - \frac{1}{2}gt'^2$

      (b) $y = (v_0/v)x - (g/2v^2)x^2$

1.9   (a) 0.04 fringe

      (b) 3.47×10^5 cm/sec

Chapter 2

2.4 $0.91c$

2.6 $c_x = v$; $c_y = c\sqrt{1 - v^2/c^2}$

2.8 15.6×10^{-6} sec

2.10   4232 cm

2.12   Star-based clock ahead of the earth-based clock by 20 years.

2.13   (b) 25.005 yrs

      (c) 3.65 days

2.14   (a) $v = 0.87c$

      (b) 50 m

Chapter 3

3.1 0.66 MeV; 2.45 MeV
3.2 (a) 0.61 MeV
     (b) 9.50 MeV
3.3 (a) 2.55 KeV, 2.57 KeV
     (b) 63.75 KeV, 78.90 KeV
     (c) 206.55 KeV, 660.02 KeV
3.7 1.96 eV
3.8 1.35 eV
3.9 (a) 67.49 MeV
     (b) 294.18 MeV, 15.48 MeV
3.10 (a) 2.068 eV/c
     (b) 12.408 KeV/c
     (c) 886.286 MeV/c
3.11 (a) 90.37 MeV
     (b) 0.142c
3.12 22°
3.13 29.87 MeV
3.15 $4m_0c^2$
3.16 $6M_0c^2$

Chapter 4

4.1 3.5×10^{18} photons/cm²-sec
4.2 $1/(4\pi r^2)$
4.6 (a) 248 Å
     (b) 1.73 Å
     (c) 0.04 Å
4.9 1.28 Å
4.10 (a) 5.36×10^{-10} cm
     (b) 10^{-14} cm
4.11 $(\hbar k)^2/2m$
4.12 (b) $E_0 = \hbar\omega/2 = (0 + \tfrac{1}{2})\hbar\omega$
     $E_1 = 3\hbar\omega/2 = (1 + \tfrac{1}{2})\hbar\omega$
     $E_2 = 5\hbar\omega/2 = (2 + \tfrac{1}{2})\hbar\omega$
4.17 0
4.18 (a) $\sqrt{2}/L$
     (b) $\tfrac{1}{2}$

Chapter 5

5.1 6565, 4863, 4342, 4103 Å
5.3 (a) $R = 1.09737 \times 10^{-3}$ Å^{-1}
     (b) $R_H = 1.09678 \times 10^{-3}$ Å^{-1}
     (c) 6561.06 Å
     (d) -6.8 eV
5.4 (a) $v = \alpha c = c/137$

5.5   (b) $\lambda_C = 3.86 \times 10^{-11}$ cm, $r_c = 2.82 \times 10^{-13}$ cm

     (c) $\lambda_\pi = 1.4 \times 10^{-13}$ cm

5.7   1468.8 eV

5.8   134.35 eV, 537.40 eV, 1209.15 eV

5.9   (a) $A_1 = \sqrt{2}/L$

     (b) $\frac{1}{2}$

5.20 2.83×10^3 gauss

5.21 7.4×10^3 gauss

5.24 (c) 7.8×10^4 gauss

Chapter 6

6.7 7×10^{-8} sec

6.8 10^8 sec^{-1}

6.11 (a) $0.75 ea_0$

     (b) 6.3×10^{-9} sec

     (c) 1.1×10^{-7} eV

     (d) $\Delta v = 2.54 \times 10^7$ sec^{-1}

6.13 (a) 10.4 MeV

6.16 (a) 1500

     (b) $\frac{1}{257}$

Chapter 7

7.4   (a) -74.8 eV

7.5   (a) $L = 1, 2, 3$

     (b) $J = 2, 3, 4$

7.6   1.84, 1.41, 1

7.7 3.62×10^5 gauss

7.8   Fe

7.9   (b) 1020 eV

7.10 251.0 eV

Chapter 8

8.1   1.98, 19.74, 0.147, 1.47 g/cm²

8.2   (b) 0.951 keV, 13.04 Å.

     (c) 0.821 keV, 15.10 Å

     (d) 0.931 keV

8.3   3.148, 1.408, 2.226, 1.817 Å

8.5   10.4, 8.26×10^7, 4.84×10^{15}, 4.21×10^{18} atoms/cm³

8.6   12.4 keV

8.8   2.3, 2.4, 2.6 keV

8.10 0.790, 0.701, 0.747, 0.667, 0.773, 0.633, keV

8.11 0.907, 9.05 cm

8.12 3.162

8.14 2.818 Å, (100); 1.987 Å, (110); 1.626 Å, (111);
      0.849 Å, (311); 0.646 Å, (331); 2.815 Å; 5.630 Å

8.16 0.111, 0.0786, 0.0642 atoms/Å²

8.17 3.0, 1.732, 1.225, 0.802 Å

Chapter 9

9.1 0.29 cm/sec
9.2 (a) 1.57×10^8 cm/sec
     (b) 1.85×10^{-9}
9.3 7.04, 5.61, 2.06 eV
9.4 (a) 0.6, 0.35 eV below conduction band
     (b) Si: 4.7×10^{-6}, 9.50×10^{-11}, 8.71×10^{-7}
         Ge: 8.74×10^{-8}, 1.42×10^{-6}, 2.92×10^{-4}
9.5 (a) 2.79×10^{-3}, 0.146, 0.313
     (b) 5.38×10^{-10}, 2.82×10^{-8}, 6.04×10^{-8}
9.6 389° K
9.10 2.4×10^4, 2.4×10^4 ohm^{-1} cm^{-1}, 0.04, 0.24
9.13 0.021
9.14 255° K
9.16 0.033, 0.041 cal mole^{-1} °K^{-1}
9.19 W: 2.55×10^{-18}, 4.46×10^{-12}; Pt: 6.44×10^{-22}, 2.12×10^{-14}

Chapter 10

10.2 0.565593 u, 8.781 MeV; 0.069756 u, 6.498 MeV; 0.098940 u, 7.680 MeV; 0.101027 u, 7.239 MeV; 0.563406 u, 8.747 MeV
10.3 4.1, 3.7, 3.3, 5.5, 4.1 fermis
10.4 41.3, 8.15 fermis
10.5 62.0 fermis
10.6 1.71×10^{-5}, 1.11×10^{-6}, 7.96×10^{-8}, 8.84×10^{-9}
10.7 alphas: 8.33, 6.94, 5.25, 15.24, 8.91 MeV
      protons: 4.17, 3.47, 2.62, 7.62, 4.46 MeV
      neutrons: 0
10.8 1.22, 1.07, 0.97 fermis
10.9 4.05, 1.35 fermis
10.10 211 MeV
10.11 0.002388 u, 1.112 MeV; 0.030377 u, 7.074 MeV; 0.008285 u, 2.572 MeV
10.12 4.2 fermis, 8.91 MeV, 7.06 fermis, 4.35 fermis
10.13 1.13×10^{-5}, 7.31×10^{-7}, 5.25×10^{-8}, 5.83×10^{-9}
10.17 227, 80.4, 18.0, 1.97 fermis
10.18 12.3 MeV

Chapter 11

11.2 2.792, 6.251 MeV
11.3 0.048 MeV
11.4 14.09, 14.13 MeV
11.5 5.858, 0.805, -0.829, -1.794 MeV; 5.053, 6.687, 7.652 MeV
11.6 β^+: 1.678 MeV; β^-: 0.572 MeV
11.7 1.884, 0.172 MeV
11.8 0.898, 0.978 MeV
11.9 2.9×10^{11}

11.10 10.73 MeV
11.11 0.647, 1.646 MeV
11.13 1.232, 1.163, 1.102 MeV
11.17 1.923, 0.806 MeV
11.18 10.272, 15.410 MeV
11.19 0.202

Chapter 12

12.1   neutrons: 26.8, 9.18, 36.0 MeV
       protons: 22.3, 6.48, 28.8 MeV
12.2   39.88115, 132.90027, 196.96080 u
12.4   2 or 3, $\frac{3}{2}$, 1 through 6
12.6   0.51
12.7   2.64, 3.30 MeV
12.9 $S_p(Z)$: 12.13, 8.33, 8.18, 10.82, 8.03 MeV
       $S_p(Z + 1)$: 10.21, 6.37, 6.02, 8.20, 7.35 MeV
       $S_p(Z - 1)$: 0.60, 1.09, 3.42, 5.78, 3.80 MeV
12.10 $\frac{1}{2}$, $\frac{7}{2}$, 0
12.11 295 barns
12.12 1.00 fermi, 0.535 fermi, 1.87

Chapter 13

13.1 23 ± 1 minutes
13.2 19.1×10^3 yr, 1.15×10^{-12} sec^{-1}
13.3 0.95×10^{-12} g
13.4   1978 sec, 1796 sec, 0.996×10^4 min^{-1}, 1.992×10^4 min^{-1}
13.5   0.152, 0.281, 0.483, 0.733, 0.862, 0.929 (10^4 min^{-1})
13.6   0.444, 1.22, 2.14, 3.36, 4.06, 4.46 (10^8 sec^{-1})
13.8   (a) 6.266, 6.102, 8.954, 5.213 MeV
       (b) 0.121, 0.112, 0.169, 0.100 MeV
13.9   (a) 0.005806 u
       (b) 4.008409 u
13.10 0.102 MeV
13.11 7.8×10^{-29}, 2.98×10^{21} sec^{-1}, 2.3×10^{-7} sec^{-1}
13.14 3956, 3506, 2070, 1071 sec^{-1}
       1.06, 11.8, 46.0, 69.8 (10^{-12} g)
13.15 1.24×10^{-4}%
13.16 5.9×10^9 yrs
13.17 $(19.0 \pm 0.8) \times 10^3$ yrs, 57×10^3 yrs

Chapter 14

14.1   12.014354 u
14.2   (a) 7.413×10^{-16} g cm/sec, 0
       (b) 7.413×10^{-16} g cm/sec, 8.61 keV

14.3   (a) 6.685 MeV
      (b) 3.836×10^{-16}, 3.573×10^{-16} g cm/sec
      (c) 0.263×10^{-16} g cm/sec, 10.8 eV
14.4 (1.73 ± 0.02) MeV
14.5   0.005162 u
14.8   (a) 1.15×10^{-16} g cm/sec
      (b) 1.15×10^{-16} g cm/sec, 65.5 eV
14.9   (a) 1.5×10^{6} cm/sec
      (b) 2.12×10^{6} cm/sec
14.10 (a) 9.51×10^{-17} g cm/sec
      (b) 9.51×10^{-17} g cm/sec, 60.8 eV
      (c) 6.83×10^{-5}
      (d) 1.11×10^{8} cm/sec
      (e) 88.9°
14.11 1.32×10^{-3} eV
14.12 ^{60}Ni   1.17 MeV: 2 through 6, E2
              1.33 MeV: 2, E2
      ^{22}Ne   1.28 MeV: 2, E2
      ^{38}Ar   1.60 MeV: 1 through 5, E1
              2.15 MeV: 2, E2
      ^{137}Ba   0.66 MeV: 4 through 7, M4
14.14 0.00138 u
14.15 β^-: 1.639 MeV; β^+: 0.897 MeV; EC: 1.89 MeV
14.16 2.93×10^{-20}
14.17 0.018 eV, 1.8×10^{5}

Chapter 15

15.1 2.93×10^{-9} amp
15.2   (a) 2.34×10^{-12} coulombs
      (b) 0.08 volts
15.3   1.77, 2.00, 2.96, 6.60 MeV
15.4   (a) 1024
      (b) 1.867
15.5   88 keV, 2.73 MeV
15.6   (a) 2.36×10^{-13} coulombs
      (b) 1.7×10^{-3} volts
15.7   820 gauss
15.8 1.3×10^{4} gauss
15.9   10, 6.67, 20.1 MeV
15.11 3.35 hp
15.12 (a) 236 cm
      (b) 28.9×10^{6} Hz
      (c) 16.57×10^{6} Hz
15.13 (a) 2.43 cm
      (b) 18.56 cm
      (c) 25.46 cm

15.14 (a) 7.43×10^4 cm
(b) 399 gauss
(c) 4.76×10^4 rev/sec
(d) 8.8×10^5 km
(e) 2.94 sec

Chapter 16

16.4 5×10^{-16} cm
16.5 2
16.6 4.7×10^{-24} sec
16.8 29.8 MeV
16.10 427 MeV, 471 MeV

Index

U

Uhlenbeck, G. E., 125
Uncertainty principle, 150–152, 158, 330, 432
Unified mass unit, 244
Unit cell, 197
   of KCl, 197–198

V

Vacancy
   defect, 207–208
Valence band, 228
Van de Graaff generator, 410, 420
Velocity addition formula, 21
Virtual process, 432
Volume energy, 305, 307

W

W particle, 434–435
Walton, E. T. S., 274
Wave equation: *see* Klein-Gordon equation and Schrödinger equation
Wave function, 76, 79. *See also* Schrödinger equation
Wavelength limit, short, 183
Weak interaction, 433–435
Weinberg, S., 435
Weisskopf, V. F., 33, 322, 371
Wilson, C. T. R., 395

Window, 385, 393
Work function, 48
Wu, C. S., 362, 363, 364

X

X-ray
   diffraction, 190 et seq
   levels, 184
   notation, 185
X-ray spectrum
   characteristic, 184
   continuous, 183–184
   of copper, 184
X-ray tube
   cold cathode, 181
   hot cathode, 182

Y

Yang, C. N., 362
Yield, nuclear reaction, 289
   for $^{12}C(d,p)^{13}C$ and $^{12}C(d,n)^{13}N$, 292
Yukawa, H., 430, 432
Yukawa potential, 431, 432

Z

Zeeman effect, 113–117